High-Pressure Shock Compression of Condensed Matter

Springer
New York
Berlin
Heidelberg
Barcelona
Budapest
Hong Kong
London
Milan
Paris
Santa Clara
Singapore
Tokyo

High-Pressure Shock Compression of Condensed Matter

J. Asay and *M. Shahinpoor* (Eds.): High-Pressure Shock Compression of Solids
A.A. Batsanov: Effects of Explosion on Materials: Modification and Synthesis Under High-Pressure Shock Compression
R. Cherét: Detonation of Condensed Explosives
L. Davison, D. Grady, and *M. Shahinpoor* (Eds.): High-Pressure Shock Compression of Solids II
L. Davison, Y. Horie, and *M. Shahinpoor* (Eds.): High-Pressure Shock Compression of Solids IV
L. Davison and *M. Shahinpoor* (Eds.): High-Pressure Shock Compression of Solids III
R. Graham: Solids Under High-Pressure Shock Compression
M. Sućeska: Test Methods for Explosives
J.A. Zukas and *W.P. Walters* (Eds.): Explosive Effects and Applications

Lee Davison Mohsen Shahinpoor
Editors

High-Pressure Shock Compression of Solids III

With 140 Illustrations

Springer

Lee Davison
7900 Harwood Avenue NE
Albuquerque, NM 87110
USA

Mohsen Shahinpoor
Department of Mechanical Engineering
University of New Mexico
Albuquerque, NM 87131
USA

Editor-in-Chief:
Robert A. Graham
Director of Research
The Tomé Group
383 La Entrada Road
Los Lunas, NM 87031
USA

Library of Congress Cataloging-in-Publication Data
High-pressure shock compression of solids III/[edited by] Lee Davison, Mohsen Shahinpoor.
p. cm. – (High-pressure shock compression of condensed matter)
Includes bibliographical references and index.
ISBN 0-387-98292-2 (hc: alk. paper)
1. Materials–Compression testing. 2. Porous materials–Mathematical models. 3. Powders–Mathematical models. 4. Materials at high pressures–Mathematical models. 5. Shock (Mechanics) I. Davison, L.W. (Lee W.) II. Shahinpoor, Mohsen. III. Series.
TA417.7.C65H552 1997
620.1′1242–DC21 97-22862

Printed on acid-free paper.

Production managed by Francine McNeill; manufacturing supervised by Thomas King.
Camera-ready copy supplied by the editors.
Printed and bound by Maple-Vail Book Manufacturing Group, York, PA.
Printed in the United States of America.

9 8 7 6 5 4 3 2 1

ISBN 0-387-98292-2 Springer-Verlag New York Berlin Heidelberg SPIN 10632891

Preface

Shock compression of condensed matter is of interest for several reasons. As an experimental technique, it provides access to the highest compressive stresses readily attainable, and does so in a way that permits accurate determination of stress and compression using only measurements of length and time. It is for this reason that it plays an essential role in studies of the equation of state of matter. A second aspect of the shock compression process is the rapidity with which the stress can be applied and removed, facilitating the study of nonequilibrium deformation, fracture, chemical composition, and other processes and states of matter.

Military applications motivated much of the early work in shock compression science, but interest in these applications is decreasing. In contrast, interest in equations of state for use in earth and planetary sciences, in high-rate deformation for improving understanding of mechanical processes in metals, ceramics, and other materials, in application of shock processing to prepare novel materials, and to study the initiation sensitivity and detonation performance of explosives is increasing.

Investigation of shock compression phenomena is based on theories of matter ranging from quantum mechanics to elementary continuum theories, including, along the way, statistical mechanics, chemistry, a broad range of continuum theories, and molecular-dynamic simulations. The continuum theories increasingly take account of aspects of the microstructure of the material studied, particularly when it is subject to design and control, as in the case of composite structural materials. Interest in synthesis of novel materials, sensitivity of explosives to initiation, and other problems has motivated work on shock-induced chemical effects.

Shock compression has traditionally been studied using plane shocks. At the continuum level, the deformation is homogeneous and precisely controlled. At the mesoscale level, that of defect structures, grains of polycrystalline metals or ceramics, particles of powdered material, etc., even the most carefully controlled deformations are chaotic. It is now clear that an understanding of deformation at this level is necessary for understanding nonequilibrium chemical and mechanical effects.

Newly developed experimental methods are providing information needed for developing an increasingly refined understanding of shock compression phenomena at continuum, mesoscale, and atomic or molecular levels. Most practical applications of results of investigations in the field of shock compression science involve numerical simulation of shock phenomena, and considerable effort is directed toward development of theories and computational methods for performing the necessary calculations.

In this third volume in the series *High-Pressure Shock Compression of Solids,* a broad range of topics of current interest is discussed, but with no attempt to provide comprehensive coverage.* The authors of the several chapters were selected for their expertise and experience in the specific subject matter they address.

The first four chapters of this volume address fundamental physical and chemical aspects of the response of matter to shock compression. The first of these chapters, prepared by Sikka, Godwal, and Chidambaram, reviews the status of knowledge of equations of state of matter at high pressure. This subject lies at the foundation of the field of shock compression science and is represented in the earliest work in the field. Nevertheless, this area continues to be one of deep and continuing interest and continual advance. Advances in related theoretical, computational, and experimental matters have contributed to this progress, as is demonstrated in the chapter. In the second chapter, Robertson, Brenner, and White describe and demonstrate a newer tool, molecular-dynamic analysis, that has proved capable of showing how interatomic forces and the organization of matter at the atomic or molecular level affect phenomena that are usually observed at the continuum level. The latter include shock

* Introduction of porosity into a material provides a means of independently varying the temperature and pressure produced by shock compression, thus providing more control than can usually be exercised over chemical reactions and phase transformations in the material. Volume IV of this series is devoted to shock compression of porous materials.

structure, phase transitions, and chemical reactions (including detonation phenomena). Progress in this field is very much dependent on continued advances in computing technology, and since this seems assured, we can look forward to molecular-dynamic investigations making many important contributions to our understanding of shock phenomena. The third chapter, by Coffey, addresses elastoplastic deformation of metals. This area, like that of equation of state, has a long history and lies at the foundation of much of the work in the area of shock compression science. Nevertheless, much of the work in this area is at the continuum level, with most of the remainder being based on classical continuum theories of dislocations. The results obtained by this means have provided very satisfactory pragmatic descriptions of observed phenomena, but have failed to provide a quantitative connection between continuum-level observations of plastic flow and the microscopic descriptions of the mechanisms believed to be responsible for these flows. The present chapter reports new work directed toward extending the microscopic descriptions by inclusion of quantum effects into the theory. The final member of this first group of chapters, prepared by Pangilinan and Gupta, demonstrates the impressive power of spectroscopic methods in producing information about molecular-level processes that occur during shock compression of condensed matter. Processes occurring at this level affect the equation of state of matter and lie at the heart of chemical investigations, but are only now being subjected to effective, direct experimental observation.

The second general subject area is that of the response of high-strength ceramic materials to shock compression. This area is covered in chapters prepared by Mashimo and by Cagnoux and Tranchet. Investigations of ceramics, including monocrystalline samples of the compounds on which they are based, are motivated by the appearance of these materials in areas of geological and planetary science, because they are often highly resistant to penetration by projectiles, and because shock processing provides a means of preparing ceramic compacts offering high resistance to abrasion. Application of ceramic materials requires knowledge of their equation of state, especially as regards phase transitions, and their deformation and fracture behavior.

The chapter by Engelke and Sheffield is a tutorial on shock-induced reaction of condensed-phase explosives. Both theory and experimental observation are covered. This is a topic of long-standing interest, but one in which advances continue to be made because of improvements in material preparation, experimental apparatus and technique, and capability for numerical simulation.

The next two chapters are devoted to continuum-mechanical investigations of shock phenomena observed at rather low stresses. The first of these, prepared by Addessio and Aidun, exhibits a continuum theory of fiber-reinforced composites that goes beyond the usual theories of homogeneous continuua in its incorporation of microstructural variables sufficient to capture the effects of layering, orientation, debonding, etc., that lie at the heart of the utility of these materials. This theory has been incorporated into a computer code capable of analyzing shock propagation phenomena, and a variety of numerical simulations have been provided. In the second of this pair of chapters, Davison has analyzed attenuation of elastoplastic pulses. This is an old, classical, problem, and the outlines of its solution are well known. Nevertheless, we have not seen an analysis presented in enough detail to exhibit the many and varied physical phenomena that make up the attenuation process.

Albuquerque, New Mexico

Lee Davison
Mohsen Shahinpoor

Contents

Contributors

F.L. Addessio
Los Alamos National Laboratory
Los Alamos, New Mexico 87545, USA

J.B. Aidun
Sandia National Laboratories
Albuquerque, New Mexico 87185-0437, USA

D.W. Brenner
Naval Research Laboratory
Washington, DC 20375-5342, USA

Jacques Cagnoux
Établissement Technique Central de l'Armement
Centre d' Études de Gramat
46500 Gramat, France

R.Chidambaram
High Pressure Physics Division
Bhabha Atomic Research Centre
Trombay, Bombay-400 085, India

Charles S. Coffey
Naval Surface Warfare Center
White Oak Laboratory
Silver Spring, Maryland 20903-5640, USA

Lee Davison
The University of New Mexico
7900 Harwood Avenue NE
Albuquerque, New Mexico 87110, USA

Ray Engelke
Los Alamos National Laboratory
Los Alamos, New Mexico 87545, USA

B.K. Godwal
High Pressure Physics Division
Bhabha Atomic Research Centre
Trombay, Bombay-400 085, India

Y.M. Gupta
Shock Dynamics Center and Department of Physics
Washington State University
Pullman, Washington 99164-2814, USA

Tsutomu Mashimo
High Energy Rate Laboratory
Faculty of Engineering
Kumamoto University
Kumamoto 860, Japan

G.I. Pangilinan
Shock Dynamics Center and Department of Physics
Washington State University
Pullman, Washington 99164-2814, USA

D.H. Robertson
Department of Chemistry
Indiana University–Purdue University at Indianapolis
Indianapolis, Indiana 43202, USA

Stephen A. Sheffield
Los Alamos National Laboratory
Los Alamos, New Mexico 87545, USA

S.K. Sikka
High Pressure Physics Division
Bhabha Atomic Research Centre
Trombay, Bombay-400 085, India

Jean-Yves Tranchet
Établissement Technique Central de l'Armement
Centre d'Études de Gramat
46500 Gramat, France

C.T. White
Naval Research Laboratory
Washington, DC 20375-5342, USA

CHAPTER 1

Equation of State at High Pressure

S.K. Sikka, B.K. Godwal, and R. Chidambaram

1.1. Introduction

The equation of state (EOS) of a system is a relationship between thermodynamic variables like pressure, p, and energy, E, with volume, V, and temperature, T. It has applications in a number of fields: condensed-matter physics, geophysics, astrophysics, plasma physics, and nuclear physics. Also, it is a vital input in hydrodynamic calculations for a wide spectrum of practical needs, e.g., in reactor safety simulations, design of fission and fusion energy producing devices, analysis of hypervelocity impacts, and weapon development.

An equation of state is measured experimentally by using static and shock wave techniques. In static experiments, the pressure on the sample is generated by squeezing it between pistons of, say, diamond in a diamond-anvil cell (DAC) and the volume measured by x-ray diffraction. The maximum static pressure achieved so far is about 5 Mbar at 300 K [1]. The compressions range up to 0.3. Melting lines have now been delineated as a function of pressure to pressures of 2 Mbar and temperatures of ~ 5000 K in laser heated DACs [2]. In shock experiments, the maximum pressure depends on the shock wave generator used. Chemical explosives and light gas guns achieve pressures up to 10 Mbar with accompanying temperatures of $\sim 10^4$K [3]. High-power lasers have been used up to 750 Mbar [4]. Using underground nuclear explosions, absolute Hugoniot EOS measurements have been carried out to 50 Mbar [5], and in impedance mismatch experiments, pressures as high as 4500 Mbar have been achieved in Al [6]. Other shock wave generators like rail guns, exploding foils, magnetic compression, etc. are either still being developed or have not yet achieved the accuracy of other methods. The maximum density ratios achieved in these shock EOS measurements are of the order of 4. The temperature of shock compressed iron has been measured directly up to 350 GPa [7].

It is obvious that experiments sample a very small portion of the pressure–volume–temperature surface. The rest of this surface has to be filled by theoretical calculations. In fact, the experimental measurements themselves depend upon theoretical EOSs above 50 Mbar. The theoretical problem is also a difficult one. This is because no single theory is able to account for all the physical phenomena on the p, V, T surface. This may be appreciated from Fig. 1.1, where we have sketched a generalized $p-T$ phase diagram for conditions where only electronic effects are important. The solid–solid phase transitions, melting, and pressure–temperature ionization (plasma) regimes are indicated. There may also be other related electronic phenomena like band crossings and corresponding reordering of valence electrons, band gap openings and closures, etc. Many of these may require a different theoretical treatment. Further, interpolations may be required. The interpolation formulas assumed may be accurate within a certain regime, but may give unacceptable results outside that domain. An illustration of this is Al, where seven different models have been employed [8]. Thus, the construction of the EOS

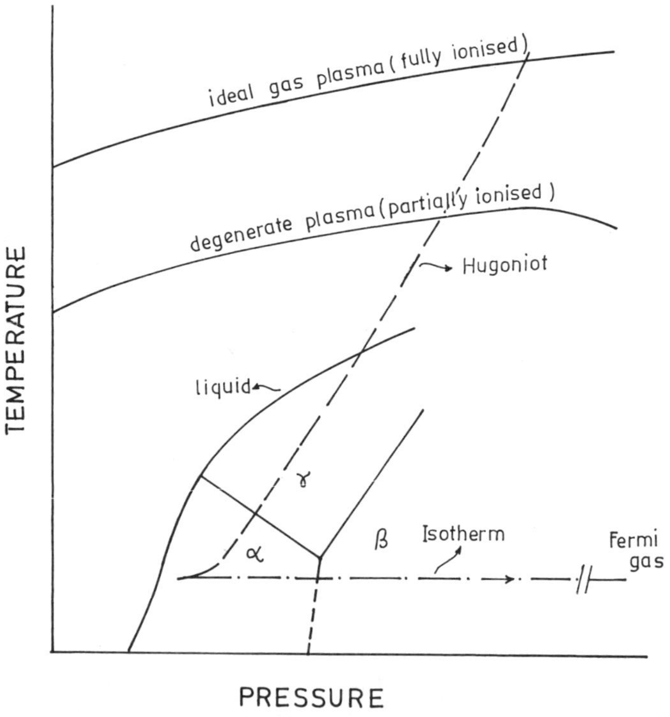

Figure 1.1. A schematic generalized p–T phase diagram. The symbols α, β, and γ denote solid phases.

of a material is a complex problem of long standing. It requires an interdisciplinary approach of subjects: solid state physics, chemistry, statistical mechanics, plasma physics, etc. For a perspective on the intricacies in the assembly of a global EOS for a material, we refer the reader to some recent books, review articles, and reports [9–15]. A handbook on the EOS of 50 materials is also available [16].

In this chapter, because of the above complexities, we shall deal only with the domain of EOS shown in Fig. 1.1. Here, during the past two decades, first-principle methods have been very successful.

1.2. General Considerations

Formally, the equation of state of a physical system can be written as

$$f(p, V, T) = 0. \tag{1.1}$$

To obtain the form of f, it is convenient to compute the Helmholtz free energy of the system, F,

$$F = E - TS, \tag{1.2}$$

where S is the entropy. The quantities p and E can be obtained as

$$p = -\left.\frac{\partial F}{\partial V}\right|_T \tag{1.3}$$

and

$$E = F + T\left.\frac{\partial F}{\partial T}\right|_V. \tag{1.4}$$

For an isotherm, T = constant. For the shock Hugoniot, which is defined as the locus of end states obtained in a series of shock experiments for the same initial state, the quantities p and E obtained from Eqs. 1.3 and 1.4 must satisfy the shock jump condition

$$E - E_0 = \tfrac{1}{2}(p + p_0)(V_0 - V). \tag{1.5}$$

It is now common practice to write the free energy as well as the pressure and the internal energy as a superposition of temperature-independent and -dependent terms:

$$E = E_c(V) + E_{lt}(V,T) + E_e(V,T) \tag{1.6}$$

and

$$p = p_c(V) + p_{lt}(V,T) + p_e(V,T). \tag{1.7}$$

Here, the symbols c, lt, and e stand for cold (0 K), lattice thermal vibrations, and electronic excitations, respectively. Implicit in this division is the assumption of Born–Oppenheimer approximation and the neglect of electron–phonon interactions.

The relative importance of these terms is shown in Fig. 1.2 for the Hugoniot of Al. At low pressures, p_c is the dominant component, accounting for about 75% of the pressure at 1 Mbar, whereas the remaining pressure is mostly due to ionic vibrations. At 18 Mbar, $p_c = 50\%$, $p_{lt} = 35\%$, and $p_e = 15\%$, whereas at 10^4 Mbar, the cold contribution is only 10 Mbar (only 0.1%). In the following section we shall give a brief description of the computational techniques that are currently in general use.

1.2.1. Solid State Theories

Density Functional Theory of the Zero-Degree Isotherm

Here, the compressed matter is modeled as a solid with electrons moving among static nuclei. The ground state energy of such a system is given by the minimum of the density functional [18]

$$E_c(n) = E_{ke}(n) + U_{i\text{-}e}(n) + U_{e\text{-}e}(n) + E_{xc}(n) + U_{i\text{-}i}\,, \tag{1.8}$$

where $n = n(\mathbf{r})$ denotes the local electron density. It is obtained from the functional variation

$$\frac{\partial E_c(n)}{\partial n} = 0 \tag{1.9}$$

under the constraint of electron number conservation

$$Z = \int n(\mathbf{r}) d\mathbf{r}^3\,. \tag{1.10}$$

In quantum mechanical calculations, the kinetic energy is given by

$$E_{ke} = \sum_k \int \psi_k^*(\mathbf{r})(-\nabla^2)\psi_k(\mathbf{r}) d^3\mathbf{r}\,, \tag{1.11}$$

where the functions ψ_k are the solutions of an effective one-electron Schrodinger equation obtained through Eqs. 1.8–1.10:

$$\left[-\nabla^2 + V(\mathbf{r})\right]\psi_k(\mathbf{r}) = \varepsilon_k \psi_k(\mathbf{r})\,. \tag{1.12}$$

Here, $V(\mathbf{r})$ is the effective one-electron potential

$$V(\mathbf{r}) = \frac{\partial U_{i\text{-}e}}{\partial n(\mathbf{r})} + \frac{\partial U_{e\text{-}e}}{\partial n(\mathbf{r})} + \frac{\partial E_{xc}}{\partial n(\mathbf{r})} \tag{1.13}$$

and

$$n(\mathbf{r}) = \sum_k |\psi_k|^2\,, \tag{1.14}$$

the sum being over all occupied electron states. For the exchange-correlation energy, E_{xc}, a local density approximation (LDA) is used.

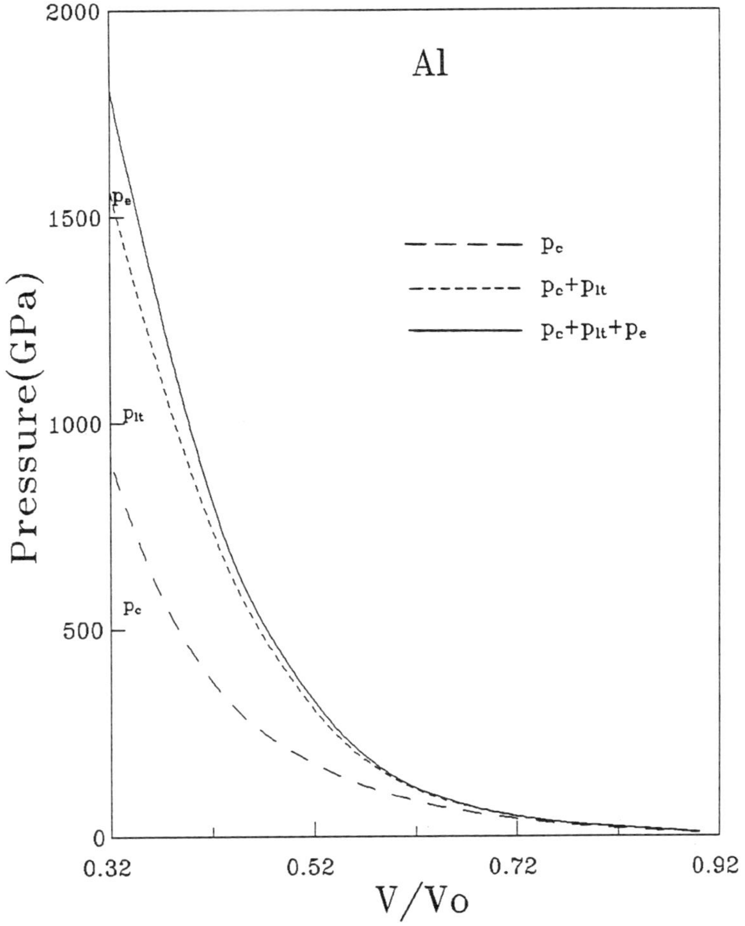

Figure 1.2. Pressure contributions to the Hugoniot of Al (data from [17]).

$$E_{xc} = \int n(\mathbf{r}) \varepsilon_{xc}(n) d^3\mathbf{r}, \tag{1.15}$$

where ε_{xc} is the contribution per electron.

The self-consistent solution of Eq. 1.12 is referred to as the band structure method. Many forms of this are available, based either on the pseudopotential concept or on the muffin-tin approximation. The former include the ab initio pseudopotential method (AP), generalized pseudopotential theory, etc., and the latter are augmented plane wave (APW), augmented spherical wave (ASW), linear muffin-tin orbital (LMTO) methods, including various linear, atomic sphere approximation (ASA) and full potential versions. However, AP and

LMTO methods are the most popular. The AP method is good for low-Z elements containing sp-electron bonding and can be extended for transition elements by use of soft-core potentials. The muffin-tin methods use radial wave functions in a spherical region close to the nuclei and different prescriptions (mostly free-electron-like) for the interstitial regions. These methods work for all types of electrons.

Now, we mention some cautions regarding the use of these methods at high compressions. The first concerns the frozen core approximation for core states, which are fully localized at a lattice point and do not contribute to the hydrostatic pressure. In this approximation, the total electron density is separated into core and valence contributions. The former is evaluated from self-consistent atomic calculations. However, at higher pressures, the core states will delocalize to form energy bands and the core part has to be redefined. For this, the muffin-tin methods are more suitable. The second difficulty is due to the derivative nature of the EOS and the increase in the size of the Brillouin zone under compression. This demands higher accuracies with consequent larger k-point sampling of the Brillouin zone, which, in turn, increases the demand on computational resources. This problem is mostly overcome with the use of super and parallel machines. The third problem is regarding the choice of crystal structures as inputs to these band structure methods, especially for phase transitions. Various recipes have been discussed by Gupta et al. [19]. In the absence of any definite knowledge of the structure, an fcc lattice may be assumed in the case of elements. It is now well established that p–V curves do not differ very much for different close-packed structures. A possible solution to this problem is the use of molecular dynamics simulations in which the crystal structures emerge naturally out of the computations. Already, there are some applications of this method [20].

Ionic Vibration Contribution

The vibrational free energy of a crystal in the quasi-harmonic-phonon approximation can be written as [21]

$$F_{\mathrm{lt}} = \tfrac{1}{2}\sum h\nu_i + k_{\mathrm{B}}T\sum \ln\left[1 - \exp\left(-\frac{h\nu_i}{k_{\mathrm{B}}T}\right)\right]. \tag{1.16}$$

Here, the quantities ν_i are classical frequencies of small-amplitude oscillations about the perfect lattice positions and are a function of the volume. The summation on i is over all normal modes of the system. Once the interatomic potentials are known, the frequencies can be obtained by diagonalizing the dynamical matrix in the standard Born–von Karman lattice dynamical theory,

$$\left\| D_{xy}(ll', \mathbf{q}) - \nu^2 \mathbf{I} \right\| = 0, \tag{1.17}$$

where the dynamical matrix element D (l and l' are the ionic labels within a unit cell, x and y are the Cartesian directions, and $\mathbf{q}$ is the wave vector) is related to the second derivative of the potential energy function

$$D_{xy}(ll', \mathbf{q}) = \nabla_{lx} \nabla_{l'x}[E_c], \tag{1.18}$$

with E_c as given by Eq. 1.8. The frequencies can also be directly computed by the so-called frozen phonon approximation. In this, the total energy is evaluated for a system with a displacement pattern according to the snapshot of the phonon movement. The frequencies are then obtained, in harmonic approximation, from the curvature of this total energy with respect to small displacements. However, this method, which involves calculations with super cells, has so far been limited to evaluations at high symmetry points of the Brillouin zone. A generalization of this to arbitrary k points has only recently been proposed by Wei and Chou [22] using the AP method. An all-electron formulation based on LDA linear-response theory has also been presented by Savrasov [23].

These direct lattice dynamics calculations have so far been employed only for a limited number of EOS determinations because, for complex crystals, these computations are very cumbersome. Therefore, the following Debye–Grüneisen approximation to Eq. 1.16 is usually employed:

$$\begin{aligned} E_{\mathrm{lt}} &= 3k_{\mathrm{B}} T D_{\mathrm{e}}(\theta_{\mathrm{D}}/T) \\ &\simeq 3k_{\mathrm{B}} T \quad \text{for} \quad T > \theta_{\mathrm{D}}, \end{aligned} \tag{1.19}$$

$$P_{\mathrm{lt}} = \frac{\gamma(V)}{V} E_{\mathrm{lt}}. \tag{1.20}$$

Here, $D_{\mathrm{e}}(\theta_{\mathrm{D}}/T)$ is the Debye function, θ_{D} is the Debye temperature, and γ is the Grüneisen parameter

$$\gamma = -\left\langle \frac{d \ln \nu_i(V)}{d \ln(V)} \right\rangle, \tag{1.21}$$

where the brackets designate an average over all of the normal modes of the solid. Either the approximation

$$\rho_0 \gamma_0 = \rho\gamma, \tag{1.22}$$

where

$$\gamma_0 = \frac{V_0 \alpha}{\kappa_0 C_V} \tag{1.23}$$

(V_0 =normal volume, α =coefficient of thermal expansion, κ_0 = isothermal compressibility, and C_V =specific heat at constant volume) has been used or γ has been evaluated from the expression

$$\gamma = \frac{t-2}{3} - \frac{V}{2}\left(\frac{d^2 p_c V^{2t/3}}{d^2 V}\right)\left(\frac{d p_c V^{2t/3}}{dV}\right)^{-1}. \tag{1.24}$$

Here $t = 0, 1, 2$, respectively, give Slater [24], Dugdale–MacDonald [25], and free-volume [26] expressions. The "cold" pressure, p_c, is as evaluated from the previous section. For a detailed discussion of the validity of this approach, and also of the neglect of anharmonic effects, see Godwal et al. [10].

Thermal Electronic Contribution

For interpretation of experimental shock wave data, Al'tshuler et al. [27] used the expressions

$$E_e = \tfrac{1}{2}\beta T^2 \tag{1.25}$$

$$p_e = \frac{\gamma_e E_e}{V} \tag{1.26}$$

based on low-temperature expansions of the Fermi–Dirac distribution function. Here, β is the electronic specific heat, taken from low-temperature measurements, and γ_e, the electronic Grüneisen parameter, is set equal to 0.5. This value falls between the value 2/3 for a free Fermi gas and the value 1/3 for an infinite atom. Some recent calculations have evaluated β and γ_e from band theory (for $T \ll T_f$ – the Fermi temperature defined as $E_f = k_B T_f$, where E_f is the Fermi energy)

$$\beta = \tfrac{1}{3}\pi^2 k_B^2 N(E_f) \tag{1.27}$$

and

$$\gamma_e = \frac{\partial \ln E_f}{\partial \ln V}, \tag{1.28}$$

where $N(E_f)$ is the density of electron states at Fermi energy, E_f, taken from LDA band structure calculations.

The LDA computations (both 0 K and finite T) have also been used directly to evaluate the electronic excitation contribution [28]. Here, E_e is determined by the effect of changes in one-electron occupancies from $n_k(0)$ to $n_k(T)$,

$$E_e = \sum_k \varepsilon_k(V)\big(n_k(T) - n_k(0)\big), \tag{1.29}$$

where the quantities ε_k denote the zero-temperature eigenvalues of Eq. 1.12. The function $n_k(T)$ is given by the Fermi–Dirac distribution function

$$n_k(T) = \left\{ \left[\exp\frac{\varepsilon_k - \mu}{k_B T} \right] + 1 \right\}^{-1}, \tag{1.30}$$

where μ is the chemical potential. The expression for p_e is

$$p_e = -\sum_k \frac{\partial \varepsilon_k}{\partial V}(n_k(T) - n_k(0)). \tag{1.31}$$

The validity of this model has been checked by McMahan and Ross [28], who find that this has excellent agreement with finite-temperature APW calculations for metallic iodine up to $T/T_f < 0.2$ [these are carried out iteratively to self-consistency using a prefactor $n_k(T)$ in Eq. 1.11 for E_{ke} in band structure calculations].

Yet another method has been used by Kerley and his co-workers [14], in which they obtain E_e and p_e as the difference between finite-temperature and zero-degree Thomas–Fermi–Dirac (TFD) theories or INFERNO, an atom-in-the-jellium model [29]. For a critique of this approach, see Bennett and Liberman [30]. The latter method has been shown to be superior as the atomic shell structure effects are automatically included in it.

1.2.2. Liquid State Theories

The liquid state methods used in EOS work are based on either perturbation approaches or computer simulations. In perturbation theories, the basic element is to expand the Helmholz free energy of a liquid about that of a reference system of known free energy, F_0:

$$F = F_0 + \{U\}_0 + \Delta F. \tag{1.32}$$

Here, U is the potential energy of the real system and the subscript 0 denotes the average of U over all reference system configurations. The quantity ΔF contains all the remaining contributions. A variational method is employed so that

$$F \leq F_0 + \{U\}_0. \tag{1.33}$$

This is the well-known Gibbs–Bogolubov inequality, which means that F is bounded. The reference systems in use include hard spheres, soft spheres, one-component plasmas, etc. The model based on hard spheres, which uses a local pseudopotential including the temperature dependence of the electron gas, is called liquid metal perturbation theory (LMPT) [31]. The one which uses the zero-degree isotherm of the solid to define an expression for the energy of a molecule in the field of its neighbors in the liquid is called the corrected rigid ion sphere model (CRIS) [32]. In both, the variational minimum is obtained by varying the hard sphere diameter. For detailed expressions, reference should be made to original papers. Both models give good agreement with experiments [33].

The computer simulation methods, Monte Carlo (MC), and molecular dynamics (MD), although used for quite some time at normal pressure, have begun to be applied at high pressures only very recently. Mention may be made of the MD calculations of melting curves of Mo by Moriarty [34], in which the interatomic potentials derived from first-principle generalized pseudopotentials were used to specify the forces for solution of Newton's equations of motion. Figure 1.3 gives a comparison of the MD and LMPT results. Another promising approach is the ab initio molecular dynamics simulations, in which the forces are directly evaluated by LDA band structure methods within the codes [35], and thus automatically include many-body interactions. Until now, the applications are very few due to excessive computational time requirements. A very recent calculation is that for the melting point of Si at normal volume [36].

1.2.3. Pressure and Temperature Ionization Effects

In the previous section, we mentioned the division of electrons in a material into core and valence states. The core electrons are bound and localized near their nuclei and the valence electrons move around and form bands. Under pressure, because of the reduction in atomic sphere radius, the wave functions of core electrons on adjacent atoms overlap, leading to a resonance broadening of atomic states into energy bands. This is called pressure ionization. Under a rise of temperature, such as in shocks, the energy gap between core and conduction levels may become $\sim k_B T$, and promotion of core electrons to the conduction band can take place. This is referred to as temperature ionization. In Al, for example, pressure ionization of the L shell occurs at 10-fold compression along the 0 K isotherm [37], whereas it begins around $V/V_0 \sim 0.3$ and $T = 10$ eV [17] along the Hugoniot. This transfer of electrons into conduction bands alters the effective ionic charge and the core radius, thus altering the ion–electron potential and the screening.

For taking into account pressure ionization, the calculations along the 0 K isotherm are straightforward and involve the redefining of the core in the band structure computations. An approximate value of the density at which this has to be done can be estimated from the

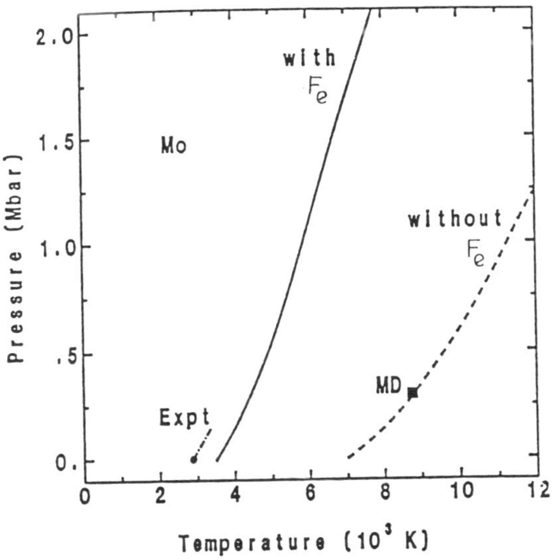

Figure 1.3. Comparison of melting curves of Mo as obtained from MD simulation and from free energy, with and without the E_e contribution, against experiment [34].

formulas given by Sikka and Godwal [38] and by More [39]. For the temperature ionization at a given volume, there are many approximate theories. These include many modified Saha equations and various finite-temperature statistical Thomas–Fermi models. In particular, the screened-Coulomb modified Saha equation (CSCP-IEEOS) of Rouse [40] has been applied by the present authors [17] to evaluate some Hugoniot points of Al and Mo, after verifying that this correctly predicts the valence charges of a number of elements at normal conditions and also the valence 2 to 3 transitions in Eu and Yb in static high-pressure experiments. The TFD theories also become reasonably accurate at these high densities [37] (see Section 1.3.10).

1.3. Some Results

Ab initio equation-of-state calculations have now been performed for scores of substances. Here, we describe a few representative examples.

1.3.1. Free-Electron Al

Al is a prototype material for EOS studies. Its EOS has been investigated over a very wide range of densities and temperatures both experimentally and theoretically. Its 300 K isotherm has been determined up to 2.2 Mbar in static pressure measurements [41] and the shock Hugoniot has been determined to 4500 Mbar [6]. The experimental and theoretical Hugoniots are compared in Fig. 1.4. The agreement is excellent at lower pressures and within the error bars of experiments at ultra-high pressures.

It is instructive to follow the condensed-matter regimes encountered along the shock Hugoniot of Al [8]. Near $p = 1.2$ Mbar, Al has a solid–liquid transition. Above 20 Mbar, the thermal pressure is greater than the cold pressure. Around 80 Mbar, the temperature rises above T_f and the electron gas ceases to be degenerate. Above 600 Mbar, the electrons and ions are weakly coupled, and near 5000 Mbar, Al becomes a fully ionized nearly ideal gas of density ratio 4. Near this density maximum on the Hugoniot, some theories predict oscillations in pressure–volume curves (Fig. 1.5), which are attributed to quantum effects of electron shell structure. The recent impedance mismatch experiments using nuclear explosions confirm this prediction [42].

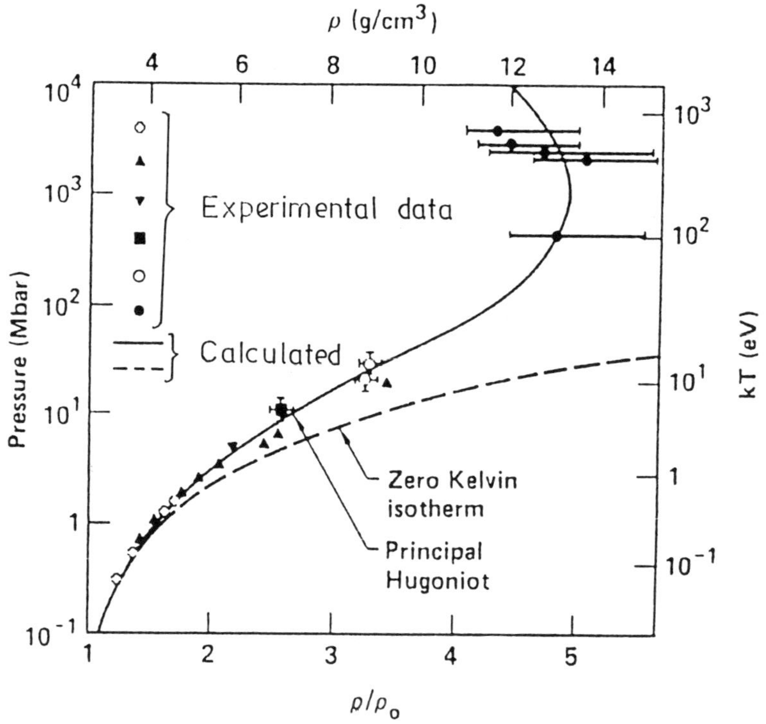

Figure 1.4. Experimental and theoretical Hugoniots of Al [8].

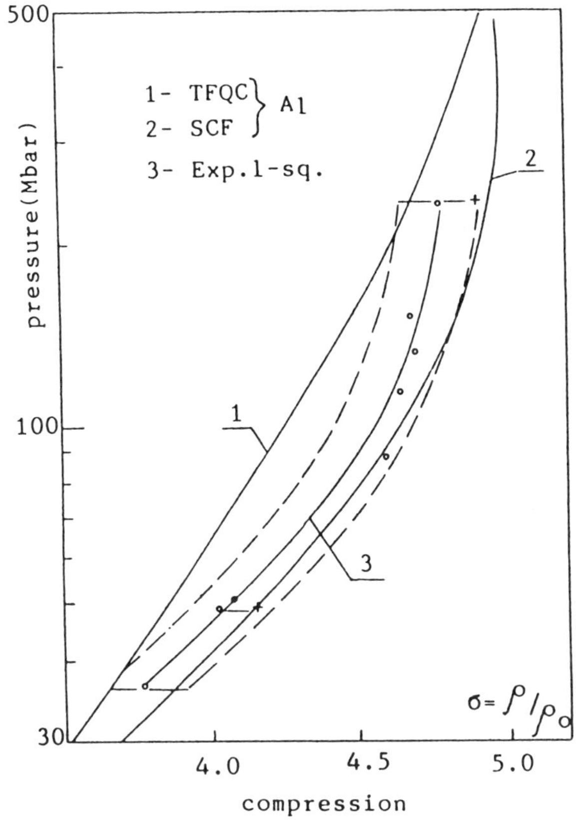

Figure 1.5. Comparison of experimental and theoretical Hugoniots of Al for higher compressions [42].

1.3.2. Discontinuities in Ti, Zr, and Hf Hugoniots

The phase diagrams of group IV B elements Ti, Zr, and Hf have been of considerable experimental and theoretical interest recently. In shock experiments, McQueen et al. [43] discovered discontinuities in their plots of shock velocity (U_s) versus particle velocity (u_p) at 17, 26, and 40 GPa, respectively. There were many speculations regarding these discontinuities. McQueen et al. assumed that these were due to an α (hcp) to β (bcc) transition, based on the observation of the β phase in the shock recovered samples of Ti. Kutsar and German [44] associated these with the $\alpha \rightarrow \omega$ (ω, a three-atom simple hexagonal structure), as they found the ω phase in shock recovered samples of Ti and Zr. Carter [45] speculated that these could be due to some electronic transitions.

In view of this disagreement, Gyanchandani et al. [46–48] resorted to total energy analysis of phase stability in these elements by the ASA–LMTO method. In Zr, they correctly found the α structure to be the most stable one at normal volume. Under compression, the first transition was found to be α → ω at 5 GPa (experimental pressure ranges 2–6 GPa [49]). In addition, a new ω → β structural transition at higher pressure (Fig. 1.6) was predicted. This was subsequently discovered in static pressure experiments at 30 GPa [50]. Thus, the cause of shock discontinuity in Zr could be attributed to this new ω → β transition.

In Ti, the calculations [47] found no ω → β transition up to very high pressures and none was detected in subsequent static pressure experiments up to 87 GPa [51]. This led Gyanchandani et al. to agree with Al'tshuler et al. [52] that the initial segment of the U_s–u_p plot of Ti is not well established. This was confirmed by later shock experiments [53]. The presence of the β phase in recovered samples of Ti was ascribed to heterogeneous heating effects.

For Hf, the calculations [48] found a direct α → β transition. However, an α → ω → β sequence was observed in experiments done simultaneously [51]. This disagreement of LMTO calculations was attributed to the Madelung correction procedure and adjustment of this predicted the α → ω phase change at 36 GPa and the ω → β transition at 55 GPa, in line with experiments. The transition pressure

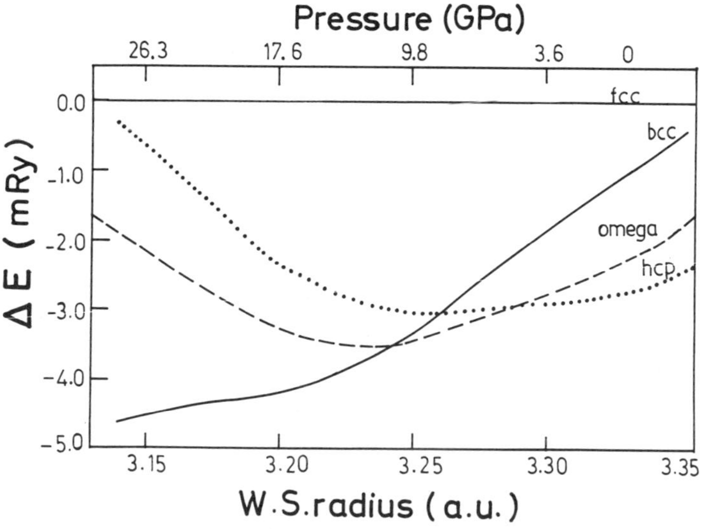

Figure 1.6. Total energy differences between Zr structures [46].

of the $\alpha \rightarrow \omega$ phase change is close to the shock discontinuity in this material.

The occurrence of the β phase in Zr and Hf also confirms the correlation between crystal structures and the *d*-electron population in the transition metal series [54]. Under compression, the d occupancy increases in these elements and they adopt the structure of the next elements. The departure of Ti from this trend can be understood in that the hard-core repulsion (which sets in earlier in Ti) does not favor the bcc structure.

1.3.3. On the 2.1 Mbar Solid–Solid Transition in Shocked Mo

This transition was seen as a discontinuity in the acoustic sound velocity at 2.1 Mbar and 4100 K in shock experiments. Together with ASA–LMTO calculations and Andersen's force theorem, this transition was concluded to be a bcc → hcp transition [55]. Subsequent total energy calculations and static pressure investigations summarized in Table 1.1 have not confirmed this. It is possible that the 2.1 Mbar shock anomaly in Mo is some other transition. Phase diagram calculations being done by Moriarty [34] using MD simulations may resolve this issue. High-temperature compression experiments in laser heated DACs may also help.

1.3.4. 5f Band Occupation in Compressed Th

Thorium crystallizes in the fcc structure at ambient conditions, and is a spd metal with an unoccupied 5f band. However, due to the broadening of this band on compression, it has a downward movement with respect to the spd bands and is expected to be populated at high pressure. Using energy dispersive x-ray diffraction with a

Table 1.1. bcc → hcp transition in shocked Mo

	p, Mbar	V / V_0
Shock data [55]	2.1	
Static data [1]	>5.8	<0.58
Hixon et al. [55]	3.2	0.62
Moriarty [56]	4.2	0.58
Sikka et al. [57]	>4.9	<0.57
Soderlind et al. [58]	5.2	0.55

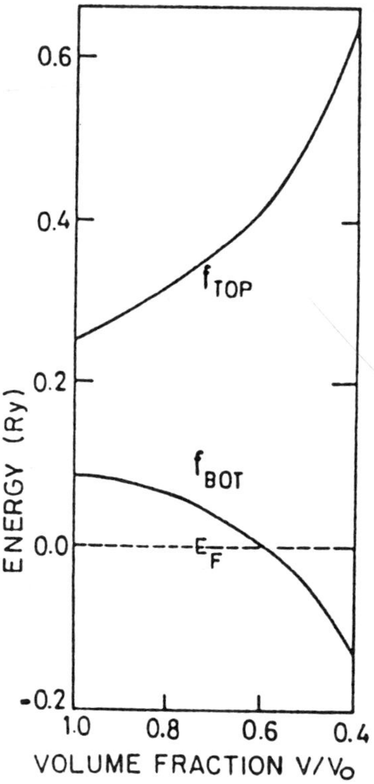

Figure 1.7. Position of 5f band relative to the Fermi energy in Th vs. V/V_0.

diamond-anvil cell, Vohra and Akella [59] generated its 300 K isotherm up to 300 GPa and found that it undergoes a fcc to bct structural transition at 80 GPa and $V/V_0 = 0.6$. Eriksson et al. [60] and Rao et al. [61] have done total energy calculations by full potential and ASA–LMTO methods, respectively, and have confirmed that this transition occurs when the bottom of the 5f band falls below the Fermi level (Fig. 1.7). The computed theoretical isotherm is compared with experimental data in Fig. 1.8. The agreement is excellent. Again, the inclusion of 5f electrons was crucial for this agreement. This can be seen from the $p-V$ curve based on a spd calculation. The shock wave data on Th has been analyzed by Gupta et al. [62] and has been again found to be consistent with spd→f electronic transfer under pressure.

1.3.5. Some Melting Curves

In general, a melting curve is established by equating solid and liquid Gibbs free energies as a function of pressure and temperature. In practice, however, it is often a difficult task because it requires finding the intersection of two nearly parallel curves. Not withstanding this difficulty, some of these calculations by first-principle methods have been attempted recently.

Figure 1.9 shows the $p-T$ phase diagram of Al, as evaluated by Moriarty et al. [63] by two theoretical treatments of cold and thermal contributions. Their prediction of melting on the shock Hugoniot, beginning at 1.2 Mbar and ending at 1.55 Mbar, is in excellent agreement with values of 1.25 and 1.5 Mbar, respectively, given by Shaner et al. [64] on the basis of sound velocity measurements.

Figure 1.10 illustrates the situation for Pb. Here, the measurements of melting temperature as a function of pressure and theoretical calculations using AP and LMTO methods and the CRIS model for liquid have been done by Godwal et al. [65]. The quantum mechanical calculations clearly point to the reliability of the laser heated DAC measurements. The predicted melting upon shock compression at 53 GPa is also in complete accord with Hugoniot sound velocity experiments [66].

Very recently, the full $p-T$ phase diagram of Mg, including the solid–solid phase transitions and the melting curve, has been calcu-

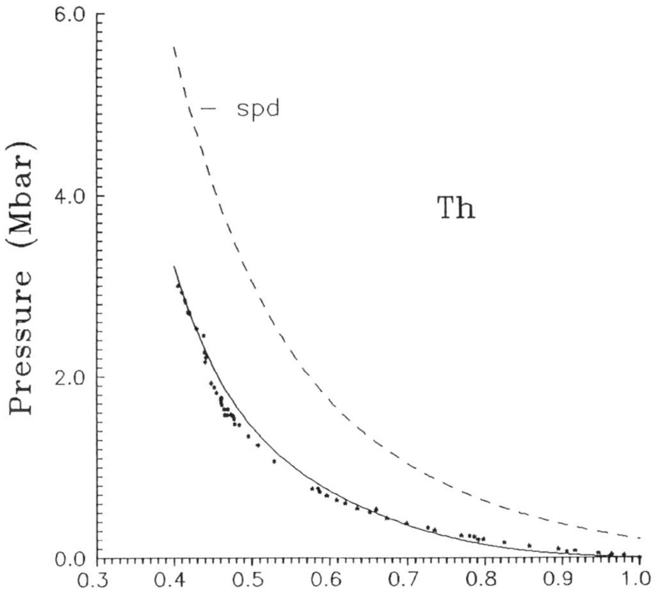

Figure 1.8. The 0 K isotherm of thorium compared to the experimental data. The curve marked spd is for the case when the f-electron contribution is supressed in the calculations [61].

lated by Moriarty and Althoff [67] up to 60 GPa and 3500 K. This is shown in Fig. 1.11. Here, the solid has been treated in quasi-harmonic approximation using GPT interatomic potentials and the liquid with the variational perturbation theory with a soft sphere reference system. The agreement of the calculated phase diagram is good with the meager experimental data available. It may be noted that the hcp→bcc transition at 50 GPa was predicted by LMTO total energy calculations prior to the experimental work [68].

It is worth recalling that sometimes semiempirical models of identifying melting have also been used in conjugation with first-principle theories. Two of these are the Lindemann law of melting [65] and the rule of constant packing fraction (~0.45) [69].

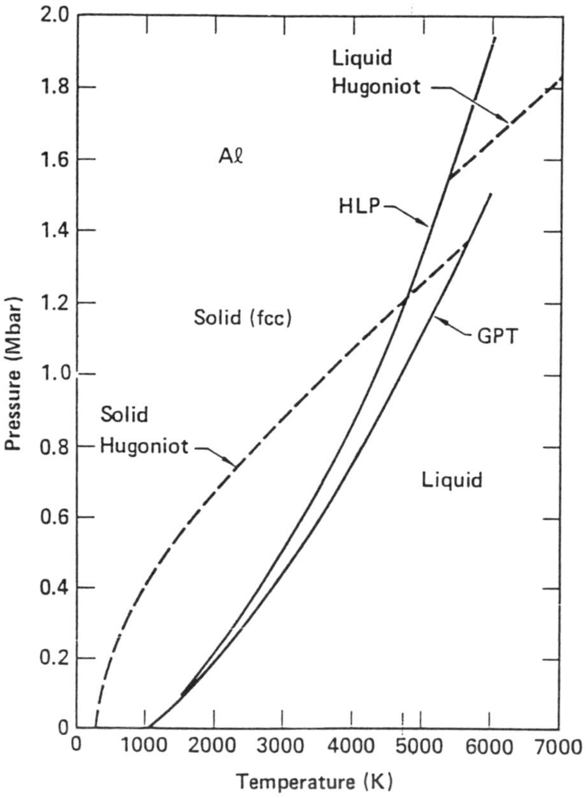

Figure 1.9. Calculated pressure and temperature along the melting line and Hugoniot curves of Al [63].

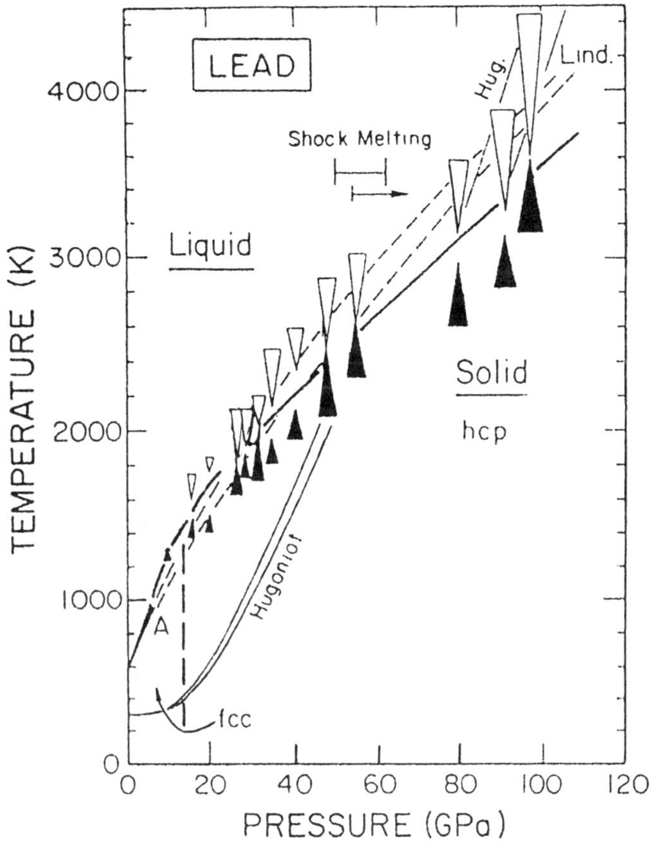

Figure 1.10. High-pressure melting curve of Pb [65]. Triangles bracket the experimental melting temperature (bold solid line). The two dotted curves are from the pseudopotential (higher curve) and LMTO calculations. Hugoniot curves are thin solid lines.

1.3.6. Crystal-to-Amorphous Transition in Quartz

The α-quartz form of SiO_2 exits at room temperature and at pressures lower than 7 GPa. At higher pressures, it persists as a metastable structure, undergoes a transition to another metastable structure before making a transition to an amorphous phase around 20 GPa (see Sharma and Sikka [70] and references therein for more details). The shock experiments also show a break in the Hugoniot around 15–20 GPa and the recovered samples contain amorphous material. There has been some controversy regarding the nature of the transition in shock waves. Some authors interpret this as an α-quartz → stishovite phase change.

Many first-principle band structure calculations and molecular dynamics simulations based on force fields derived from quantum mechanical studies have now been done to understand the nature (driving mechanism, change of bonding, coordination, etc.) of this crystal-to-amorphous transition. Chelikowsky et al. [71] have employed the pseudopotential method and Pomponio and Continenza [72] used the full-potential LAPW technique to accurately reproduce the pressure-induced variations of V/V_0, Si–O–Si and O–Si–O angles, and the interpolyhedral O–O distance in α-quartz. The latter distance has a value of 2.7 Å at 15 GPa, the shortest distance known in silicates. This is regarded as the cause of the instability of the α-quartz structure because the compression beyond will be very costly in energy [73]. The calculated $p-V/V_0$ curve from MD simulations [74] is compared with experimental data in Fig. 1.12. The agreement with the data in the crystalline phase is excellent [75]. At higher pressures, the $p-V$ behavior of the simulated amorphous phase is very close to that of the shocked α-quartz [76]. The figure also shows that the unloading path in the calculations is much steeper, which is again in conformity with release wave measurements in the shock experiments on x-cut quartz [77].

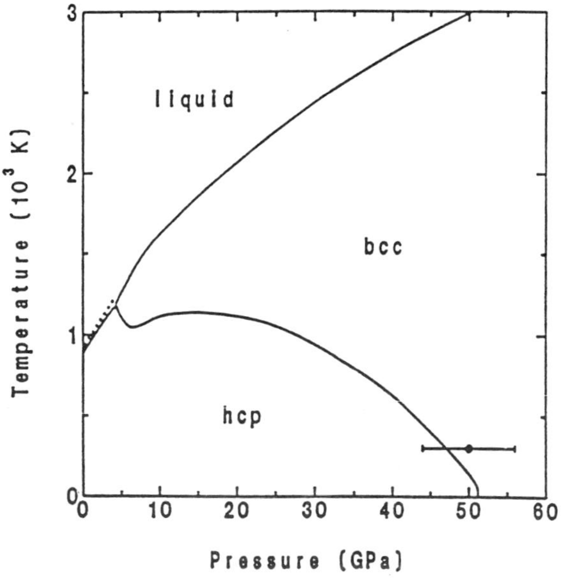

Figure 1.11. Calculated p–T phase diagram of Mg compared with experimental data [67].

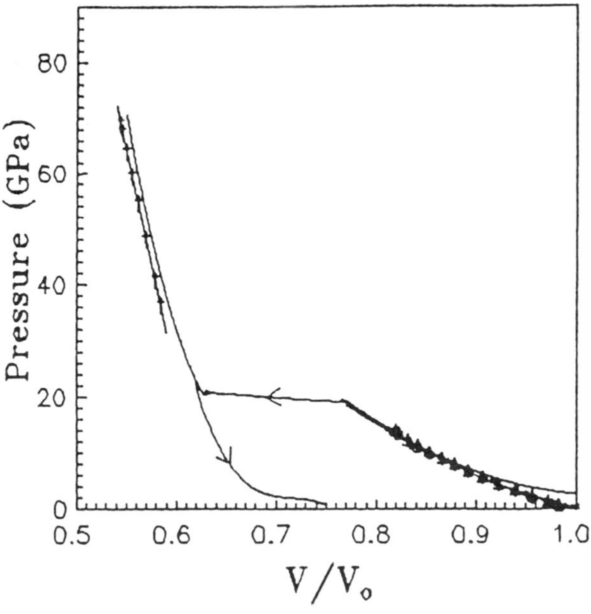

Figure 1.12. V/V_0 vs. pressure for quartz SiO_2 [74]. The experimental data (Δ) for α-quartz are from Ref. 75 and the shock data (↑) are from Ref. 76.

1.3.7. Metallization of CsI

CsI has been an archetypical material for metallization studies in both static and shock high-pressure investigations. Self-consistent nonrelativistic APW calculations on CsI [78] suggested that the insulator-to-metal transition would occur in this compound in the pressure range 100 ± 10 GPa. This is in accord with the optical absorption measurements [79] done later. A calculation of the shock Hugoniot, in which the contribution of the thermally excited electrons across the band gap was accounted for by a Boltzman-like expression, was also in good agreement with the experimental data. However, the 300 K $p-V$ data coincided with the Hugoniot curve. This implied that the Grüneisen parameter γ is effectively zero along the Hugoniot, a conclusion inconsistent with conventional understanding. This discrepancy between static and shock measurements has recently been resolved by Mao et al. [80]. They discovered that CsI undergoes a continuous transition from B2 to an hcp-like phase, stable to at least 302 GPa. This intermediate orthorhombic phase was found to be different from the previous structural assignments and about 10% denser. With this new static isotherm, EOS results of static and shock experiments could be reconciled with theoretical calculations (see Fig. 1.13).

1.3.8. Metallic Hydrogen

Theoretical studies have been carried out on dense solid hydrogen for over a half a century [81]. These predict that, under pressure, the insulating molecular phase will undergo a band overlap transition to a molecular metallic phase before dissociating to form a monoatomic-metallic solid. The metallic phase of hydrogen has importance in the evolution and composition of heavy planets and is expected to be a high-temperature superconductor. However, there is a wide variation

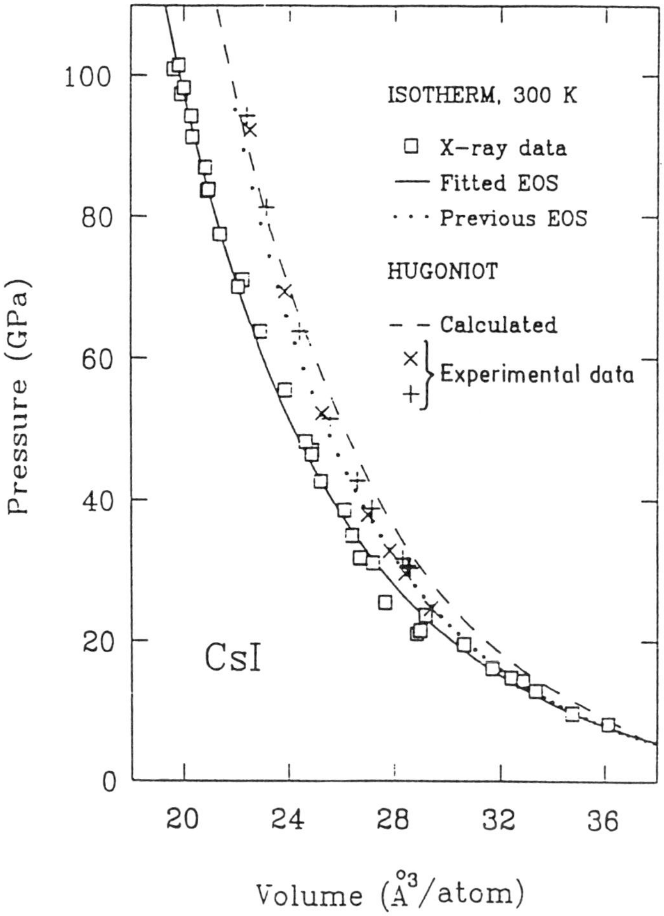

Figure 1.13. Pressure–volume relation for CsI [80].

in the predicted pressures for these transitions, depending upon the structure, molecular orientation, and approximations in the theory employed. For example, LDA calculations of a *c*-axis ordered hcp phase predict band gap closure at 40 GPa [82], whereas quasi-particle GW calculations [83] give the value 151 GPa. Lowering the symmetry increases this band gap at 150 GPa [84]. Similarly, for the upper transition, the predictions range from 2.5 to 6 Mbar [85].

The current experimental situation can be summarized as follows. Hydrogen, in the molecular form (H_2), solidifies at 14 K in the hcp structure ($\rho/\rho_0 = 1$). The molecular structure has been shown to be stable up to 42 GPa ($\rho/\rho_0 = 5.8$) by x-ray diffraction [86]. Raman spectra [87] suggest that this hexagonal order persists up to 150 GPa ($\rho/\rho_0 = 9$). At this pressure, the vibron frequency shows a discontinuity. The nature of this new phase is controversial, although its pressure is in the range of predictions for insulator-to-metal molecular phase transition. Further compression to 250 GPa led to the disappearance of Raman spectra and the optical properties changed drastically [88]. However, because of the small sample size at these pressures, there are large uncertainties in the experiments.

The static $p-V$ equation of state up to 42 GPa is shown in Fig. 1.14. It has provided important data for determination of intermolecular potentials and helped discriminate among some theoretical models.

1.3.9. Validation of Empirical EOS Forms

Static Pressure Data

A wide variety of different expressions are employed for fitting of experimental EOS data. These are mostly two-parameter equations of state. The parameters are usually the ambient-pressure volume (V_0), the bulk modulus (K_0), and its first pressure derivative (K_0'). In this section, we will examine the question which of these forms is physically more reasonable.

We consider here the equations proposed by Birch [89], Rose et al. [90], and Holzapfel [91]. The Birch (B–M) EOS is based on Eulerian finite strain formalism and is given by

$$p = 1.5\ K_0 X^{-7}(1 - X^2)\{1 + \tfrac{3}{4}(K_0' - 4)(X^{-2} - 1) + \cdots\}, \qquad (1.34)$$

when truncated at the third order of energy in strain. Here, $X = (V/V_0)^{1/3}$ is the linear compression. The universal EOS (UEOS) of Rose et al. is

$$p = 3K_0 X^{-2}(1 - X)\exp[\tfrac{3}{2}(K_0' - 1)(1 - X)]. \qquad (1.35)$$

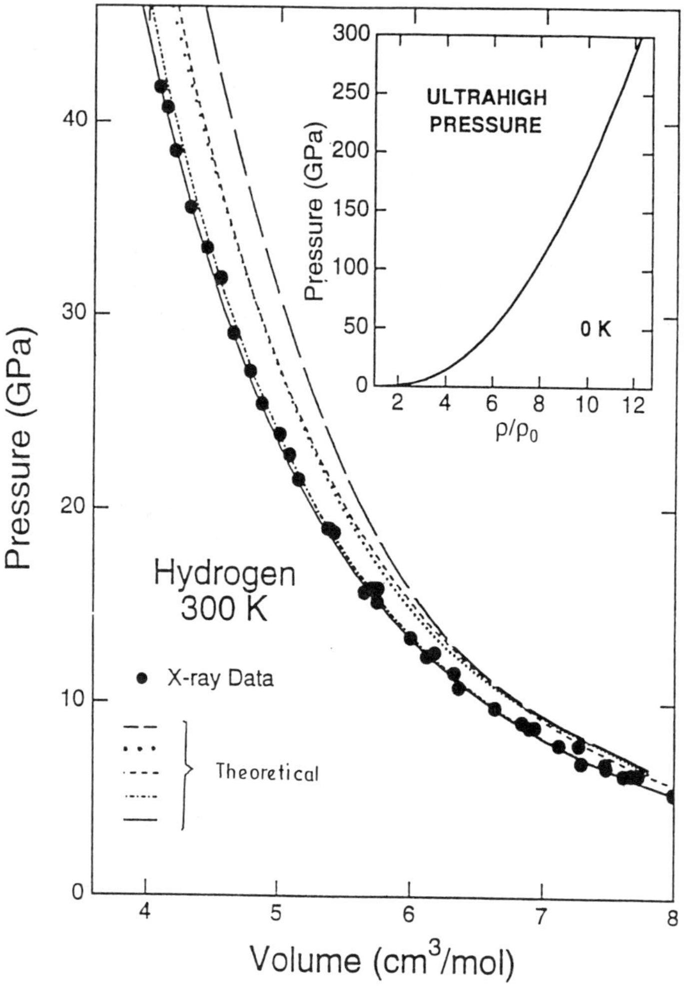

Figure 1.14. Pressure–volume relation for solid hydrogen up to 42 GPa [86]. Inset is the extrapolation near 300 GPa using UEOS form parameters determined up to 42 GPa.

This is based on the Taylor expansion of the energy of the condensed system as a function of normalized interparticle distance (a^*)

$$E^*(a^*) = -(1 + a^* + \cdots)\exp(-a^*). \tag{1.36}$$

Holzapfel has recently proposed the form (H02)

$$p = 3K_0 X^{-5}(1 - X)\exp[c_0(1 - X)], \tag{1.37}$$

which is also a zeroth-order approximation to one which approaches the correct asymptotic Fermi gas limit. The coefficient c_0 is defined in terms of K_0 and the Fermi gas pressure p_{FG0} as

$$c_0 = \ln p_{FG0} - \ln 3K_0, \tag{1.38}$$

with $p_{FG0} = a_{FG}(Z/V)^{5/3}$ and $a_{FG} = 2.3366$ TPa Å^5.

For tests here, we employ the data of 300 K isotherms (EOS1) of Al, Cu, and Pb up to 1 TPa, given by Nellis et al. [92] for use in calibration of diamond cell experiments conducted in the Mbar range of pressure. They generated these by combining shock data with first-principle theory. Figures 1.15 and 1.16 show the comparison for Al and Cu. Ultrasonic values for K_0 and K_0' are employed (see Sikka [93] for details). The B-M form fails above moderate compressions. For Al, both UEOS and H02 fit very well, and for Cu, these have excellent agreement up to $V/V_0 = 0.535$ and a pressure of 4 Mbar. The reason for departure for Cu at higher compressions is d$\rightarrow$s electron transfer as noted by Sikka [93]. Inclusion of higher-order terms is required to take this into account. This kind of deviation of these EOS forms is a general occurrence in presence of electronic transitions [94].

Some comments may also be made regarding the application of the uniform interstitial electron gas model for EOS of metallic elements. It was noted by Moruzzi et al. [95] that the bulk modulus continues to scale as r_s^{-5}, as for free electrons, provided r_s is determined from the electron density in the interstitial region (the volume excluding the muffin-tin spheres). For K_0', a dependence on r_s has also been deduced [96] from its expression in H02,

$$K_0' = A + B \ln\left(\frac{Z}{Z_B}\right), \tag{1.39}$$

where Z_B is the bonding charge determined by the product of the interstitial electron density and the volume per atom. This is shown in Fig. 1.17. A correlation is indicated, in spite of large uncertainties in experimental K_0' values.

Shock Wave Data

It was recognized by Rice et al. [97] that the shock velocity (U_s) and particle velocity (u_p) in shock wave experiments can be related as

$$U_s = C_0 + su_p + s'u_p^2 + \cdots . \tag{1.40}$$

The coefficient s' can be neglected when a substantial phase change (structural or electronic with a large volume change) in a material

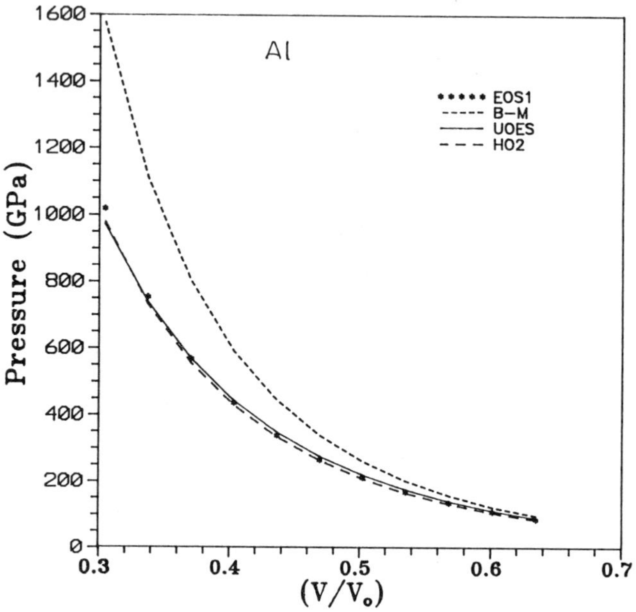

Figure 1.15. The 300 K isotherm of Al from different empirical EOS forms against the tabular data of Nellis et al. [92].

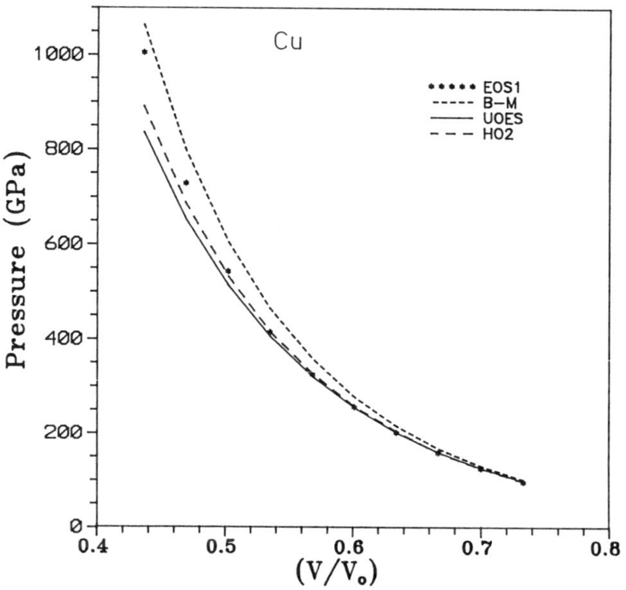

Figure 1.16. The 300 K isotherm of Cu from different empirical EOS forms against the tabular data of Nellis et al. [92].

does not occur. Here, C_0 is the bulk sound speed at ambient pressure and s is related to the repulsive forces (see Fig. 5 of Sikka et al. [98]). These can be expressed in terms of K_0 and K_0' as

$$K_0 = \rho_0 C_0^2, \tag{1.41}$$

$$K_0' = c(4s - 1), \tag{1.42}$$

where $c \cong 1.5$ for elemental solids [99]. For a relationship of C_0, s, and s' to B-M and UEOS, see Jeanloz and Grover [100] and Gupta et al. [62]. Their connection to the uniform interstitial electron gas is the same as that for K_0 and K_0'.

It may be pointed out that the linear $U_s - u_p$ relation holds up to higher pressures in shock wave data than the above isothermal equations in static pressures. Moriarty [56] has noted that for Mo, because of s→d electron transfer, the UEOS, fitted in the linear form

$$\ln H = \ln K_0 + \tfrac{3}{2}(K_0' - 1)(1 - X), \tag{1.43}$$

breaks down along 0 K at 2 Mbar. But in shock waves, the lower-pressure initial C_0 and s parameters fit up to 20 Mbar (Fig. 1.18). The reason for this, probably, is that the rise in temperature in dynamic experiments inhibits the electronic s→d transfer. Similar behavior has been noted by Gupta et al. [62] for Th and U.

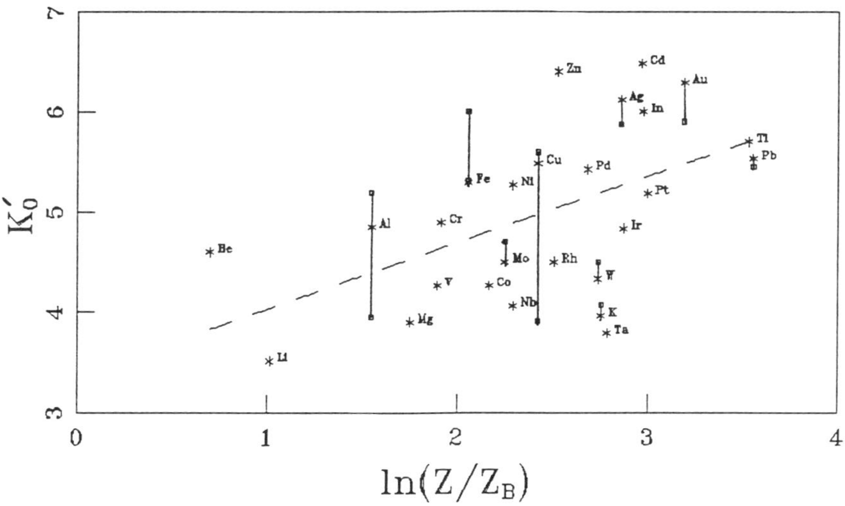

Figure 1.17. A correlation of K_0' to Z/Z_B [96]. The vertical lines show the range in the experimental values of K_0' for some elements. The dashed curve is the best fit.

1.3.10. Tests of Statistical Thomas–Fermi Models

The basic ingradient of these theories is again the density functional formalism. However, they differ from quantum mechanical calculations in the way the kinetic energy (E_{ke}) is treated. Here, one expands the kinetic energy in terms of derivatives of $n(\mathbf{r})$

$$E_{ke}(n) = \int d^3\mathbf{r}\, n(\mathbf{r}) \left[\tfrac{5}{3}(3\pi^2 n)^{5/3} + \tfrac{1}{4}\sigma |\nabla n / n|^2 + \cdots \right]. \quad (1.44)$$

The first term in square brackets is the energy of the free Fermi gas and the second one is the gradient correction which accounts for nonlocal quantum behavior of electrons. The coefficient σ has a value 1 in the Weizsacker derivation [102] and 1/9 in the Kirzhinits formulation [103]. By a variational procedure, one obtains the following basic equation for Thomas–Fermi theories [39, 104].

$$E_f = (3\pi^2 n)^{2/3} - \frac{2}{\pi}(3\pi^2 n)^{1/3} + \frac{1}{4}\sigma\left(\left| \frac{\nabla n}{n} \right|^2 - 2\frac{\nabla^2 n}{n} \right) + V_{xc}(\mathbf{r}) + V(\mathbf{r}), \quad (1.45)$$

where $V(\mathbf{r})$ is the electrostatic potential

$$V(\mathbf{r}) = -2\frac{Z}{\mathbf{r}} + 2\int \frac{n(\mathbf{r})}{|\mathbf{r} - \mathbf{r}'|} d^3\mathbf{r}. \quad (1.46)$$

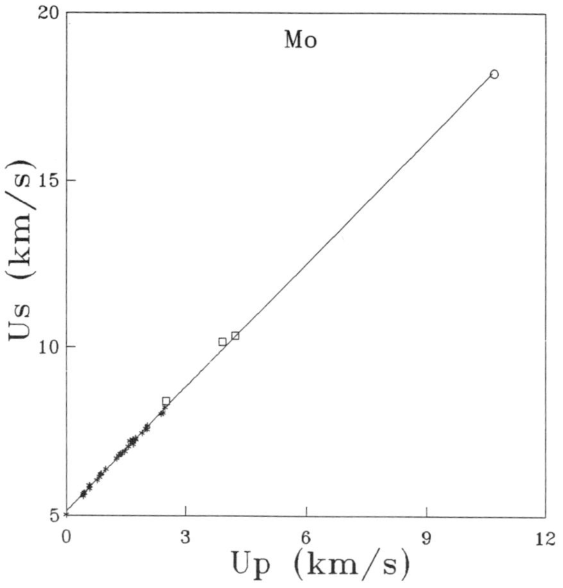

Figure 1.18. U_s–u_p shock data for Mo. Data: * from Ref. 43, □ from Ref. 101, O from Ref. 5. The solid curve is fit to the lower-pressure data from Ref. 43.

The Thomas–Fermi–Dirac theory is obtained by setting $\sigma = 0$ and the TF theory by omitting the exchange-correlation term as well. The model in which Eqs. 1.45 and 1.46 are solved self-consistently is called the quantum-statistical model (QSM) of Kalitkin and Kuzmina [105]. A finite-temperature version was presented by Perrot [106]. A comparison of QSM with APW calculations on Al was done by More [104]. A very extensive work is that of Vehn and Zittel [37], who reported ASW and QSM results at 0 K on Li, He, Be, Al, and K. Their motivation was to clarify some ambiguities regarding the occurrence of discontinuous anomalies due to pressure ionization as reported by Kirzhnits et al. [107], Zink [108], and others. Figure 1.19 shows such a comparison for Al. They find that pressure ionization is continuous, in agreement with the continuity theorem [39]. However, an oscillatory behavior of the cold-pressure curves was obtained due to the pressure-induced electron redistribution from bands of low angular momentum to bands of higher angular momentum (e.g., 3s, 3p → 3d band reordering in Al near $\rho / \rho_0 = 10$ in Fig. 1.19). This is because different partial waves react differently to compression, as they have different spatial extension and ultimately approach the asymptotic distribution for a uniform Fermi gas.

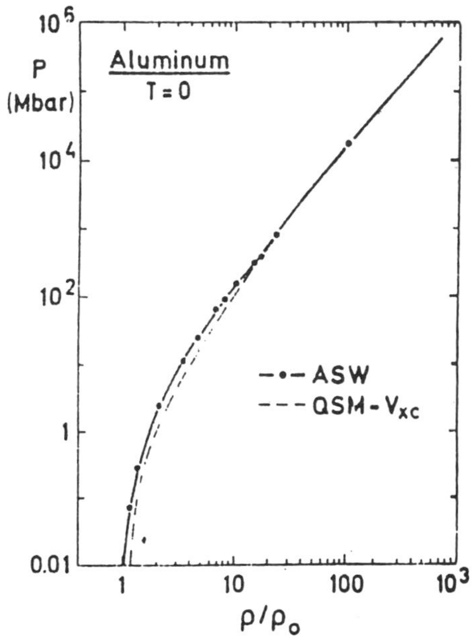

Figure 1.19. Cold pressure of Al as a function of compression: ASW (solid with dots) and QSM (dashed line) [37].

1.4 Summary

The present chapter discusses the current capabilities of the ab initio theories of equations of state. It emerges that these quantum-mchanical calculations, based only on the knowledge of atomic numbers of the elements in a material, are able to accurately reproduce the experimental data and have sometimes yielded reliable predictions and also led to reinterpretation of experiments. The examples described also show how these theoretical techniques have provided understanding of the electronic and structural changes on the equation-of-state surfaces under compression from a microscopic point. It is also now clear that these theories, which earlier were limited to 0 K computations, are able to handle fairly well the temperature-induced effects. This has brought the calculations of the full $p-T$ phase diagrams within reach of first-principle methods.

References

[1] A.L. Ruoff and H. Luo, in *Recent Trends in Recent Trends in High Pressure Research* (ed. A.K. Singh), Oxford, New Delhi, pp. 779–781 (1992).

[2] R. Boehler, *Nature* **363**, pp. 534–536 (1993).

[3] T.J. Ahrens, in *High Pressure Shock Compression of Solids* (eds. J.R. Asay and M. Shahinpoor), Springer-Verlag, New York, pp. 75–114 (1993).

[4] R. Cauble, D.W. Pillion, T.J. Hoover, N.C. Holmes, J.D. Kilkenny, and R.W. Lee, *Phys. Rev. Lett.* **70**, pp. 2102–2105 (1993).

[5] C.E. Ragan III, M.G. Silbert, and B.C. Divon, *J. Appl. Phys.* **48**, pp. 2860–2870 (1977); R.F. Trunin, *Phys. Usp.* **37**, pp. 1123–1145 (1994).

[6] A.S. Vladimirov, N.P. Voloshin, V.A. Nagin, A.V. Petrovtsev, and V.A. Simonenko, *JETP Lett.* **39**, pp. 82–85 (1984).

[7] C.S. Yoo, N.C. Holmes, M. Ross, D.J. Webb, and C. Puke, *Phys. Rev. Lett.* **70**, pp. 3931–3934 (1993).

[8] D.A. Young, J.K. Wolford, F.J. Rogers, and K.S. Holian, *Phys. Lett.* **108A**, pp. 157–160 (1985).

[9] E. Eliezer and R.A. Ricci, *High-Pressure Equations of State: Theory and Applications,* North-Holland, Amsterdam, (1991).

[10] B.K. Godwal, S.K. Sikka, and R. Chidambaram, *Phys. Rep.* **102**, pp. 121–197 (1983).

[11] A.V. Bushman and V.E. Fortov, *Sov. Phys. Usp.* **26**, pp. 465–496 (1983).

[12] M. Ross, *Rept. Prog. Phys.* **48**, pp. 1–52 (1985).

[13] J.R. Asay and G.I. Kerley, *Int. J. Impact Engng.* **5**, pp. 69–99 (1987).

[14] G.I. Kerley, technical report SAND88-2291, Sandia National Laboratories, Albuquerque, NM (1991).

[15] M. Ross and D.A. Young, *Ann. Rev. Phys. Chem.* **44**, pp. 61–87 (1993).

[16] K.S. Holian, technical report LA-10160 MS, Los Alamos National Laboratory, Los Alamos, NM (1984).

[17] B.K. Godwal, S.K. Sikka, and R. Chidambaram, *Phys. Rev.* **B20**, pp. 2362–2365 (1979); *Phys. Rev. Lett.* **47**, pp. 1144–1147 (1981).

[18] P. Hohenberg and W.Kohn, *Phys. Rev.* **B136**, pp. 864–871 (1964); W. Kohn and L.J. Sham, *Phys. Rev.* **A140**, pp. 1133–1138 (1965).

[19] S.C. Gupta, J.M. Daswani, S.K. Sikka, and R. Chidambaram, *Curr. Sci. (India)* **65**, pp. 399–406 (1993).

[20] M.S. Somayazulu, S.M. Sharma, and S.K. Sikka, *Phys. Rev. Lett.* **73**, pp. 98–101 (1994).

[21] M. Born and K. Huang, *Dynamical Theory Of Crystal Lattices*, Oxford Univ. Press, London, (1954).

[22] S. Wei and M.Y. Chou, *Phys. Rev. Lett.* **69**, pp. 2799–2802 (1992).

[23] S.Yu. Savrasov, *Phys. Rev. Lett.* **69**, pp. 2819–2822 (1992).

[24] J.C. Slater, *Introduction to Chemical Physics,* McGraw-Hill, New York (1939), Chapter XIV.

[25] J.S. Dugdale and D.K.C. McDonald, *Phys. Rev.* **89**, pp. 832–834 (1953).

[26] Y.Ya. Vashchenko and V.N. Zubarev, *Sov. Phys. Solid State* **5**, pp. 653–655 (1963).

[27] L.V. Al'tshuler, S.B. Kormer, A.A. Bakanova, and R.F. Trunin, *Sov. Phys. JETP* **11**, pp. 573–579 (1960).

[28] A.K. McMahan and M. Ross, *Phys. Rev.* **B15**, pp. 718–725 (1977).

[29] D.A. Liberman, *Phys. Rev.* **B20**, pp. 4981–4989 (1979).

[30] B.I. Bennett and D.A. Liberman, in *Shock Compression of Condensed Matter—1991* (eds. S.C. Schmidt, R.D. Dick, J.W. Forbes, and D.G. Tasker), Elsevier, New York, pp. 49–52 (1992).

[31] M. Ross, *Phys. Rev.* **B21**, pp. 3140–3151 (1980).

[32] G.I. Kerley, *J. Chem. Phys.* **73**, pp. 487–494 (1980).

[33] B.K. Godwal, S.K. Sikka, and R. Chidambaram, *J. Phys.* **29,** pp. 93–101 (1987).

[34] J.A. Moriarty, in *High Pressure Science and Technology—1993* (eds. S.C. Schmidt, J.W. Shaner, G.A. Samara, and M. Ross), American Institute of Physics, NewYork, pp. 233–236 (1994).

[35] R. Car and M. Parrinello, *Phys. Rev. Lett.* **55**, pp. 2471–2474 (1985).

[36] O. Sugino and R. Car, *Phys. Rev. Lett.* **74**, pp. 1823–1826 (1995).

[37] J. Meyer-ter-Vehn and W.Zittel, *Phys. Rev.* **B37**, pp. 8674–8688 (1986).

[38] S.K. Sikka and B.K. Godwal, *Phys. Rev.* **B35**, pp. 1446–1447 (1987).

[39] R.M. More, *Adv. Atomic Molec. Phys.* **21**, pp. 305–356 (1985).

[40] C.A. Rouse, *Prog. High Temp. Phys. Chem.* **4**, pp. 139–191 (1971).

[41] R.G. Greene, H. Luo, and A.L. Ruoff, *Phys. Rev. Lett.* **73**, pp. 2075–2078 (1994).

[42] V.A. Simonenko, in *Shock Compression of Condensed Matter—1991* (eds. S.C. Schmidt, R.D. Dick, J.W. Forbes, and D.G. Tasker), Elsevier, NewYork, pp. 41–47 (1992).

[43] R.G. McQueen, S.P. Marsh, J.W. Taylor, J.N. Fritz, and W. Carter, in *High Velocity Impact Phenomena* (ed. R. Kinslow), Academic Press, New York, pp. 293–417 (1971).

[44] A.R. Kutsar and V.N. German, in *Proceedings of the 3rd Int. Conf. on Ti*, Moscow (1976).

[45] W.J. Carter, in *Metall. Effects of High Strain Rate* (eds. R.W. Rohde, B.M. Butcher, R. M. Holland, and C.H. Karnes), Plenum, New York, pp. 171–184 (1973).

[46] J.S. Gyanchadani, S.C. Gupta, S.K. Sikka, and R. Chidambaram, *High Press. Res.* **4**, pp. 472–474 (1990).

[47] J.S. Gyanchandani, S.C. Gupta, S.K. Sikka, and R. Chidambaram, *J. Phys. Condens. Matter* **2**, pp. 301–305 (1990).

[48] J.S. Gyanchandani, S.C. Gupta, S.K. Sikka, and R. Chidambaram, *J. Phys. Condens. Matter* **2**, pp. 6457–6459 (1990).

[49] S.K. Sikka, Y.K. Vohra, and R. Chidambaram, *Prog. Mater. Sci.* **27**, pp. 245–310 (1982).

[50] H. Xia, S.J. Duclos, A.L. Ruoff, and Y.K. Vohra, *Phys. Rev. Lett.* **64**, pp. 204–207 (1990).

[51] H. Xia, G. Parthsarthy, H. Luo, Y.K. Vohra, and A.L. Ruoff, *Phys. Rev.* **B42**, pp. 6736–6738 (1990).

[52] L.V. Al'tshuler, A.A. Bakanova, I.P. Dudoladov, E.A. Dynin, R.F. Trunin, and B.S. Chekin, *Zh. Prikl. Mekh. Tekh. Fiz.* **2**, pp. 3–34 (1981).

[53] G.T. Gray III, C.E. Morris, and A.C. Lawson, in *Proc. 7th. Inter. Conf. on Ti*, TMS, Pittsburgh (1993).

[54] D.J. Pettifor, *J. Phys.* **C3**, pp. 367–377 (1970).

[55] R.S. Hixson, D.A. Boness, J.W. Shaner, and J.A. Moriarty, *Phys. Rev. Lett.* **62**, pp. 637–640 (1989).

[56] J.A. Moriarty, *Phys. Rev.* **B45**, pp. 2004–2014 (1992).

[57] S.K. Sikka, B.K. Godwal, and R.S. Rao, *High Press. Res.* **10**, pp. 707–709 (1992).

[58] P. Soderlind, R. Ahuja, O. Eriksson, B. Johansson, and J.M. Willis, *Phys. Rev.* **B49**, pp. 9365–9371 (1994).

[59] Y.K. Vohra and J. Akella, *Phys. Rev. Lett.* **67**, pp. 3563–3566 (1991).

[60] O. Eriksson, P. Soderlind, and J.M. Willis, *Phys. Rev.* **B45**, pp. 12588–12591 (1992).

[61] R.S. Rao, B.K. Godwal, and S.K. Sikka, *Phys. Rev.* **B46**, pp. 5780–5782 (1992).

[62] S.C. Gupta, N. Suresh, and S.K. Sikka, in *High Pressure Science and Technology–1993* (eds. S.C. Schmidt, J.W. Shaner, G.A. Samara, and M. Ross), American Institute of Physics, New York, pp. 183–185 (1994).

[63] J.A. Moriarty, D.A. Young, and M. Ross, *Phys. Rev.* **B30**, pp. 578–588 (1984).

[64] J.W. Shaner, J.M. Brown, and R.G. McQueen, in *Proc. of IX AIRAPT International Conf.*, North-Holland, New York, p. 137 (1983).

[65] B.K. Godwal, C. Meade, R. Jeanloz, A. Garcia, A.Y. Liu, and M.L. Cohen, *Science* **248**, pp. 462–465 (1990).

[66] D.A. Boness, J.M. Brown, and J.W. Shaner, in *Shock Waves in Condensed Matter—1987* (eds. S.C. Schmidt and N.C. Holmes), Elsevier, New York, pp.115–118 (1988).

[67] J.A. Moriarty and J.D. Althoff, *Phys. Rev.* **B51**, pp. 5609–5615 (1995).

[68] J.A. Moriarty and A.K. McMahan, *Phys. Rev. Lett.* **48**, pp. 809–812 (1982).

[69] N.W. Ashcroft and J. Lekner, *Phys. Rev.* **145**, pp. 83–90 (1966).

[70] S.M. Sharma and S.K. Sikka, *Prog. Mater. Sci.* (in press).

[71] J.R. Chelikowsky, H.E. King, Jr., N. Troullier, J.L. Martins, and J. Glinnemann, *Phys. Rev. Lett.* **65**, pp. 3309–3312 (1990).

[72] A.Di. Pomponio and A. Continenza, *Phys. Rev.* **B48**, pp. 12558–12565 (1993).

[73] S.K. Sikka and S.M. Sharma, *Current Sci.* **63**, pp. 317–320 (1992).

[74] M.S. Somayazulu, S.M. Sharma, S.K. Sikka , N. Garg, and S.L. Chaplot, in *High Pressure Science and Technology—1993* (eds. S.C. Schmidt, J.W. Shaner, G.A. Samara, and M. Ross), American Institute of Physics, New York, pp. 815–818 (1994).

[75] D.B. McWhan, *J. Appl. Phys.* **38**, pp. 347–352 (1967).

[76] J. Wackerle, *J. Appl. Phys.* **33**, pp. 922–937 (1962).

[77] L.C. Chhabildas, in *Shock Waves in Condensed Matter—1985* (ed. Y.M. Gupta), Plenum, New York, pp. 601–605 (1986).

[78] J. Aidun, M.S.T. Bukowinski, and M. Ross, *Phy. Rev.* **B29**, pp. 2611–2622 (1984).

[79] R. Reichlin, M. Ross, S. Martin, and K.A. Goettel, *Phys. Rev. Lett.* **56**, pp. 2858–2860 (1986).

[80] H.K. Mao, Y. Wu, R.J. Hemley, L.C. Chen, J.F. Shu, L.W. Finger, and D.E. Cox, *Phys. Rev.* **64**, pp. 1749–1752 (1990).

[81] E. Wigner and H.B. Huntington, *J. Chem. Phys.* **3**, pp. 764–770 (1935); D.E. Ramaker, L. Kumar, and F.E. Harris, *Phys. Rev. Lett.* **34**, pp. 812–814 (1975).

[82] T.W. Barbee III, A. Garcia, M.L. Cohen, and J.L. Martins, *Phys. Rev. Lett.* **62**, pp. 1150–1153 (1989).

[83] H. Chacham and S.G. Louie, *Phys. Rev. Lett.* **66**, pp. 64–67 (1991).

[84] E. Kaxiras, J. Broughton, and R.J. Hemley, *Phys. Rev. Lett.* **67**, pp. 1138–1141 (1991).

[85] H. Nagara and T. Nakamura, *Phys. Rev. Lett.* **68**, pp. 2468–2471 (1992).

[86] J. Hu, H.K. Mao, J.F. Shu, and R.J. Hemley, in *High Pressure Science and Technology—1993* (eds. S.C. Schmidt, J.W. Shaner, G.A. Samara, and M. Ross), American Institute of Physics, New York, pp. 441–444 (1994); H.K. Mao and R.J. Hemley, *Rev. Mod. Phys.* **66**, pp. 671–692 (1994).

[87] R.J. Hemley and H.K. Mao, *Phys. Rev. Lett.* **61**, pp. 857–860 (1988).

[88] H.K. Mao and R.J. Hemley, *Science* **244**, pp. 1462–1465 (1989).

[89] F. Birch, *J. Geophys. Res.* **83**, pp. 1257–1266 (1978).

[90] J.H. Rose, J.R. Smith, F. Guinea, and J. Ferrante, *Phys. Rev.* **B29**, pp. 2963–2969 (1984).

[91] W.B. Holzapfel, in *Molecular Solids under Pressure* (eds. R. Pucci and G. Piccitto), North-Holland, Amsterdam, pp. 61–88 (1991).

[92] W.J. Nellis, J.A. Moriarty, A.C. Mitchell, M. Ross, R.G. Dandrea, N.W. Ashcroft, N.C. Holmes, and G.R. Gathers, *Phys. Rev. Lett.* **60**, pp. 1414–1417 (1988).

[93] S.K. Sikka, *Phys. Lett.* **135A**, pp. 129–131 (1989).

[94] S.K. Sikka, *Phys. Rev.* **B38**, pp. 8463–8464 (1988).

[95] V.L. Moruzzi, J.F. Janak, and A.R. Williams, in *Calculated Electronic Properties of Metals*, Pergamon, NewYork, (1978).

[96] N. Suresh, S.C. Gupta, and S. K. Sikka, to be published.

[97] M.H. Rice, R.G. McQueen, and J.M. Walsh, *Solid State Phys.* **6**, pp. 40–63 (1958).

[98] S.K. Sikka, B.K. Godwal, and R. Chidambaram, *Bull. Mater. Sci.* **7**, pp. 377–386 (1985).

[99] D.J. Steinberg, *J. Phys. Chem. Solids* **43**, pp. 1175-1182 (1982).

[100] R. Jeanloz and R. Grover, in *Shock Waves in Condensed Matter—1987* (eds. S.C. Schmidt and N.C. Holmes), Plenum, New York, pp. 69–72 (1988).

[101] A.C. Mitchell, W.J. Nellis, J.A. Moriarty, R.A. Heinie, N.C. Holmes, R.E. Tipton, and G.W. Repp, *J. Appl. Phys.* **69**, pp. 2986–2984 (1991).

[102] C.F. von Weizsaker, *Z. Phys.* **96**, p. 431 (1935).

[103] D.A. Kirzhnits, *Sov. Phys. JETP* **5**, pp. 64–71 (1957).

[104] R.M. More, *Phys. Rev.* **A19**, pp. 1234–1246 (1979).

[105] N.N. Kalitkin and L.V. Kuzmina, *Sov. Phys. Solid State* **13**, pp. 1938–1942 (1972).

[106] F. Perrot, *Phys. Rev.* **A20**, pp. 586–594 (1978).

[107] D.A. Kirzhnits, Yu.E. Lozovik, and G.V. Shpatakovskaya, *Sov. Phys. Usp.* **18**, pp. 648–672 (1976).

[108] W. Zink, *Phys. Rev.* **176**, pp. 279–284 (1968).

CHAPTER 2

Molecular Dynamics Analysis of Shock Phenomena

D.H. Robertson, D.W. Brenner, and C.T. White

2.1. Introduction

The effects of shock waves are evident in such common occurrences as the thunderclap following a lightening strike or the craters visible on the Moon [1]. Shock waves can result from or cause various physical and chemical processes. For example, shock waves can cause chemical reactions which subsequently couple with the shock wave to generate an explosive self-sustaining detonation (see, for example, Ref. 2). The ability of chemical reactions to couple with shock waves has direct implications on the strength and sensitivity of explosives and, therefore, their usefulness and safety in industrial and defense applications. Likewise, shock waves can also cause physical changes such as a transition to a high-pressure phase in a material [3]. This polymorphic phase transition can couple with the shock to produce a split shock wave profile [4]. These phase changes might place the material in a state that remains stable upon the release of pressure, such as the transition of graphite to diamond (lonsdaleite) [5]. Shock waves can therefore transform a material into a state that has more desirable properties or economic value and at the same time produce interesting phenomena such as shock wave splitting.

These physical or chemical changes associated with the passage of a shock wave in condensed matter occur on the atomic scale of the material. A more detailed understanding of the atomic-scale behavior of material under these extreme shock-loading conditions could assist in the formulation and production of new stronger, more stress-resistant materials and safer more reliable explosives. The atomic-level details of shock-related phenomena are difficult to probe experimentally and by their nature are unavailable from continuum calculations. However, it is just the time (subpicosecond) and length (subnanometer) scales of the shock wave which make experimental studies difficult that make shock-related phenomena amenable to direct study by molecular dynamics (MD) simulations.

Various MD simulations in Lennard–Jones systems have shown that atomic-scale simulations can reach sufficient sizes to reproduce continuum behavior of nonreactive planar shock waves [1,6,7]. However, to model shock-related phenomena such as split shock waves or detonations, potentials capable of supporting distinct phases with significant volume changes or chemical reactions are necessary. This prevents the use of Lennard–Jones-type potentials or force-field approaches which lack such key ingredients. Additionally, these potentials must permit the rapid computation of simulations containing tens of thousands of atoms so that it can be of a size and duration to allow comparison with continuum theory and its predictions. This precludes semiempirical quantum chemical or tight-binding models that allow for reactivity but are currently too intensive computationally to allow for simulations of tens of thousands of atoms on current computers in a reasonable amount of time.

The atomic-scale simulations presented herein will use empirical bond-order potentials that allow for a chemically reasonable description of reactivity in addition to supporting high-pressure phases of the material. These chemically reactive potentials have allowed the atomic-level investigation of such processes as the initial steps of diamond chemical vapor deposition [8], detonations [9,10], adhesion, friction and tribochemistry [11–13], initial stages of fullerene formation [14], shock-induced phase changes [15], and hypervelocity impacts [16]. In this chapter we focus on the use of these potentials in MD simulations probing the atomic-level details of shock phenomena such as shock wave splitting and chemically sustained shock waves.

2.2. Model and Methods

2.2.1. Empirical Bond-Order (EBO) Model

The model potentials used in these shock simulations are based on the formalism first introduced by Abell [17] and subsequently used by Tersoff [18] to model silicon. We have tailored this formalism to treat chemically reactive diatomic molecules [9,10,15]. Within this formalism, the energy of a collection of N atoms is given by

$$E = \sum_{i}^{N} \sum_{j>i}^{N} \left\{ f_c(r_{ij}) \left[V_R(r_{ij}) - \overline{B}_{ij} V_A(r_{ij}) \right] + V_{vdW}(r_{ij}) \right\},$$

where the molecular bonding portion of this potential consists of an attractive term, V_A, and a repulsive term, V_R, both modeled by exponentials. Additionally, a Lennard–Jones term is added to describe the van der Waals interactions that allow the individual molecules to condense into a stable molecular crystal.

The bond-order term in the above equation, $\bar{B}_{ij} = (B_{ij} + B_{ji}) / 2$, is a many-body function that modifies the attractive term according to the local environment. This empirical parameter models the underlying effects of the valence or bonding electrons. For an isolated diatomic molecule, $\bar{B}_{ij}$ is unity and the combination of V_R and V_A reduces to a generalized Morse potential [19]. For more highly coordinated structures, $\bar{B}_{ij}$ is no longer unity, but decreases with the increasing number and strength of competing bonds that are attached to atoms i and j. This decrease in $\bar{B}_{ij}$ reflects the finite number of valence electrons and orbitals available to participate in bonding. For our diatomic model, $\bar{B}_{ij}$ is tailored to favor a single bond, but $\bar{B}_{ij}$ can also be parameterized to favor tetrahedrally coordinated semiconductors [18] as well as highly coordinated metals [17].

2.2.2. AB Model

The EBO potential has been parameterized to model a series of diatomic molecular solids. The functions and parameters used in these models have been presented elsewhere [9,10,15]. Herein, we will focus on results obtained using the parameters described in Ref. 8 for the energetic material and Ref. 14 for the split shock wave studies. These diatomic molecules are composed of two types of atoms, A and B. This allows for the formation of both homogeneous (A_2 and B_2) and heterogeneous (AB) diatomic molecules. These molecules are parameterized to have reasonable physical and chemical properties such as bond length, bond energies, vibrational frequencies, and barriers to chemical reaction [9]. The parameterization of the nonbonding component of the potential is set so that the diatomic molecules will condense into a molecular solid with reasonable properties such as speed of sound and cohesion energies [9].

The energetics of the parameterization is set so that the homonuclear species (either A_2 or B_2) have bond strengths of 5 eV, whereas the heteronuclear molecules (AB) have a binding energy of 2 eV. This allows for chemical reactions such as $A + AB \rightarrow A_2 + B$ to occur with an exothermicity of 3 eV. The exothermicity of this reaction is similar to the reaction $N + NO \rightarrow N_2 + O$ which is thought to be important in the detonation of NO. Using this parameterization, if the initial molecular solid is composed of AB molecules, then this models an energetic material that can undergo exothermic reactions to form the more stable A_2 or B_2. However, if the initial molecular solid is composed of A_2 or B_2 molecules, then the system can still undergo reactions, but there is no net exothermicity. This parameterization allows shock wave studies to be performed on both energetic and nonenergetic materials, depending on whether the initial material is AB or A_2/B_2, respectively.

The potential energy surface for the collinear $A + AB \rightarrow A_2 + B$ reaction is plotted in the left of Fig. 2.1. The dashed line shows the minimum energy path for this reaction. This reaction has an early barrier that is typical in highly exothermic reactions. The plots to the right in Fig. 2.1 show the potential energy along the possible collinear reaction paths for exothermic energy release and the two non-exothermic exchange reactions. The barrier to reaction for the exothermic reaction (solid line) is 0.11 eV, which is similar to but greater than a similar reaction $F + H_2 \rightarrow FH + H$ [20]. This A_2 model also supports a high-pressure phase under hydrostatic loading similar to that seen in I_2 or Br_2 [15]. Therefore, the flexibility of this EBO potential allows for a reasonable chemical description of energetic or nonenergetic molecular solids that are capable of supporting chemical reactions and the existence of high-pressure phases.

2.2.3. Methods

These simulations are performed in either two (2D) or three (3D) dimensions with the system initially at low temperature to allow the

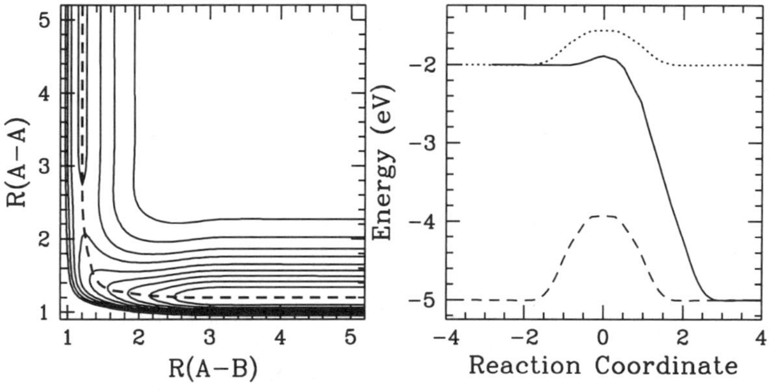

Figure 2.1. *Left:* Potential energy (PE) surface for collinear gas-phase reaction of $A + AB \rightarrow A_2 + B$. Contours are drawn at −4.5, −4.0, −3.5, −3.0, −2.5, −1.95, −1.5, −1.0, −0.5. Dashed line is the reaction's minimum energy path (MEP). *Right:* PE along the MEP for the reaction $A + AB \rightarrow A_2 + B$ (solid), $A + A_2 \rightarrow A_2 + A$ (dashed), and $A + BA \rightarrow AB + A$ (dotted).

formation of a stable molecular crystal. These equilibrated molecular crystals for both 2D and 3D are shown in Fig. 2.2. The material is taken as semi-infinite in the direction of the shock propagation; i.e., material is added in front of the propagating shock wave as necessary. Either periodic or free boundary conditions are enforced in the dimension(s) perpendicular to the direction of shock propagation, depending on the simulation being performed. Shock waves are introduced into the system by either impacting the crystal's free edge with a portion of the material or by driving it with a constant-velocity piston (constant-velocity quadratic potential). Once the boundary and initial conditions have been set for the simulation, the motion of the atoms is integrated using Hamilton's equations of motion and a Nordsiek predictor–corrector method [21].

2.3. Nonenergetic A_2 Piston-Driven Simulations

First we discuss results from simulations of shock waves generated in the nonenergetic A_2 material by driving the edge of the crystal with a constant-velocity piston.

2.3.1. Two-Dimensional (2D) Simulations

Figure 2.3 presents snapshots of the atomic positions at 2 ps for 2D simulations driven with piston velocities of 2.0, 5.0, and 8.0 km/s, from top to bottom, respectively. These snapshots have been offset to align the piston positions. The shock front positions are visible as sharp changes in density between adjacent regions. The piston and shock waves are propagating from left to right. Figure 2.4 plots the

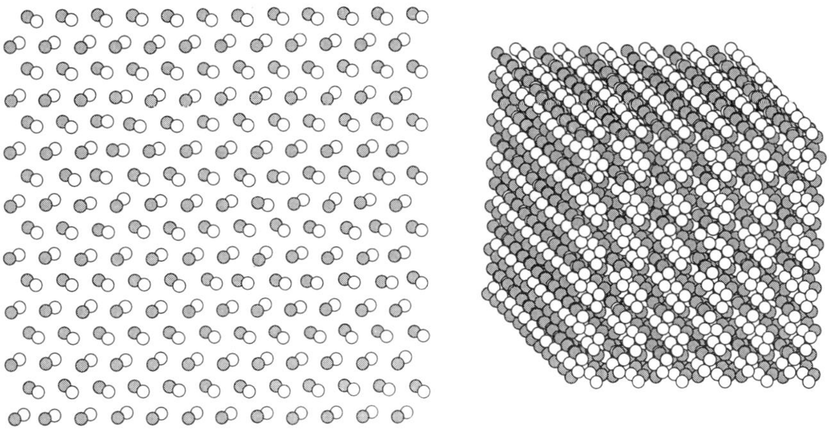

Figure 2.2. Crystal structures of equilibrated AB molecular solid in both two dimensions (*left*) and three dimensions (*right*).

servable due to an elastic–plastic transition. During this transition the unshocked material undergoes first an elastic deformation as the elastic shock passes and subsequently passes through a second plastic shock accompanying the disruptive plastic deformations of the material. The periodicity in the density graph in Fig. 2.5 results from the incommensurate mapping of the lattice parameters of the molecular crystal onto the grid used for averaging the density. This periodicity continues through the elastic region but is lost as plastic deformations destroy the symmetry of the molecular crystal. Visibly, the transition from unshocked material to elastic wave is very difficult to observe in the shapshots of the atomic positions, whereas the plastic wave is much more distinct.

At higher piston velocities, the plastic wave overruns the elastic precursor and becomes the single wave. Then, as in the 2D simulation, when the piston velocity reaches a value at which the system can enter a dissociative phase, a split shock wave again appears with the leading wave being the plastic wave and the trailing wave being the shock associated with the transition into the dissociative phase.

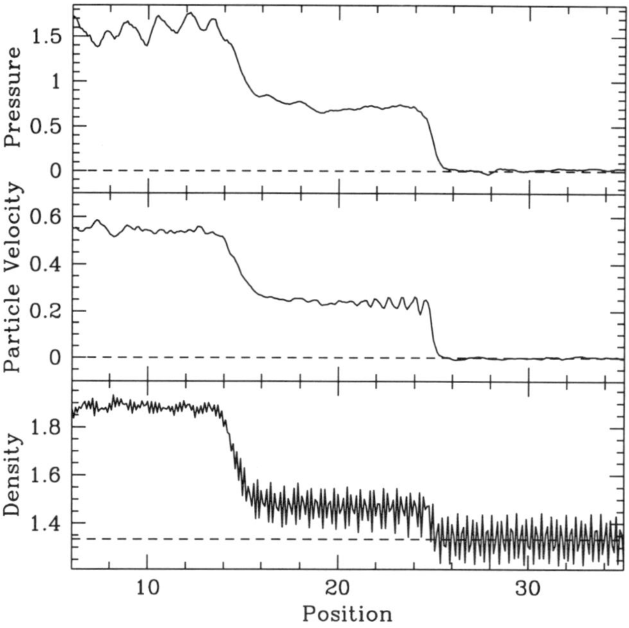

Figure 2.5. Sectional averages for 3D simulation with a piston velocity of 0.55 km/s. From top to bottom are pressure (GPa), particle velocity (km/s) and density (g/cm3). Dashed lines are preshock values.

This shock behavior involving elastic, plastic, and dissociative shock waves is summarized in Fig. 2.6, which plots the shock front velocities at varying piston velocities. At piston velocities below 0.25 km/s, only an elastic shock wave is present. In the region of 0.25 to 0.90 km/s, both an elastic and plastic shock waves coexist. When the piston velocity reaches the range of 0.90 to 4.0 km/s, again only a single plastic shock wave is evident. Over the range of 4.0 to 6.8 km/s, the dissociative phase is present and a split shock wave is present. Finally, at a piston velocity greater than 6.8 km/s, only the single dissociative shock wave is present in the system.

2.3.3. Discussion of A_2 Piston-Driven Simulations

An extensive analysis of this system for the 2D simulations and its associated Hugoniot has been presented earlier [15]. This analysis holds for the 3D simulations, which show results similar to the 2D simulations. The differences observed in the 3D simulations are that the shock velocities are lower, a more distinct elastic wave is present,

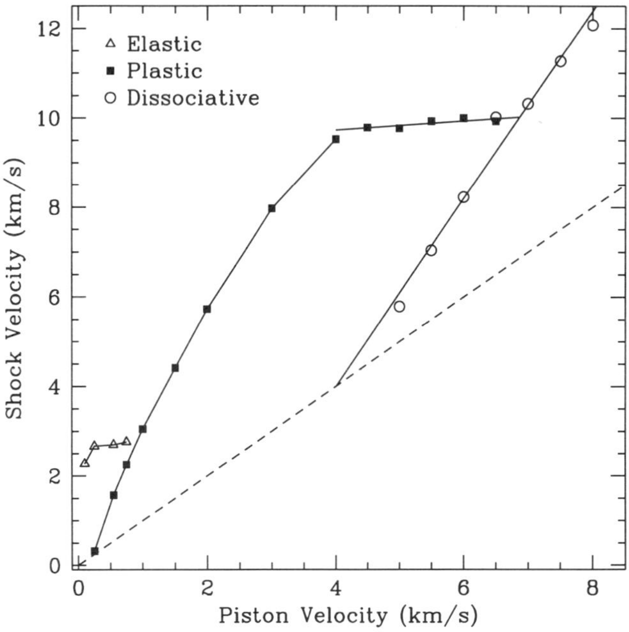

Figure 2.6. Plot of shock velocities as a function of the driving-piston velocity for 3D simulations. The open triangles, filled squares, and open circles correspond to the shocks of the elastic, plastic, and dissociative shock waves, respectively. The dashed line is the piston velocity.

and the region in which the plastic and dissociative shock wave coexist occurs over a smaller range of piston velocities (4–6.8 km/s as compared to 3.0–7.0 km/s for 2D [15]). Both simulations show the behavior that when the piston velocity generates a final state of the shock wave in which the split shock wave structure is present, the initial shock wave travels at a near-constant velocity which is defined by the characteristics of the phase transition and initial state. As seen in Fig. 2.6, any additional piston velocity above that required to generate the split shock wave increases only the velocity of the second shock wave. At sufficiently high piston velocities, the velocity of the second wave can exceed the leading shock velocity, collapsing the split shock wave down to a single dissociative shock wave. Additionally, in the low-piston-velocity region, as the piston velocity slows then the elastic wave velocity approaches the material's speed of sound (1.8 km/s), as expected from theory.

These simulations give excellent quantitative agreement with continuum theory and predictions for shock-induced polymorphic phase transitions [15] as well as showing the coexistence of elastic and plastic waves at lower final pressures. We have also used this model to show that hypervelocity impacts can induce dissociative phase transitions in target materials [16]. Molecular dynamics coupled with reactive many-body potentials are able to model complex behavior in shock-induced phase transitions and hold promise for further investigations into the atomic-scale behavior of material under extreme shock-loading conditions.

2.4. Energetic Chemically Sustained Shock Waves

If, instead of A_2, the energetic AB diatomic molecules are used in the simulations, then the possibility exists for exothermic chemical reactions. Simulations initiated by either a constant-velocity piston or flyer-plate impact are capable of generating exothermic reactions, releasing energy and possibly developing a self-sustaining shock wave.

2.4.1. Flyer-Plate Impact in 2D Energetic Material

Figure 2.7 presents a snapshot of the atomic positions for 2D simulations both before and after the impact of a 6-km/s flyer plate against the edge of an energetic material. The top snapshot in Fig. 2.7 shows the incoming flyer plate (*left*) which is moving from left to right to impact the edge of the quiescent energetic material. The bottom snapshot shows the results for the simulation 12.5 ps after impact. The positions are shifted so that the front of the propagating shock wave is in view. The reactants (white/dark heteronuclear diatomic molecules) can be seen to the right of Fig. 2.7. These reactants are

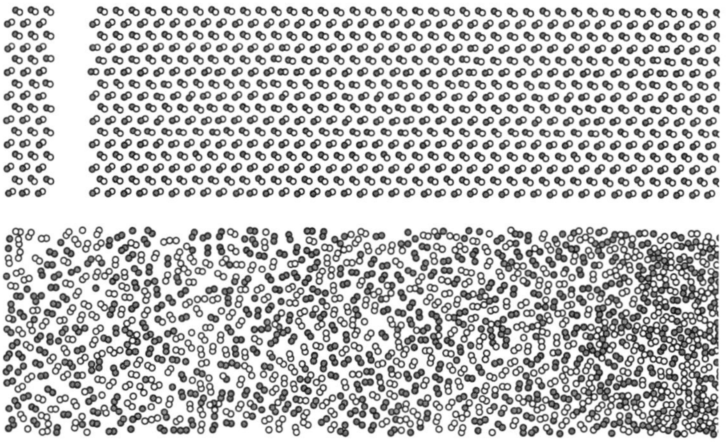

Figure 2.7. *Top*: Snapshot of initial conditions for 2D simulation initiating a chemically sustained shock wave. The impact flyer plate is at the left. *Bottom*: Snapshot of chemically sustained shock wave resulting from impact of a 6-km/s flyer plate at 12.5 ps after impact. The types of atoms are denoted by light and dark symbols. The shock is propagating from left to right.

compressed and heated as they progress through the shock front to ultimately form the more stable products (white/white and dark/dark homonuclear molecules) with an accompanying release of energy into the system. These products can then expand and cool as the shock front continues to propagate through the system.

Profiles of the sectional averages of the pressure, particle velocity, and density for the snapshot in Fig. 2.7 are plotted in Fig. 2.8. The dashed lines in these plots are the initial values of these quantities. From these plots, it can be seen that the initial rise in the values of these properties is very abrupt on the order of molecular dimensions. These properties then relax, with the pressure and density approaching zero as the products that are formed expand to the rear of the shock wave. Analysis of these properties with respect to the Rankine–Hugoniot relations [22] shows that these simulations rapidly stabilize near the shock front and then progressively satisfy the Rankine–Hugoniot relations over a longer distance behind the shock front as the simulation progresses [9]. Additionally, the behavior of this shock wave–supported through chemical reactions–is consistent with that predicted by the Zel'dovich–von Neumann–Doering (ZND) model for detonations [2].

2.4.2. Comparison of 2D and 3D Simulation Results

Results obtained from a simulation in which a three-dimensional core of energetic material is impacted with a flyer plate is shown in

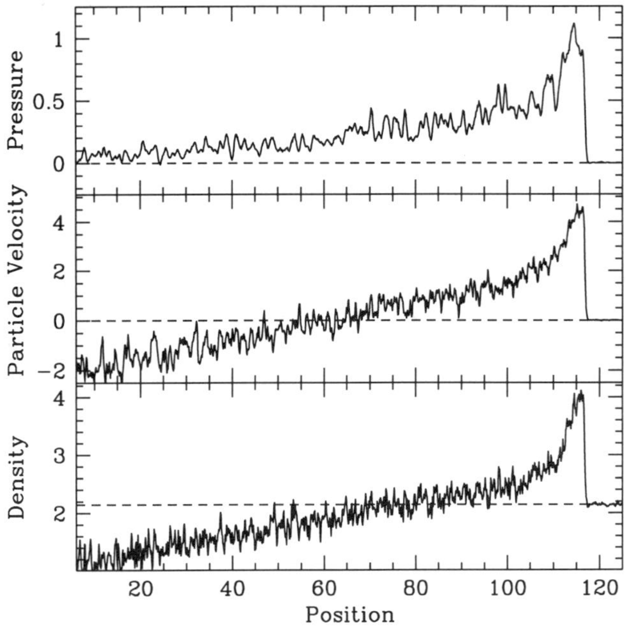

Figure 2.8. Profiles for the pressure (eV/Å^2), particle velocity (km/s), and density (amu/Å^2) for the snapshot in the bottom of Fig. 2.7. The position (nm) is relative to the initial edge of the unreacted energetic material.

Fig. 2.9, which plots the profiles of the pressure, particle velocity and density at 12.5 ps after impact. These results are very similar to those from the 2D simulations. The profiles in Fig. 2.9 show less fluctuations than those in Fig. 2.8 due to a greater amount of spatial averaging in the 3D results (more material per unit in the shock dimension). Additionally, the profiles for the 3D simulations show a more well-defined initial peak at the front of the shock. This profile which is better defined in the 3D simulations is partially visible in the density in Fig. 2.8 from the 2D simulations.

If one compares the shock front positions for both the 2D and 3D profiles in Figs. 2.8 and 2.9, it can be seen that shock front in the 2D simulation has moved further over the same period of time following impact. This is shown in Fig. 2.10, which plots the shock front positions with respect to the time since impact. This figure shows that the shock front velocities quickly reach constant but differing values as indicated by straight but diverging lines. The slopes of the lines in Fig. 2.10 correspond to velocities of 9.3 and 7.2 km/s for the 2D and 3D simulations, respectively. Therefore, both the 2D and 3D simulations rapidly reach a constant velocity in which the shock wave propagation is sustained by the continual release of energy as the reactants are converted into products.

The data presented above shows that the 2D and 3D simulations have qualitatively similar results in terms of shock profiles and the ability to support chemically sustained shock waves. The main difference in that, in the 3D simulations, the chemically sustained shock wave travels at a slower rate (closer to real experimental values) than in the 2D simulations. One other significant difference is the number of atoms which must be treated in 3D as compared to 2D. For the results presented above, only 10000 atoms were needed to carry a 5-nm-wide 2D simulation to a time of 12.5 ps, whereas 33000 atoms were necessary for a narrower core of 3 × 3-nm material in the 3D simulations. This difference in number of atoms as well as the added dimension (high number of neighbors) for the 3D simulations makes them much more computationally expensive and longer to perform than the 2D simulations. For this reason, in the remainder of this chapter, we will focus on results from 2D simulations. As

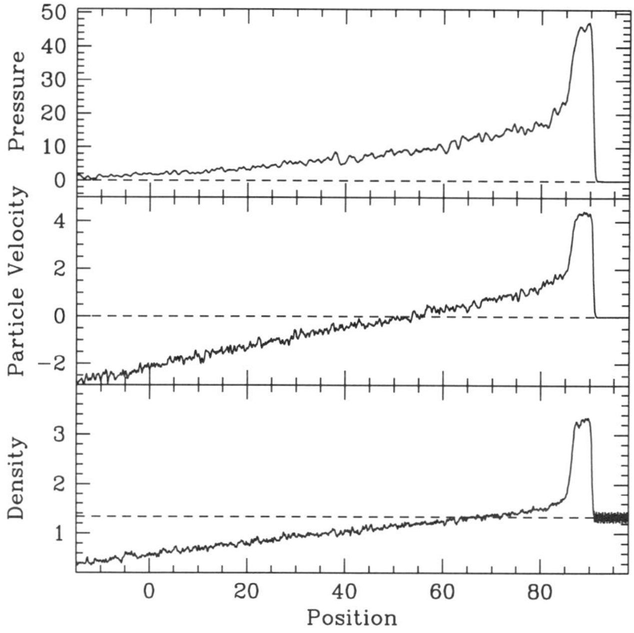

Figure 2.9. Profiles for the pressure (GPa), particle velocity (km/s) and density (g/cm3) for a 3D simulation at 12.5 ps after impact. The position (nm) is relative to the initial edge of the unreacted energetic material. The dashed lines are the unshocked values for these properties.

cannot be initiated no matter how large the flyer plate. Thus, this material has a threshold for initiation. This is necessary if this model is to be useful in investigating the effect that defects, voids, or other modifications to the molecular solid or model parameterization will have on the sensitivity of the energetic material and provide insight into initiation at the atomic level. Studies probing the effects of defects and voids on shock wave propagation and initiation are currently under way.

Independent studies using this model substantiate this initiation behavior [23]. Moreover, the analysis presented therein shows that the results from this model are both consistent with thermal explosion theory (with chemically reasonable parameters) as well as the behavior of real systems of explosives.

2.4.4. Critical Diameter Studies

The previous simulations enforced periodic boundary conditions perpendicular to the shock propagation direction. This simulated an infi-

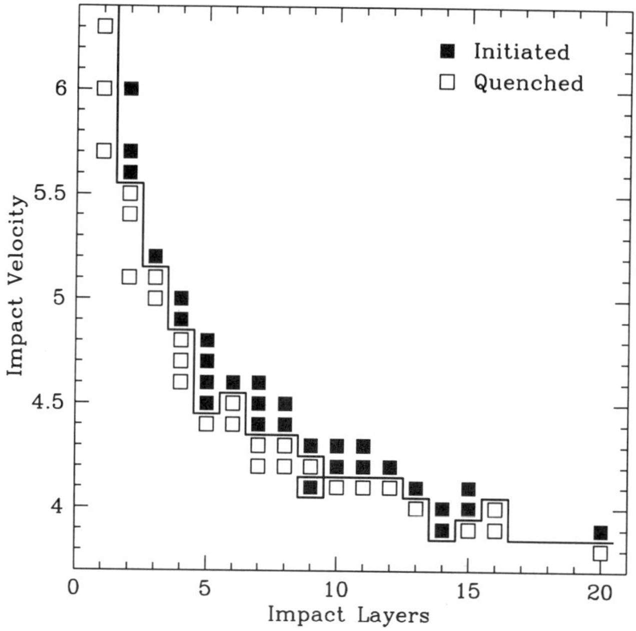

Figure 2.12. Results from 2D simulations in which both the number of layers and velocity of the flyer plate were varied. Solid squares are simulations in which a chemically sustained shock wave was initiated, and open squares are those which did not. The solid line follows the interface between these two regions.

nite amount of material in this dimension. If this boundary condition is removed, then the material has free edges through which expansion can occur. Real explosives have a critical diameter below which a detonation will not sustain itself [2]. By using free-boundary conditions in our simulations, we can study this model's critical diameter.

Figure 2.13 shows the results for a variety of simulations in which the energetic material has a finite width. From this figure, it can be seen that up to a width of 12.5 nm the simulations fail to initiate. Once the energetic material's width is greater than 12.5 nm, a chemically sustained shock wave can be initiated and sustained in the system. These shock waves in the finite-width simulations continue to propagate at constant velocity once initiated similar to the periodic boundary simulations. However, the shock wave velocity is dependent on the energetic material's width and the chemically sustained shock wave travels at slower velocities for narrower widths. As the width increases, then the value of the shock velocity asymptotes to the value for the periodic boundary (infinite width) simulations.

Removing the periodic boundary conditions perpendicular to the direction of shock propagation allows the shock front to develop curvature. This is shown in Fig. 2.14 which displays a snapshot of the

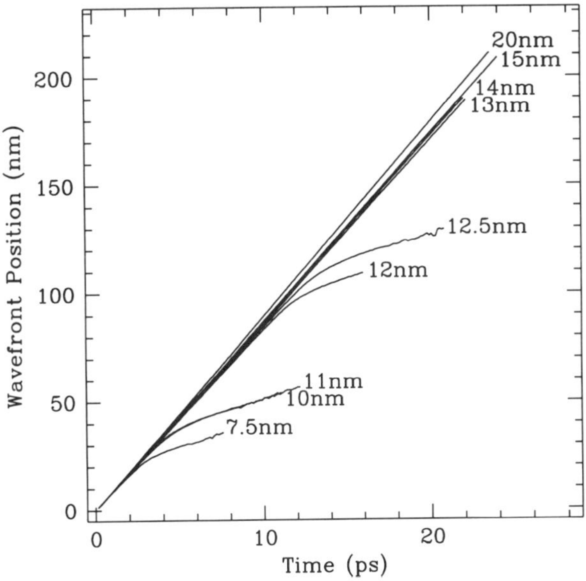

Figure 2.13. Shock front position versus time from 2D simulations in which the material had finite widths of 17.5, 10, 11, 12, 12.5, 13, 14, 15, and 20 nm.

15-nm-wide simulation at 24 ps after impact from the flyer plate. This snapshot has curvature in the shock front together with expansion of the products out the material's sides. A pie-shaped region of higher density behind the shock front that results from the expansion waves approaching from either side of the material is visible in Fig. 2.14. This particular snapshot also shows a slight defect in the shock front curvature in the upper half. This may be due to some slight irregularity in the propagation of the shock wave, but it is not of sufficient magnitude to disrupt or terminate the self-sustaining nature of the shock wave and the chemically sustained shock wave is stable against these types of imperfections or irregularities.

2.4.5. Chemically Sustained Shock Waves Conclusions

These simulations show behavior consistent with that of real explosives and ZND theory. The 2D and 3D simulations show very similar results. This model does not always initiate; but when it does, the prop-

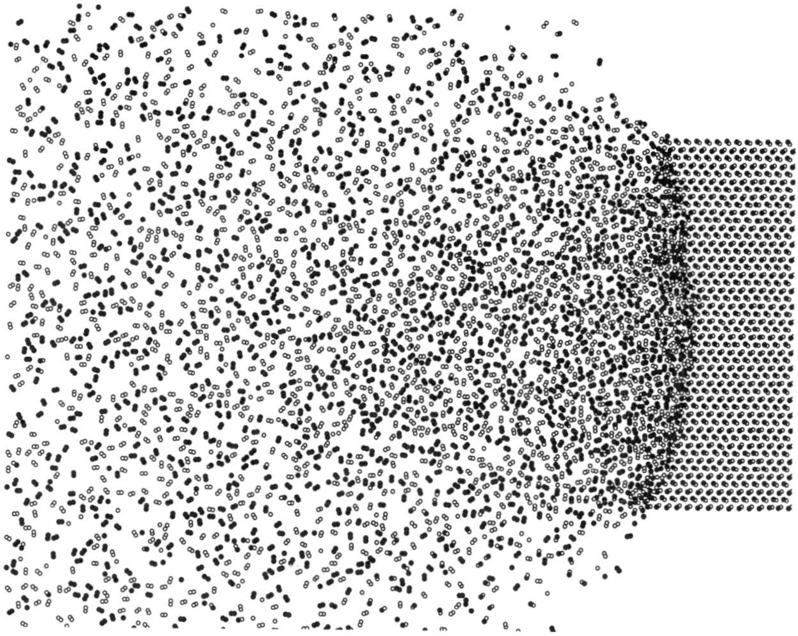

Figure 2.14. Snapshot of the atomic positions for the 15-nm-wide simulation at a time after impact of 24 ps. The atoms type are denoted by black and white; the shock is propagating from left to right.

erties of the resulting chemically sustained shock wave are independent of initial conditions. Independent studies confirm that initiation in this model is consistent with thermal explosion theory and real energetic materials. Finite-width simulations show that this model has a critical diameter below which it will not form a stable chemically sustained shock wave. This model shows great promise for the ability of MD simulations and chemically reasonable potentials to probe how the atomic-level details of an energetic material affects the macroscopic details of its sensitivity, initiation, and propagation characteristics.

2.5. Conclusions

Chemically reactive potentials with the possibility for chemical reactions and high-pressure phases have been used in MD simulations to study coupling of shock waves to chemical and physical properties. Piston-driven shock wave simulations of a nonenergetic material show the presence of a split shock wave depending on the piston velocity. These results are in agreement with the behavior expected from experiments and continuum theory. Simulations using an energetic material are shown to develop stable, constant-velocity, chemically sustained shock waves. These chemically sustained shock waves show initiation and critical diameter results consistent with continuum theory and real energetic materials. Although the chemically sustained shock waves shown in Figs. 2.8 and 2.9 have a very simple profile, these are not the only profiles supported by these EBO potentials. Other simulations with different EBO parameterizations can support a chemically sustained shock wave with a split shock wave structure similar to that discussed in Sec. 2.3 [10,24].

The MD simulations with reactive many-body potentials have been shown to be able to probe the atomic-level details of shock-related phenomena. These simulations have been able to reproduce such complex behavior as shock wave splitting induced by a polymorphic phase transition as well as suggest the possibility for chemically sustained shock waves to include polymorphic phase transitions and thereby show interesting unanticipated behavior. Additionally, the finite-width simulations showing a 12.5-nm critical diameter in this model indicates a lower bound of nanometers for the reaction-zone lengths in some detonating systems—much shorter than experiment has yet been able to measure. Further studies, more complex models, and faster computers should increase the ability to probe these atomic-level details of shock-related phenomena by MD techniques.

Acknowledgments

This work was supported in part by the ONR through NRL and the ONR Physics Division.

References

[1] B.L. Holian, in *Macroscopic Simulations of Complex Hydrodynamic Phenomena* (eds. M. Mareschal and B.L. Holian), Plenum Press, New York, pp. 75–85 (1992).

[2] W.C. Davis, *Sci. Am.* **256,** pp. 106–112 (1987); Ya. B. Zel'dovich and A.S. Kompaneets, *Theory of Detonation*, Academic, New York (1960); W. Fickett and W.C. Davis, *Detonation*, University of California Press, Berkeley (1979); W. Fickett, *Introduction to Detonation Theory*, University of California Press, Berkeley (1985).

[3] G.E. Duvall and R.A. Graham, *Rev. Mod. Phys.* **49**, pp. 523–579 (1977).

[4] D. Bancroft, E.L. Peterson, and S. Minshall, *J. Appl. Phys.* **27**, pp. 291–298 (1956).

[5] P.S. DeCarli and J.C. Jamieson, *Science* **133**, pp. 1821–1822 (1961).

[6] B.L. Holian, W.G. Hoover, W. Moran, and G.K. Straub, *Phys. Rev.* **22**, p. 2798 (1980).

[7] A.N. Dremin and V.Yu. Klimenko, *Prog. Astronaut. Aeronaut.* **75**, p. 253 (1981).

[8] D.W. Brenner, *Phys. Rev. B.* **42**, p. 9458 (1990).

[9] D.W. Brenner, D.H. Robertson, M.L. Elert, and C.T. White, *Phys. Rev. Lett.* **70**, pp. 2174–2177 (1993).

[10] C.T. White, D.H. Robertson, M.L. Elert, and D.W. Brenner, in *Macroscopic Simulations of Complex Hydrodynamic Phenomena*, (eds. M. Mareschal and B.L. Holian), Plenum Press, New York, pp. 111–123 (1992).

[11] J.A. Harrison, C.T. White, R.J. Colton, and D.W. Brenner, *Phys. Rev. B* **39**, p. 1453 (1989).

[12] J.A. Harrison, C.T. White, R.J. Colton, and D.W. Brenner, *MRS Bulliten*, **18**, p. 50–53 (1993).

[13] J.A. Harrison and D.W. Brenner, *J. Am. Chem. Soc.* **116**, pp. 10399–10402 (1994).

[14] D.H. Robertson, D.W. Brenner, and C.T. White, *J. Phys. Chem.* **96**, pp. 6133–6135 (1992).

[15] D.H. Robertson, D.W. Brenner, and C.T. White, *Phys. Rev. Lett.* **67**, pp. 3132–3135 (1991).

[16] C.T. White, S.B. Sinnott, J.W. Mintmire, D.W. Brenner, and D.H. Robertson, *Int. J. Quant. Chem.* **28**, pp. 129–137 (1994).

[17] G.C. Abell, *Phys. Rev. B* **31**, p. 6184–6196 (1985).

[18] J. Tersoff, *Phys. Rev. B* **37**, p. 6991–7000 (1988).

[19] J.N. Murrell, S. Carter, S.C. Farantos, P. Huxley, and A.J. Varandas, *Molecular Potential Energy Functions*, Wiley, Chichester (1984).

[20] G.C. Fettis, J.H. Knox, and A.F. Trotman-Dickenson, *J. Chem. Soc. London*, p. 1064 (1960).

[21] C.W. Gear, *Numerical Initial Value Problems in Ordinary Differential Equations*, Prentice-Hall, Englewood Cliffs, NJ, p. 148 (1971).

[22] Ya.B. Zel'dovich and Yu.P. Raizer, *Physics of Shock Waves and High-Temperature Hydrodynamics Phenomena*, Academic Press, New York (Vols. 1 & 2) (1966).

[23] P.J. Haskins and M.D Cook,.*High-Pressure Science and Technology —1993* (eds. S.C. Schmidt, J.W. Shaner, G.A. Samara, and M. Ross), American Institute of Physics, New York, pp. 1341–1344 (1994).

[24] D.H. Robertson, *High-Pressure Science and Technology — 1993* (eds. S.C. Schmidt, J.W. Shaner, G.A. Samara, and M. Ross), American Institute of Physics, New York, pp. 1345–1348 (1994).

CHAPTER 3

Mechanisms of Elastoplastic Response of Metals to Impact

C.S. Coffey

3.1. Introduction

The elastoplastic response of metals to impact and shock has been the subject of a great many investigations. Many, if not most, of these were developed for mild impact conditions where thermal effects are important. The author's own peculiar situation has led him to examine the elastoplastic response of metals from the perspective of high-amplitude impacts and shocks. In this regime, thermal effects, although important, are secondary to the effects of the impact or shock. Thermal effects are necessary to properly account for plastic flow behavior for low-amplitude impacts and shocks and to provide a more correct picture of plastic flow at higher stress levels. Much of what follows arose from efforts to account for the energy dissipation and localization that occurs in crystalline solids during shock or impact. These processes are related to plastic deformation since all occur due to the creation and motion of dislocations.

Within the solid, the elastoplastic deformation of a metal must have its basis at the atomic or molecular level. The goal here is to set this out in a correct and usable way. Unfortunately, enough is still uncertain about some of the microscopic properties involved in plastic flow that some amount of "calibration" will be necessary and may always be necessary. As much as possible, the causes of these uncertainties will be identified as they are encountered in the text. It will be assumed that plastic flow within a crystalline solid occurs by the creation and motion of dislocations and that the reader is familiar with the basic concepts of dislocations [1–6]. Plastic deformation due to shock or impact is a dynamic process and, although interesting, the purely static aspects of dislocations are not really relevant to this situation. It will also be assumed that the classical picture of dislocations is adequate to describe the dynamic behavior of dislocations in simple metals to at least first order. This picture will likely have to be modified somewhat to extend it to describe dislocation motion in molecular crystals [7]. Because the dislocation motion responsible for plastic flow involves movement of atoms or molecules, quantum mechanics is the appropriate means of describing the elastoplastic response of crystalline solids to impact or shock. Equally important is the need to carry this microscopic description to the macroscopic level to make it accessible for general use. Finally, while conditions leading to crystal failure are developed, no attempt is made to extend the analysis into the failure regime.

3.2. Dislocation Motion

It is appropriate to start by considering the means by which dislocations move through a crystalline solid because this contains the essence of plastic deformation. The conventional development of dislocation theory identifies the shear wave speed as the maximum, or nearly the maximum, velocity attainable by moving dislocations [5,6]. This was confirmed by the classic experiments on LiF by Johnston and Gilman [8]. Similar confirming results have since been obtained for a number of materials and these have been summarized by Kim and Clifton [9]. These experimental results describe the dislocation velocity over the entire range of applied shear stress, an achievement which has not yet been possible with most theoretical approaches.

Classical Picture

It is generally recognized that the core of the dislocation resides in a potential well formed by the displaced atoms or molecules that surround the core region. Because of the break in the lattice pattern, this potential well is considerably distorted and reduced from the regular array of potential wells that characterize the solid. These distorted potential wells extend along the line of the dislocation. The dislocation core vibrates in the usual manner within the confines imposed by this distorted well. In the conventional dislocation models, movement occurs when the energy of the dislocation core exceeds the height of the potential barrier that confines it, thus allowing the dislocation to move into the adjacent potential well [6]. Often, this is envisioned as occurring due to thermal fluctuations which temporarily contribute sufficient energy to lift the core over the Peierls potential barrier of the distorted well and into the adjoining well.

A problem arises here because the time required for such a fluctuation to occur has been shown to be of the order of 10^{-10} to 10^{-11} s [10]. If these are truly fluctuations, the energy will vary with time and the dislocations will not always be able to overcome the potential barrier. This will diminish the average dislocation velocity. However, in the presence of a strong shock or impact, a dislocation moving at or near the shear wave speed of 3 to 5 km/s in a lattice with an atomic or molecular spacing of about 3×10^{-10} m requires only 10^{-13} s or less to overcome the potential well barrier and travel one lattice spacing. It cannot afford to wait for a fluctuation to occur, especially at low and moderate temperatures. Further, a dislocation may move many tens and even hundreds of atomic or molecular spacings during a shock or impact, so many such fluctuations must occur over macroscopic distances. On this basis, it does not seem likely that thermal fluctuations can account for dislocation motion in response to strong shocks or impacts.

Quantum Tunneling

In an effort to determine the processes by which dislocations move through the lattice of a crystal, we have proposed that dislocation motion occurs by quantum-mechanical tunneling [11]. In an early article, Weertman examined dislocation motion by this process [12]. The dislocation core was treated as a string of single atoms and, generally, the conditions of low shear stress and low temperatures applied. It was pointed out that, at low shear stress levels, the tunneling probability depends approximately on the quantity $\exp(-U^{1/2})$, where U is the height of the Peierls barrier potential. From this it was concluded that the tunneling probability was very small and that, at normal temperatures, thermal fluctuations were the likely means by which dislocations were able to surmount the Peierls barrier and transit into the next potential well.

However, at higher levels of applied shear stress, which is the situation of interest here, it is necessary to consider the more exact quantity $(U - E)^{1/2}$ in order to determine the tunneling probability [11,13]. The quantity E is the energy of the tunneling particle and is made up of both applied stress and thermal components. For low-energy dislocations, this reduces to the above approximation. But as $E \rightarrow U$, as will occur at high shear stress levels or high temperatures, the probability of tunneling becomes significant and consequently must be taken into account when determining plastic flow.

The dislocation model that we propose follows from the single particle model of Weertman. The core of an edge dislocation, for example, is viewed as composed of a string of segments, each made up of the atoms or molecules at the termination of the extra half-plane. Each core segment is joined to its nearest neighbors along the line of the dislocation by the full lattice potential. However, in the slip direction(s), each core segment resides in the distorted and retential barrier and will underestimate the tunneling probability of the high-energy dislocations that reside near the top of the potential barrier.

Beginning with the Schrodinger equation for the quasi-particle dislocation core segment of mass m located an idealized potential well, the probability, T, that the core segment will tunnel through the barrier along the slip plane is*

$$T = \left[1 + \frac{U^2 m}{2E\hbar^2} S(U-E)\right]^{-1}, \tag{3.1}$$

where

$$S(U-E) = \frac{\hbar^2}{2m(U-E)} \sinh^2\left[\frac{2m}{\hbar^2}(U-E)\right]^{½} W \tag{3.2}$$

when $E < U$, and

$$S(U-E) = \frac{\hbar^2}{2m(E-U)} \sin^2\left[\frac{2m}{\hbar^2}(E-U)\right]^{½} W \tag{3.3}$$

when $E > U$.

For the regime where detectable tunneling and plastic flow begin, $E << U$, the approximation can be made that

$$T \cong \frac{2E\hbar^2}{mU^2}[S(U-E)]^{-1}. \tag{3.4}$$

For the situation where E approaches U,

$$T \approx \frac{2\hbar^2}{mW^2}\frac{E}{U^2}, \tag{3.5}$$

which holds independently of the sign of $U-E$. Most important, when $E >> U$, the tunneling probability approaches unity, $T \to 1$, and the dislocations have energies greater than the reduced potential barrier and are no longer constrained by the lattice, so that some form of lattice failure will have occurred.

3.2.2. The Dislocation Energy

Assume local equilibrium, so that the energy density of a dislocation must be the same as the local energy density of the host lattice.

---

* Except for a slight change in notation these results are taken directly from Ref. 14, p. 97.

During shock or impact, the energy will have two components: one due to the shear stress imposed by the shock or impact and the other from the shear stresses arising from the thermal background. Even for high-level shocks, most of the solid will behave elasticitically as a Hooke's-law solid and most of the shear stress energy of the dislocation core will be elastic energy. The average energy with which a segment of the dislocation core approaches the potential barrier is approximately

$$E \approx \frac{\tau \gamma V}{2} + \frac{k\theta}{2}, \tag{3.6}$$

where τ and γ are the shear stress and strain, θ is the temperature, and k is Boltzmann's constant. The effects of thermal fluctuations about the mean temperature have been ignored. The nonlinear contribution of the dislocation core region to the energy has also been neglected. For a mostly elastic solid, $\gamma = \tau / G$, where G is the shear modulus, and the energy of a dislocation core segment approaching the potential barrier becomes

$$E \approx \frac{\tau^2 V}{2G} + \frac{k\theta}{2}. \tag{3.7}$$

3.2.3. Dislocation Velocity

The dislocation velocity is determined by observing that a dislocation core segment residing in a potential well has both thermal and applied stress-induced motions. These cause the core segment to vibrate back and forth between the walls of the potential well. Occasionally, the segment will tunnel through the barrier and transit into the adjacent well where it continues to oscillate between the walls of the new potential well until tunneling again occurs and the dislocation segment moves into the next potential well. Since most often the potential barriers prevent dislocation motion, it is useful to consider only the average velocity. If n is the number of times that a dislocation segment approaches a potential barrier and T is the probability of tunneling through the barrier, then nT is the average number of potential barriers through which the segment tunnels. With each successful tunneling event, the dislocation segment moves a lattice distance b. To simplify things, assume that a sufficiently strong thermal or shear stress exists so that, for the most part, the dislocation segment does not tunnel back into a previously occupied well. This simplification is based on the observation that a dislocation segment is accelerated in the direction of the shear stress so that its kinetic energy is greatest when it approaches a potential barrier in this direction. Since tunneling depends strongly on the kinetic energy with which the segment approaches the barrier, it will occur preferentially in the direction of the shear stress. For this case, the average velocity of a segment is approximately

$$v \cong \frac{nTb}{t}, \tag{3.8}$$

where t is the time interval. Because energy is not conserved during tunneling, the actual tunneling time is infinitesimally small, determined by the uncertainty principal. It will be assumed that the time to complete each successful tunneling event is just the time for the lattice to rearrange itself to accommodate the new position of the dislocation core segment. This occurs at the local sound speed, v_0, so that the time required to accomplish tunneling one lattice spacing is $\Delta t \cong b / v_0$. The appropriate time interval over which to average the dislocation velocity is just $t = n\,\Delta t$, so that the average segment velocity becomes

$$v = T v_0. \tag{3.9}$$

Since a dislocation is made up of a string of these core segments, all of which experience the same shear stresses, Eq. 3.9 is the average velocity with which the dislocation moves.

This form gives the correct limits for the average dislocation velocity both at very low shear stress levels, $\tau \sim 0$, where $T \to 0$ and $v \to 0$, and at high shear stress levels, $\tau \gg 0$, where $T \to 1$ and $v \to v_0$. Gilman, in his work with LiF, developed an empirical expression for the tunneling coefficient, $T \approx \exp(-\tau_0 / \tau)$, where τ_0 is a characteristic shear stress of the order of the yield stress [15]. The instantaneous dislocation velocity is discontinuous since, for $T < 1$, not every attempt by the dislocation to tunnel through the potential barrier will be successful. When tunneling does not succeed, dislocation motion will be arrested until the dislocation is able to penetrate the barrier after repeated trials.

3.2.4. Dislocation Sources

Given the possible complexities of the lattice and its defect structures, there appears to be a plethora of potential dislocation sources and much has been written about several of these. Here, we take a more practical approach and develop a simple generic source of characteristic length l_0, where l_0 is the effective source length averaged over all the source types active in the crystal [16].

The number of moving dislocations can be estimated by assuming that the dislocations are created by N_s sources whose dislocations intercept a unit surface area of the crystal. To each of these sources is assigned a probability, $p_c(\tau)$, of creating a new dislocation pair at a shear stress τ and a characteristic source size, l_0, beyond which the newly created dislocation pair must propagate before the back stress due their presence is sufficiently reduced to allow the source to create another dislocation pair. The rate at which a single generic source can create dislocations is [16]

$$\frac{dn}{dt} = 2p_c(\tau)\frac{v}{l_0}. \tag{3.10}$$

The factor of 2 accounts for the simultaneous creation of a pair of dislocations. If τ does not depend strongly on time but resembles a step function, $\tau(t) \approx \tau u(t)$, then integrating Eq. 3.10 gives

$$n(\tau, t) \approx 2T(\tau)\, p_c(\tau)\frac{v_0}{l_0}\, t, \tag{3.11}$$

where the time t is measured from the first application of the shock or impact.

It has been observed that dislocations are often concentrated in band-like regions after deformation by shock or impact [8]. If it is assumed that the dislocations in a shear band are created by a single dislocation source, the rate of increase in the width of a band, dw/dt, is just

$$\frac{dw}{dt} = \frac{D}{N_0}\frac{dn}{dt}, \tag{3.12}$$

where D is the average separation distance between active slip planes in the band and N_0 is the number of dislocations in a pileup on an active slip plane [6]. Assuming some typical values, $D \approx 10^{-9}$ m, $N_0 \approx 10^2$ to 10^3 dislocations in a static pileup, and $dn/dt \approx 10^{11}$ dislocations/s, the rate of increase in width of a shear band during a strong shock is approximately 0.1 to 1 m/s.

3.3. Plastic Strain Rate

The plastic strain rate, $d\gamma_p/dt$, is the quantity that is most often measured experimentally. It is convenient to use the Orowan relation to write the plastic strain rate as [4]

$$\frac{d\gamma_p}{dt} = Nvb, \tag{3.13}$$

where N is the number of moving dislocations per unit area of surface, v is the average dislocation velocity, and b is the Burgers length. Combining Eqs. 3.9, 3.11, and 3.13, the plastic strain rate from N_s dislocation sources that intercept a unit area is approximately

$$\frac{d\gamma_p}{dt} \approx 2[T(\tau)]^2 \frac{v_0^2\, b}{l_0} p_c(\tau) N_s\, t. \tag{3.14}$$

The plastic deformation rate given in Eq. 3.14 admits two time regimes of interest. These are delineated by the time required for a newly created dislocation to move away from its source, at a velocity v, and encounter an obstacle which forces it to stop. Let L_0 be the average distance between the dislocation source and an obstacle such as a grain boundary. Then, at early times after the application of a shock or impact when $t < L_0/v$, the number of moving dislocations increases with time and the plastic strain rate is given by Eq. 3.14. However, at later times when $t \geq L_0/v$, the moving dislocations will begin to encounter obstacles and their motion will be halted. A nearly steady state is reached when the number of newly created dislocations is offset by the number of recently stopped dislocations. The number of moving dislocations approaches a constant determined by substituting the time $t = L_0/v$ into Eq. 3.14. For this steady state case, $t > L_0/v$, the plastic strain rate becomes

$$\frac{d\gamma_p}{dt} \approx \frac{2T(\tau)\, v_0\, b\, p_c(\tau) N_s\, L_0}{l_0} \tag{3.15}$$

and is only implicitly dependent on time through the shear stress, $\tau(t)$.

Over the range of possible shear stresses, the plastic strain rate displays three regimes. Two of these are asymptotic regimes. One is at low values of shear stress where plastic deformation and dislocation movement are barely detectable and where $T \approx 0$ and $d\gamma_p/dt \approx 0$. The other asymptotic regime is the high shear stress region where the shear stress approaches and exceeds the levels associated with rapid plastic flow of the crystal and the tunneling coefficient approaches unity, $T \approx 1$. In this high shear stress regime, the plastic strain rate can be very large but changes very slowly with increasing shear stress.

Between these asymptotic regions is a transition regime where the shear stress is of the order of or greater than the yield stress of the crystal and the energy of the tunneling particle increases to become a substantial fraction of the energy of the potential barrier. Over this relatively narrow range of shear stress, the tunneling coefficient changes rapidly from near zero to near unity as $E \to U$. The plastic strain rate increases rapidly as well. Because of the quantum mechanical nature of the tunneling process, the plastic strain rate will exhibit a damped oscillatory behavior as the applied shear stress approaches the flow shear stress and $T \to 1$ [13]. The shear stress amplitude at which this transition regime occurs is determined by the condition that the energy of the tunneling particle approaches the height of the potential barrier, $E \to U$. For most mild steels, this will only be of the order of a few hundred MPa.

Together, the asymptotic regimes and the transition region determine the main features of the plastic strain rate as a function of the applied shear stress for any crystalline solid. This may be altered somewhat if, at high shear stress levels, it becomes possible to tunnel through the distorted Peierls barrier in different directions, which requires higher tunneling energies and results in different slip planes becoming active, or if the number of dislocation sources were to change.

3.4. Comparison with Experiments

It is not the intention here, nor was it in our previous work, to obtain a best curve fit for the experimental data. Rather, the intent is (and was) to compare the theoretical predictions with the major features of the data. Too many quantities such as the shape, height, and width of the distorted Peierls barrier and the effective mass of the dislocation core segments are currently only imprecisely known. Also, at best, a rectangular potential barrier is only an approximation to the real Peierls barrier. It is inappropriate to expect an exact comparison between the predictions of the theory and the observations of experiment. This sort of comparison belongs to the realm of calibration experiments.

A large number of experiments have been conducted to examine the relation between the applied shear stress and the rate of plastic deformation. Perhaps the most notable of these is the work by Campbell and Ferguson [17] who examined the plastic deformation rate in a mild steel for different applied shear stress levels over a wide range of temperatures. Our initial comparisons with these data estimated the energy stored in the dislocation core and the effective mass of the core. Here, the rectangular Peierls barrier is maintained, but to take into account the decreasing width of the real potential barrier with increasing energy, the width of the barrier is given the form $W = W_0[1 - \exp(-U / E)]$. The quantity W_0 is the width at the base of the barrier. To estimate the barrier height, observe that at the melt temperature, under conditions of zero applied shear stress, the dislocations must be able to move freely about the solid. On this basis, the height of the potential barrier was estimated by setting $\tau = 0$, $\theta = \theta_{melt}$, and choosing U to give $T \approx 0.5$.

In our earlier work, a comparison was made between the experimental results of Campbell and Ferguson and the plastic strain rate predictions given above. Here, in Fig. 3.1, the calculated* plastic strain rate is shown as a function of the applied shear stress for two temperatures, $\theta = 300$ K and $\theta = 700$ K, to illustrate the asymptotic behaviors of the strain rate for low and high shear stress, the transition region connecting the asymptotic regimes and the quantum mechanical oscillations associated with the transition region.

The predicted asymptotic behavior in strain rate with applied shear stress that occurs at $d\gamma_p/dt \approx 5 \times 10^4$ s^{-1} at all temperatures is due to the tunneling probability approaching unity as $E \rightarrow U$. This behavior is evident in the experiments of Campbell and Ferguson and has been observed in other experiments on steel, copper, and β brass [18–21]. The conventional macroscopic constitutive approach cannot account for this asymptotic behavior of the plastic strain rate at high shear stress levels.

At low shear stress levels and temperatures $\theta < 500$ K the plastic strain rate is nearly independent of temperature. Initially, this was attributed to a restricted functioning of the dislocation sources at low stress levels. Our present realization is that for these conditions, the dislocations lie deep in the potential well, far below the top of the reduced Peierls barrier, and have relatively large effective mass so that tunneling is extremely unlikely.

A characteristic of the tunneling process is the damped oscillations that occur in the tunneling probability as $T \rightarrow 1$ [13]. From Eq. 3.13 or 3.14 these damped oscillations will occur in the plastic strain

---

* The material constants were taken as $b = 3 \times 10^{-10}$ m, $W = 0.5 \times 10^{-10}$ m, $l_0 = 3 \times 10^{-8}$ m, $v_0 = 5.5 \times 10^3$ m/s, $G = 7.5 \times 10^{11}$ dynes/cm^2, $\rho = 8.6 \times 10^3$ kg/m^3, $U = 0.7 \times 10^{-13}$ ergs. The number of dislocation sources per unit area is determined in conjunction with the mean free path distance, L_0, that a dislocation travels between its source and a blocking obstacle. It will be assumed that there is only one source per mean free path, so that $N_s L_0^2 \approx 1$. Taking $L_0 = 10^{-4}$ m, the approximate crystal size, then $N_s \approx 10^8$ m^{-2}. The mass of a dislocation core segment is taken as $m \approx 10^4$ electron masses, and the volume, V, of the lattice that is perturbed by the presence of the moving dislocation core is taken as a cylindrical disk centered at the dislocation core and of height b and radius $7b$ so that $V \approx 150b^3$. For simplicity, it will be assumed that the amplitude of the applied shear stress is large enough so that the source can readily create dislocations allowing $p_c(\tau) \rightarrow 1$.

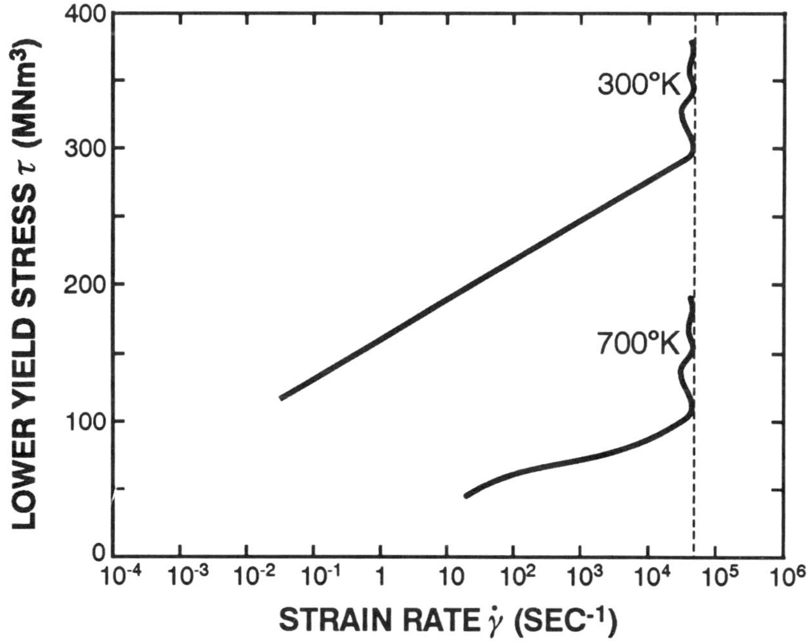

Figure 3.1. Plastic deformation rate as a function of the applied shear stress in mild steel at temperatures of 300 K and 700 K. Shown are the two asymptotic regimes, $T \approx 0$ and $T \approx 1$, and the transition region where τ is approximately a few hundred MPa, $d\gamma_p / dt \approx 10^4$ s^{-1}. The predicted quantum oscillations are also shown. The results are in reasonably good agreement with the experiments of Campbell and Ferguson [17].

rate as well. Campbell and Ferguson observed similar damped oscillations in their experimental results but attributed them to the excitation of a "breathing mode" resonance in their test apparatus [17]. Although this may indeed have been the case, these damped oscillations could also have been a manifestation of quantum-mechanical tunneling.

These results can be extended to other materials. The prediction* of the dislocation velocity in lithium fluoride crystals determined by Eq. 3.9 compares reasonably well with the experimental measure-

---

* Using the following nominal parameter values for lithium flouride to calculate both the dislocation velocity and the plastic strain rate as a function of shear stress: $G = 5.3 \times 10^{11}$ dynes / cm^2, $U = 0.22 \times 10^{-13}$ ergs, $W = 1.5 \times 10^{-10}$ m, $v_0 = 3.6 \times 10^3$ m/s, $b = 3 \times 10^{-10}$ m, $l_0 = 3 \times 10^{-9}$ m, $\rho = 2.6 \times 10^3$ kg / m^3, $m = 6\rho b^3$, $V = 40 b^3$. No real effort was made to optimize these calculations.

ments of Johnson and Gilman on LiF at several temperatures [8]. Again, the use of a simple square well potential overestimates the dislocation velocity at low shear stress levels.

Finally, these results apply to screw dislocations as well as edge dislocations since the only differences between these is the potential barrier that each type of dislocation must tunnel through and, perhaps, the effective mass. This will only alter the tunneling probability, T, for mid-range shear stresses. The asymptotic behavior at very high- and very low-amplitude shear stresses will be the same in both cases which is in keeping with the experimental observations of dislocation velocities [8].

3.5. High-Amplitude Shock Loading

High-amplitude shock loading offers an unique insight into plastic deformation because the energy with which the dislocations approach the reduced Peierls barrier nearly equals the barrier height $E \rightarrow U$. For this case the tunneling probability can be approximated by Eq. 3.5 and the plastic strain rate at early times, Eq. 3.14, simplifies to

$$\frac{d\gamma_p}{dt} = \frac{2}{U^4}\left(\frac{\tau^2 V}{2G} + \frac{k\theta}{2}\right)^2 \left(\frac{2\hbar^2 v_0}{mW^2}\right)^2 \frac{b}{l_0} p_c(\tau)\, N_s\, t. \tag{3.16}$$

When the energy imparted to the dislocation segments by the applied shear stress greatly exceeds their thermal energy, as in the case of a high-amplitude shock, the thermal term in Eq. 3.16 can be ignored. Thus, in the high-amplitude shock limit at early times, the average plastic strain rate is proportional to the fourth power of the applied shear stress. Swegle and Grady [22] have observed that within a few tens of nanoseconds after the onset of a high-amplitude shock, the plastic strain rate in a number of different materials varied as the fourth power of the stress level. Choosing representative numerical values* for the constants appearing in Eq. 3.16 and applying a shear stress $\tau = 1$ GPa gives $d\gamma/dt \approx 10^5$ s^{-1} at $t = 10^{-8}$s, which is in the range of the experimental data.

3.6. Elastic and Plastic Waves in Shocks

It is well known that the leading edge of a shock wave is composed of an elastic precursor which travels at the local sound speed, v_0, and a

---

* $U = 0.5 \times 10^{-13}$ ergs, $v_0 = 3 \times 10^3$ m/s, $W = 0.5 \times 10^{-10}$ m, $b/l_0 = 10^{-2}$, $G = 7 \times 10^{11}$ dynes/cm^2, $m = 10^4 m_e \approx 10^{-23}$ g, and $N_s = 10^8$ sources/m^2.

plastic component which moves at a speed $\upsilon(\tau)$ that is dependent on stress wave amplitude. Experimentally, it has been observed that $\upsilon(\tau) \to \upsilon_0$ for large amplitude shocks. The reason for this is clear from Eq. 3.9. Since the maximum value of $T(\tau, \theta)$ approaches unity for high-amplitude shocks, the plastic wave velocity can never exceed the local sound wave speed. A measurement of the ratio $\upsilon(\tau)/\upsilon_0$ gives an experimental determination of the average tunneling probability, $\langle T(\tau)\rangle$, averaged over all of the crystal orientations present in the shocked sample.

3.7. Electroplastic Effects

The other quantities in the plastic strain rate relations of Eq. 3.14 or 3.15 that can be altered experimentally are the height and width of the potential energy barrier. This can be achieved by superimposing an electric field on the crystal as it is undergoing deformation [11]. Physically, this amounts to slightly altering the charge distribution around the atoms or molecules in the lattice and, so, slightly changing their interaction with their neighbors. Let $U = U_0 - \alpha E_A$, where E_A is the applied electric field and U_0 is the potential barrier height in the absence of the electric field. The quantity α is the electric dipole moment and measures the polarization strength at the potential barrier minimum where tunneling takes place. The width of the potential barrier can be approximated in the same way by letting $W = W_0 + \beta E_A$, where W_0 is the width of the barrier in the absence of the electric field.

In order to compare with the available experimental data, consider the ratio of the plastic strain rate when the electric field is applied to the plastic strain rate without the electric field

$$\frac{(d\gamma_P/dt)_{E_A}}{(d\gamma_P/dt)_0} = \frac{T(\tau, E_A)}{T(\tau, E_A = 0)}. \tag{3.17}$$

Here, Eq. 3.15 has been used, since the data to be cited shortly do not appear to have been obtained from fast shock experiments. In the limit where significant plastic deformation occurs, the probability of tunneling may be approximated by Eq. 3.5, and Eq. 3.17 reduces to

$$\frac{(d\gamma_P/dt)_{E_A}}{(d\gamma_P/dt)_0} = \left[\left(1-\alpha\frac{E_A}{U_0}\right)^2\left(1+\beta\frac{E_A}{W_0}\right)^2\right]^{-1}. \tag{3.18}$$

Troitskii and Likhtman [23] and Conrad et al. [24,25] have measured the plastic strain rate in several metals during deformation, both with and without an applied electric field, and have found the ratio of the two to have a nearly cubic dependency over a range of five orders of magnitude in strain rate. Equation 3.18 also predicts a dependency on the applied electric field; however, more complete descriptions of the effect of the electric field on the lattice potential and the effect of joule heating on the plastic deformation rate are needed.

Equation 3.18 predicts a similar enhanced plastic flow for non-conducting insulator and semiconductor crystals when they undergo deformation in the presence of an electric field, and there appears to be experimental evidence for such behavior [26,27]. In a recent series of exploratory experiments, the electroplastic effect was demonstrated in large single crystals of NaCl [28].

3.8. Impediments to Dislocation Motion and Crystal Failure

Although the actual mechanisms of crystal fracture and failure are yet to be understood, it is likely that plastic deformation must be reduced or eliminated for brittle failure to occur. Two means of reducing plastic deformation of a crystalline solid are to increase its defect content and to lower its temperature. Both serve to slow or block the movement of dislocations.

In the tunneling picture, it is easy to see how this might arise. Defects, in the form of other dislocations, twins, or impurities, form barriers to dislocation motion and plastic flow by distorting the lattice in the vicinity of the defect. To pass a blocking obstacle, a mobile dislocation must tunnel through the distorted potential barrier created by the defect. It is likely that the local distortion of the host lattice by the blocking obstacles will present a barrier of both greater height and width than the more easily surmounted reduced potential barriers encountered by the dislocations along the unencumbered preferred crystalline slip systems. Gilman has investigated the case of point defects and has demonstrated that mobile dislocations passed these obstacles by tunneling [29]. Dislocation tunneling through obstacles posed by a "forest" of immobile dislocations has been investigated by Glen [30] and Mott [31].

Often, dislocations are pinned against potential barriers and are unable to move until their energy approaches the barrier energy. Of particular interest is the case where the potential of a blocking obstacle, U_B, exceeds the potential barrier height, U_0, of a lattice otherwise free of obstacles, $U_B > U_0$. For example, a dislocation encountering a grain boundary or similar obstacle must tunnel through the blocking potential, U_B, in order to continue its motion. If the applied shear stress is insufficient to cause tunneling, other dislocations on the same slip plane will pile up behind it. This increases the shear stress on the lead dislocation by N times the applied shear stress, where N is the number of dislocations in the pile-up [6]. The energy of the lead dislocation is approximately

$$E \approx \frac{(N\tau)^2}{2G} + \frac{K\theta}{2}. \tag{3.19}$$

As this energy approaches the energy level needed to tunnel through the blocking potential, the lead dislocation and many of the immediately following dislocations may no longer be bound by the potential of the host lattice since, for these lead dislocations, the condition will arise where $E > U_0$ as $E \to U_B$. Consequently, in the local regions where $E > U_0$, the material will fail, and the behavior of these unconstrained dislocations may not be predicted by the standard dislocation picture. Rather, a new physical description will apply which accounts for the failure of the crystal lattice to restrict the motion of these unbounded dislocations and the sudden and possibly uncontrolled deformation that must follow.

At low temperatures, brittle failure frequently occurs during shock or impact loading. The plastic strain rate expression of Eq. 3.19 provides some insights into how this may happen. At low temperatures, say $\theta \leq 250$ K, the thermal contribution to the tunneling energy decreases rapidly, so that achieving a dislocation velocity or plastic strain rate equivalent to that at a higher temperature requires a significant increase in the applied shear stress. Equation 3.19 shows that more dislocations will be needed in a pile-up to overcome a blocking obstacle than would be required at higher temperatures. Thus, at low temperatures, as $E \to U_B$, many more dislocations in the pile-up will have energy in excess of the energy of the potential wells of the host lattice than would be the case at higher temperatures. This suggests that failure will be more catastrophic at low temperatures than at high temperatures since many more high energy dislocations will be involved.

3.9. Energy Dissipation by Moving Dislocations

The energy dissipated during rapid deformation is also a part of the elastoplastic response of a material to shock or impact. The moving dislocations responsible for plastic flow are also responsible for energy dissipation due to the lattice displacements produced as the dislocations move through the solid. To understand how energy dissipation and localization occurs during plastic flow, the interaction of a moving dislocation with the lattice can be described by the interaction Hamiltonian $H_I = \vec{\sigma} \cdot \vec{\varepsilon} / 2$. [32]. Expressing the strain, $\vec{\varepsilon}$, in terms of the lattice displacement produced by the dislocation and the stress field, $\vec{\sigma}$, as the stress generated by a classical edge dislocation [6], the interaction Hamiltonian becomes [32]

$$H_I = \frac{\Delta X}{4\pi(1-\nu)} \frac{Gb}{r} \left[(\sin\phi + \cos\phi) \left(\frac{\cos\phi}{d_2} - \frac{\sin\phi}{d_1} \right) \right]. \qquad (3.20)$$

The shear modulus, G, provides a measure of the coupling between the dislocation and the lattice. To simplify matters, it will be assumed that G is a constant, independent of crystal slip directions and pressure effects. The quantity b is the Burgers length, ν is the Poisson ratio, r is the distance from the dislocation core, ϕ is the angle between the shear stress and the strain, and d_1 and d_2 are the lattice spacings. Through second quantization, the lattice displacement, ΔX, transforms directly to the appropriate quantum mechanical expression [33].

Treating a moving dislocation as a superposition of plane waves, the energy dissipation rate for N moving edge dislocations is [32,34, 35]

$$\frac{dE}{dt} = N\Gamma G^2 \int (n_q + 1) dq + N \sum_j^l \hbar\omega_{j,j-1} \sum_{f,u} \left| \sum_{l=1}^{l} \frac{\langle f | H' | l \rangle \langle l | H' | u \rangle}{E_l - E_u - \hbar \sum_{j=1} \omega_{j,j-1}} \right|^2 - K \frac{dT}{dx} \qquad (3.21)$$

where

$$\Gamma = \frac{1}{32\pi^3 \rho} \left(\frac{Rb}{1-\nu} \right)^2 \frac{1}{v_0 d^2}.$$

In this expression, R is the radius of the dislocation core, ρ is the mass density, v_0 is the shear wave speed and n_q is the phonon number density of wave vector q. For simplicity, it was assumed that the lattice spacings were nearly equal in all directions, $d \approx d_1 \approx d_2$. The quantity H' is the interaction Hamiltonian coupling the lattice perturbations due to the moving dislocation with the internal vibrational modes present in molecular crystals. The last term in Eq. 3.21 describes the thermal conduction of energy from the shear zone. For most shock or impact situations, this term will only be important when the thermal gradient is large or for relatively long impact times.

A dislocation moving with a speed v encounters and perturbs the lattice potential v/d times per second. This perturbation generates a spectrum of phonons centered at the frequency $\omega_0 = 2\pi v/d$. The dislocation velocity can be written as $v = Tv_0$. Here we will use the empirical expression that $v = v_0 \exp(-\tau_0/\tau)$, where τ_0 is a characteristic shear stress of the material [15]. In order to evaluate the energy dissipation rate, the approximation is made that the phonons generated by the moving dislocations can be treated as being concentrated at the center frequency, ω_0 [35].

When these approximations are introduced into Eq. 3.21, the rate of energy dissipation becomes

$$\frac{dE}{dt} \approx \frac{4\pi\Gamma G^2 N}{d} e^{-\tau_0/\tau} + N\sum_{j}^{l} \hbar\omega_{j,j-1} \sum_{f,u} \left| \sum_{l=1} \frac{\langle f|H'|l\rangle\langle l|H'|u\rangle}{E_l - E_u - \hbar\sum_{j=1}^{l}\omega_{j,j-1}} \right|^2 - K\frac{dT}{dx}, \tag{3.22}$$

where $\omega_{j,j-1}$ is contained in the phonon spectrum centered about $\omega_0 = (2\pi v_0/d)\exp(-\tau_0/\tau)$.

The first term of Eq. 3.22 arises from the first-order transitions and is responsible for dissipation in atomic crystalline solids where no internal molecular modes exist. It is also important in molecular solids subjected to impact and low-level shock.

The higher-order terms in Eq. 3.22 are mainly of interest for molecular crystals, where they are responsible for energy dissipation by rapid shear deformation arising from relatively large amplitude shocks. For these terms to be important, resonance must occur. Resonance takes place due to excitation of the internal molecular vibrational modes by the optical phonons generated by high-velocity dislocations and contained in the spectrum centered about ω_0. At resonance, $\omega = \omega_{l,l-1} = (E_l - E_{l-1})/\hbar$, where E_l and E_{l-1} are the energies of the adjacent l and $l-1$ vibrational levels. Typically, $\omega_{l,l-1} \approx 10^{13}$ rad/s. For many molecular crystals, the maximum dislocation velocity is approximately $v_0 \approx 2 \times 10^3$ m/s and the intermolecular spacing is about $d \approx 10^{-9}$ m, so that $\omega_0 \approx 10^{13}\exp(-\tau_0/\tau)$ rad/s. Thus, resonance requires that the dislocation velocity approach its maximum value, $v \to v_0$, which can only happen in relatively high-amplitude shocks at shear stress levels where $\tau \gg \tau_0$ and the tunneling coefficient approaches unity.

A further simplification is possible. Many molecular solids of interest are composed of molecules containing 20 or more atoms so that numerous internal molecular vibrations are possible. The phonon spectrum generated by the rapidly moving dislocations and centered at ω_0 is broad and has sufficient number density so that it can be assumed that there will always be a vibrational ladder to excite and that all but perhaps the most widely spaced vibrational energy levels will be resonantly excited. At resonance, the transition probability between adjacent vibrational levels approaches unity. Only for the more widely spaced lowest vibrational levels are nonresonant conditions likely to occur. Thus, the energy dissipation–molecular excitation rate in molecular crystals subjected to rapid shear deformation by a strong shock will be determined by the product of only a few nonresonant lowest-level transition probabilities. This results in an extremely fast ($\approx 10^{-11}$ to 10^{-12} s) multiphonon process in which energy is transferred directly from the rapidly moving dislocations to excite the internal vibrational modes and even to dissociate the molecules in the shear/plastic deformation regions of the host crystal. This is believed to be the process responsible for initiating detonation in explosive crystals.

3.10. Conclusions

Plastic deformation of metals, and crystals in general, by shock or impact occurs due to the creation and motion of dislocations. This involves the relative displacement and excitation of the atoms or molecules which are microscopic processes and requires a quantum mechanical treatment. The analysis developed here suggests that dislocations propagate in crystalline solids by quantum mechanical tunneling. The appropriate tunneling regime for plastic deformation occurs at higher shear stress levels where the tunneling particle is forced against the restraining Peierls barrier and its energy begins to approach the barrier height. This is still very much a quantum mechanical regime. The effective mass of the dislocation core segments associated with the tunneling particles can be quite large, but, in this high-energy regime as $E \rightarrow U$, the size of the mass is not as important as at lower energies and tunneling will still be very probable. This permits the derivation of the elastoplastic response of the solid when subjected to shock or impact at arbitrary temperatures and even in the presence of an electric field.

Energy dissipation during plastic deformation occurs due to the lattice perturbations produced by the moving dislocations. This energy can be localized in shear bands that contain concentrations of dislocations, probably created by a single or at most just a few dislocation sources.

Finally, it is suggested that crystal failure can occur because of the inability of the host lattice to constrain those dislocations that pile up behind blocking obstacles and have energies that exceed the depth of the potential wells of the host lattice.

Acknowledgments

This work was supported by the Office of Naval Research and by the Naval Surface Warfare Center (NSWC) Independent Research Funds. The author wishes to thank Dr. D.H. Liebenberg of the ONR and Dr. C.W. Dickinson of NSWC for their support and encouragement. He also wishes to acknowledge and thank Dr. J.J. Gilman of the University of California at Los Angeles, Dr. R.N. Lee of NSWC, and Mr. M.S. Coffey of the College of William and Mary for their help and insights into this problem.

References

[1] G.I. Taylor, *Proc. Roy. Soc. A* **145**, p. 362 (1934).

[2] M. Polanyi, *Z. Phys.* **89**, p.660 (1934).

[3] J.M. Burgers, *Proc. Kon. Ned. Akad. Wetenschap.* **42**, pp. 293, 378 (1939).

[4] E. Orowan, *Z. Phys.* **89**, pp.605, 634 (1934).

[5] J. Frenkel, *Z. Phys.* **37**, p. 572 (1926).

[6] J. Hirth and J. Lothe, *Theory of Dislocations*, McGraw-Hill Book Co., New York, pp. 701–706 (1968).

[7] J. Sharma, in *Shock Compression of Condensed Matter—1995* (eds. S.C. Schmidt and W.C. Tao). American Institute of Physics, New York (1996).

[8] W.G. Johnston and J.J. Gilman, *J. Appl. Phys.* **33**, p. 129 (1959).

[9] K.S. Kim and R.J. Clifton, *J. Mater. Sci.* **19**, p. 1428, (1984).

[10] U.F. Kocks, A.S. Argon, and M.F. Ashby, in *Prog. Mater. Sci.* **19** (1975).

[11] C.S. Coffey, *Phys. Rev.* **B49**, p. 208 (1994).

[12] J. Weertman, *J. Appl. Phys.* **29**, p. 1685 (1958).

[13] A. Messiah, *Quantum Mechanics*, Vol. 1. Wiley, New York, p.77 (1961).

[14] V.L. Indenbom, B.V. Petukhov, and J. Lothe, in *Elastic Strain Fields and Dislocation Mobility* (eds. V. L. Indenbom and J. Lothe), North-Holland, Amsterdam, p. 489 (1992).

[15] J.J. Gilman, *Aust. J. Phys.* **13**, p. 327 (1960).

[16] C.S. Coffey, *J. Appl. Phys.* **66**, p. 1654 (1989).

[17] J.D. Campbell and W.G. Ferguson, *Phil. Mag.* **21**, p. 63 (1970).

[18] A.R. Rosenfield and G.T. Hahn, *Trans. Am. Soc. Metals* **59**, p. 962 (1966).

[19] P.S. Follansbee and U.F. Kocks, *Acta Met.* **36**, p. 81 (1988).

[20] O. Vohringer, "Deformation Behavior of Metallic Materials," International Summer School on Dynamic Behavior of Materials, ENSM, Nantes (1989).

[21] W.G. Ferguson, Ph.D. Thesis, University of Auckland, New Zealand (1964).

[22] J.W. Swegle and D.E. Grady, *J. Appl. Phys.* **58**, p. 692 (1985).

[23] O.A. Troitskii and V.I. Likhtman, *Dokl. Akad. Nauk. SSSR*, **148**, p. 332 (1963).

[24] H. Conrad and A.F. Sprecher, *Dislocations in Solids* (ed. F.R.N. Nabarro), Elsevier Science, New York, (1989), Chapter 43.

[25] H. Conrad, A.F. Sprecher, W.D. Cao, and X.P. Lu, *J. Miner. Metals Mater. Soc.* **42**, p. 28 (1990).

[26] A.N. Kulichenko and B.I. Smirnov, *Sov. Phys. Solid State* **23**, p. 595 (1981).

[27] A.N. Kulichenko and B.I. Smirnov, *Sov. Phys. Solid State* **26**, p. 570 (1984).

[28] C.T. Stanton, C.S. Coffey, and F. Zerilli, *J. Mater. Res.* **10**, p. 258 (1995).

[29] J.J. Gilman, *J. Appl. Phys.* **39**, p. 6086 (1968).

[30] J.W. Glen, *Phil. Mag.* **1**, p. 400 (1956).

[31] N.F. Mott, *Phil. Mag.* **1**, p. 568 (1956).

[32] C.S. Coffey, *Phys. Rev.* **B24**, p. 6984 (1981).

[33] C. Kittel, *Quantum Theory of Solids*, J. Wiley & Sons, Inc., New York, (1963).

[34] C.S. Coffey, *Phys. Rev.* **B32**, p. 5335 (1984).

[35] C.S. Coffey, in *Structure and Properties of Energetic Materials* (eds. D.H. Liebenberg, R.W. Armstrong, and J.J. Gilman), Material Research Society, Boston, Volume 296, p. 63 (1993).

CHAPTER 4

Molecular Processes in a Shocked Explosive: Time-Resolved Spectroscopy of Liquid Nitromethane

G.I. Pangilinan and Y.M. Gupta

4.1. Introduction

Scientific studies of shock-induced chemical decomposition in condensed high explosives and the subsequent buildup and propagation of detonation waves comprise a significant element of shock wave research. High explosives play a dual role in this challenging field: They are commonly used to produce intense shock waves to permit an examination of condensed matter at extreme conditions; they can also be used to gain a fundamental understanding of the atomic/molecular processes governing shock-induced chemical changes. In this chapter, we focus our attention on the latter aspect.

Historically, shock wave research on condensed explosives, similar to the work on nonreacting condensed materials, has focused on continuum and thermodynamic investigations [1,2]. This is due to several factors: the strong emphasis on applications, the enormous experimental difficulties encountered in examining explosives, and the fact that relating continuum investigations of reactive flow problems to molecular mechanisms associated with structural and chemical changes is a scientifically difficult problem. The books by Fickett and Davis [1] and Chéret [2] provide a comprehensive background and summary of past work on detonation-related studies in explosives. With regard to nonlinear wave propagation, the liberation of chemical energy and associated kinetics provide interesting phenomena related to the growth of shock waves [3].

It is useful to consider the shock initiation problem in condensed explosives to occur sequentially as follows. First, the passage of a plane shock wave results in an "excitation" of the explosive molecules. This is followed by structural and molecular changes associated with the onset of chemical reaction in the excited molecules and, finally, by release of the chemical energy which is coupled to the propagating shock wave. The last step results in a macroscopically observable pressure increase. The development of time-resolved stress and/or particle velocity measurements in the seventies provided considerable information on chemical energy release rates in a variety of high explosives [4]. These data were used to develop phenomenological approaches for modeling chemical reactions and coupling of the chemical energy to the propagating shock wave [5]. Although the benefits of continuum measurements and analyses are well recognized, they cannot be used to discern fundamental molecular processes associated with the onset of chemical reactions.

In the last fifteen years, there has been a growing interest in understanding the role of molecular and electronic structure on the sensitivity of explosives to a variety of stimuli. On the theoretical side, developments in quantum chemical calculation, crystal structure analysis, and molecular dynamics provide approaches to probe fundamental issues. Similarly, experimental methods (optical spectroscopy) are needed to directly probe molecular changes associated with the initial and subsequent stages of shock-induced chemical reactions. The development and use of such spectroscopic probes for examining time-dependent atomic and molecular changes in shocked condensed materials has been a major activity at Washington State University. In this chapter, we present a comprehensive summary of our spectroscopy effort on an insensitive high explosive (liquid nitromethane) to indicate the scientific issues that can be addressed by such studies. To facilitate the discussion in later sections, a very brief background regarding spectroscopic probes is presented in the next section.

4.2. Optical Spectroscopy Probes

The need to examine atomic/molecular processes in shocked condensed materials led to the development of optical spectroscopy, over the past fifteen years, at several organizations: Commissariat à l'Énergie Atomic (France), Livermore, Los Alamos, and Sandia National Laboratories, the Naval Research Laboratory (Washington, D.C.), and our laboratory at Washington State University. Both electronic spectroscopy (UV-vis absorption [6–9], luminescence [10], reflection [11,12]) and vibrational spectroscopy (coherent anti-Stokes Raman scattering (CARS) [13,14], spontaneous Raman [15–20], infrared absorption [21]) methods were developed. A noteworthy aspect of the work at Washington State University has been the emphasis on the development of time-resolved measurements for both electronic and vibrational spectroscopy. Despite the difficulty associated with collecting time-resolved data, such measurements are necessary for understanding mechanisms governing shock-induced chemical reactions.

Several points are worth emphasizing. Although optical spectroscopic methods have been employed routinely for decades in physics and chemistry, their use in shock wave experiments pose special challenges due to the single-event nature of these experiments, extreme pressures and temperatures, and need for temporal synchronization to achieve nanosecond resolution. Because spectroscopic methods provide information at the molecular level, they are inherently more sensitive than stress and/or particle velocity measurements, which can provide information only after significant chemical energy has been released. The spectroscopic probes, therefore, should be viewed as a complement to, but not a substitute for, continuum measurements. They sample different stages of a complex, time-dependent problem. Without a good knowledge of the continuum parameters (shock amplitude, pulse duration, temperature when applicable, etc.), the spectroscopic data have limited value. In the following paragraphs, a very brief description of the underlying principles used in our electronic and vibrational spectroscopy is presented. More detailed accounts may be seen elsewhere [22,23].

4.2.1. UV-Visible Absorption Spectroscopy

Absorption spectroscopy over the UV-visible range is commonly used to probe the electronic energy levels (and, hence, the electronic structure) of molecules [22]. Because chemical reactions represent a drastic change in the electronic structure and due to the relative ease in performing these experiments, time-resolved absorption experiments were the first to be used to probe shock-induced chemical changes [6]. Light of intensity I_0 passes through a thickness L of a sample and is attenuated to a value I due to absorption (and/or scattering) in the sample. The change in transmittance can be related to absorbance in the material which, in turn, is related to the electronic energy levels for the material [23].

For most organic compounds like nitromethane, which are transparent at visible wavelengths, the absorbance rises sharply in the UV where the energy gaps between the highest occupied and lowest unoccupied levels are located [22]. Due to various experimental considerations, it is difficult to probe absorbance at wavelengths shorter than 250–260 nm in shock wave experiments. Hence, only the lowest energy absorption band can be probed in these measurements.

Since the pioneering work by Duvall and co-workers [6] on CS_2, time-resolved absorption (and later reflection) experiments have been used extensively to probe shock-induced chemical changes in CS_2. The comprehensive studies on CS_2 [7–9,17,24,25] and the more cursory examination of other materials [26] have established the usefulness of these measurements to probe shock-induced chemical changes, including an accurate determination of the threshold conditions. Despite the many benefits of using electronic spectroscopy, precise determination of molecular changes associated with shock-induced chemical reactions is difficult. The lack of sharp transitions in condensed materials, coupled with the very limited number of electronic transitions that can be examined, makes the data analysis difficult.

4.2.2. Raman Spectroscopy

Raman spectroscopy is a convenient method to probe intermolecular and intramolecular vibrations using visible light. In a typical Raman scattering experiment, incident light of frequency ω_{inc} is inelastically scattered to frequencies $\omega_{inc} \pm \omega_v$, where ω_v represents the frequency of molecular vibrations. Typically, the Stokes frequencies $(\omega_{inc} - \omega_v)$ are recorded because of intensity considerations.

The intensities of Raman lines are governed by three interactions: i) coupling of the incident radiation with the electrons in the material, ii) modulation of the electronic states due to oscillations of the nuclei about their equilibrium positions; and iii) transition to the electronic ground state with the emission of the scattered radiation [27]. Raman scattering is inherently a weak process, where the intensity of the scattered light (at $\omega_{inc} - \omega_v$) is typically 7–9 orders of magnitude lower than that of the incident light. Hence, considerable care is needed to eliminate the incident light (at ω_{inc}). It is, however, particularly suited for obtaining a molecular description of the onset of chemical changes because of several features. First, vibrations arising from different parts of the molecule can be simultaneously monitored. Second, at incident frequencies away from resonance, intensities are linearly dependent on the number of molecules being investigated. Furthermore, an inhomogeneous distribution of molecular energies at high pressure can be observed through the spread (width) in the vibrational frequencies. Finally, the vibrational frequencies, which typically increase with pressure due to anharmonic interactions, reflect the intermolecular and intramolecular forces in the material.

Time-resolved Raman scattering has proven valuable in the microscopic characterization of structural and symmetry changes in the shocked state. Raman work on quartz has emphasized the role of nonhydrostatic stresses on compression [19] and tension [20] at the molecular level. In diamond, the Raman work [18] has shown, through a splitting of the triply degenerate vibrational mode, the lowering of symmetry due to uniaxial strain loading. Our current time-resolved Raman spectroscopy builds on these initial investigations. We also bring attention to the extensive studies by Moore, Schmidt, and co-workers on shocked nitrogen and other diatomic molecules. Through the use of coherent anti-Stokes Raman scattering (CARS), they were able to obtain very strong signals [13,14].

4.3. Shock Response of Nitromethane and Sensitized Nitromethane

To date, the most comprehensive optical spectroscopy effort on a shocked, condensed explosive has been the work on liquid nitromethane at Washington State University (W.S.U.). In this section, we present a summary of this ongoing effort with an emphasis on the results that have provided an insight into the molecular processes governing shock-induced chemical decomposition. Other noteworthy examples of comprehensive spectroscopic investigations of shock-induced molecular changes can be seen in the Los Alamos work on liquid diatomics [13,14], and the W.S.U. work on liquid CS_2 [7–9,17, 24,25].

Due to the cost and difficulty associated with spectroscopic measurements in shocked energetic materials, the proper choice of materials is important. Nitromethane is an insensitive high explosive well suited for investigation in an academic institution. Due to its relatively simple structure, it has served as a good prototype for investigating chemical decomposition of nitrocompounds and is one of the best studied energetic materials. By examining a liquid in contrast to a solid, we avoid many complexities: sample preparation, role of defects both at the atomic and the mesoscopic scale, and effects due to nonhydrostatic deformation. These issues will be considered in future studies on energetic solids. There exists a large body of spectroscopic studies at ambient [28–31] and at static high pressures [32–35] and continuum studies under shock loading [36–38] on nitromethane. At single-shock pressures of 8.0 GPa, reaction as measured by light emission in streak camera records or changes in particle velocities in VISAR measurements is observed to occur on microsecond time scales [38]. Other continuum studies have shown that nitromethane is sensitized upon addition of small amounts of amines, including ethylenediamine ($NH_2CH_2CH_2NH_2$) [39]. There existed, prior to our work, some spectroscopic data on shocked nitromethane

[40–42], but most of these data constituted exploratory efforts and the results were not always in agreement.

Potential mechanisms for shock-induced chemical decomposition of pure and sensitized nitromethane have been proposed by Engelke et al. [43–45], Bardo [46], Cook and Haskins [47,48], and Constantinou [49]. Engelke proposed that the nitromethyl aci-ion ($H_2C = NO_2^-$) is responsible for shock-induced decomposition of pure nitromethane and its sensitization by amine addition. Engelke suggested that the aci-ion concentration is enhanced at high hydrostatic pressures [45], or upon addition of small amounts of amines [44]. Bardo, noting the volume compression under shock loading, emphasized the importance of bimolecular interactions and proposed a head-to-tail bimolecular association as the first step in the shock decomposition of pure nitromethane [46]. The effects of the molecular environment were further emphasized in relating the molecular structure of nitromethane with its sensitivity [50]. In 1989, Cook and Haskins [47] questioned the role of the aci-ion in the sensitization of nitromethane and proposed, through quantum mechanical calculations, a hydrogen-bonded complex between a nitromethane molecule and an amino group that lowers the activation energy for CN bond scission. This reaction is shown to be autocatalytic and exothermic. In 1992, Constantinou [49], based on slow thermal decomposition measurements, proposed that the sensitization is indeed due to CN weakening; but the CN weakening is caused by a charge transfer complex, and not hydrogen bonding. The thesis work by Constantinou provides a comprehensive account of the thermal decomposition of pure and sensitized nitromethane at ambient pressure and serves as an excellent background for this work.

Experimental measurements are necessary to evaluate the different molecular models postulated for shock decomposition. The objective of our nitromethane work was to address the following broad issues:

i) What are the molecular changes associated with shock-induced decomposition in pure and sensitized nitromethane?

ii) How does amine addition sensitize nitromethane? Are the molecular processes in sensitized nitromethane similar to those in pure nitromethane with lowered pressure and temperature thresholds, or are new molecular processes introduced?

iii) How are the spectroscopic and continuum studies related?

iv) Is the initiation of chemical reaction controlled by pressure or temperature?

4.3.1. Experimental Method

Details of the experimental methods used in our spectroscopic studies are provided in published articles and reviews [24,25,51–53]; only a summary is presented here. Briefly, optical spectroscopy under shock loading can be considered as a "pump and probe" experiment where the pump is the propagating shock. Although there exist many techniques for producing shock waves, flyer-plate impact methods provide the best control for producing pulses of known amplitude and duration. Because of the single-event nature of the experiments and because the pump is a large amplitude mechanical shock wave produced by impact, these experiments require considerable care. Figure 4.1 shows a typical experimental setup for our time-resolved Raman spectroscopy measurements in shocked liquid nitromethane. In our impact experiments, thin liquid samples (ranging from 10 to 300 μm thick, depending on the spectroscopic method), are subjected to stepwise loading (SWL) due to shock reverberation between two optical windows (typically, a-cut sapphire). By varying the impactor velocity and window thickness, shock wave amplitudes and durations are controlled precisely.

Raman measurements are obtained continuously at predetermined time intervals (typically 50 ns) while the sample is being subjected to uniaxial strain loading. The typical duration of our experiments is 1 μs, a time interval limited by the arrival of release waves from the lateral boundaries of the cell at the center of the sample. Light from a laser is sent to the sample and scattered light is spectrally dispersed by the spectrometer, temporally dispersed by the electronic streak camera, and recorded on a two-dimensional CCD detector; the data consist of intensity–wavelength–time output. Other electronic and optical components are indicated in the figure. Light transmission to and from the sample is carried out using optical fibers, an important feature for these destructive experiments. Using appropriate trigger pins and time delays, the laser and the streak camera are carefully synchronized with the impact event and the arrival of the shock wave at the sample. For higher time resolution, this time synchronization becomes an increasingly difficult and important problem.

Our time-resolved, UV-visible absorption experiment is a single-pass transmission measurement so that the incident light from a flash lamp is introduced through the impactor [51]. Light collection and synchronization use an arrangement very similar to that shown in Fig. 4.1, although the specific components are different.

Two aspects of the optical spectroscopy experiments are worth noting. First, because a shock wave is a propagating disturbance, both

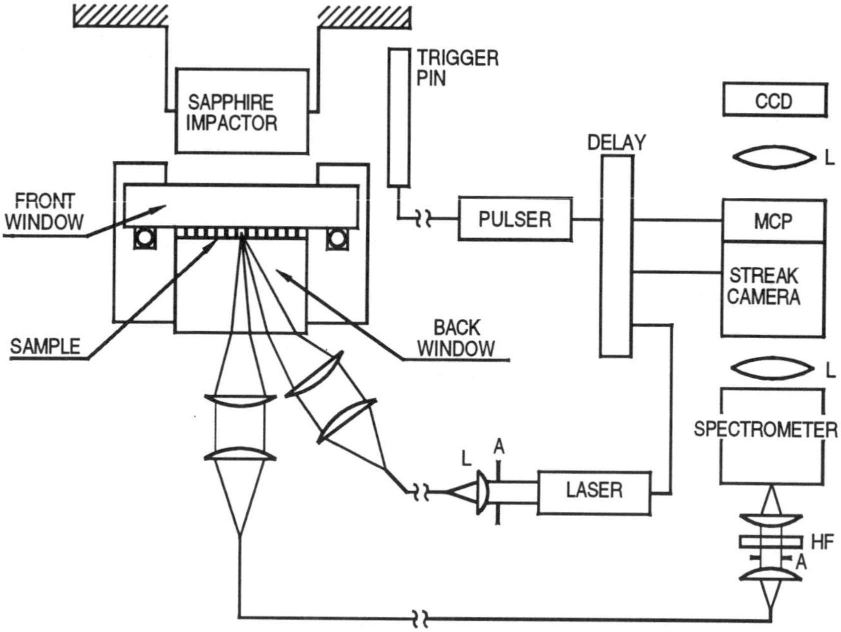

Figure 4.1. Experimental configuration for obtaining time-resolved Raman measurements in shocked nitromethane. L: lens; A: aperture; HF: holographic filter; MCP: microchannel plate intensifier; CCD: charge coupled device detector. Light transmission to and from the sample is carried out using optical fibers [52].

spatial and temporal variations need to be considered for the physical quantities of interest. It is for this reason that time-resolved continuum measurements using Lagrangian probes (stress and particle velocity histories at a fixed material location) have proven immensely beneficial for reactive flow studies [3]. Ideally, one would want the highest spatial and temporal resolution in shock wave experiments. In our spectroscopy experiments, we address this requirement by using thin samples and nanosecond resolution. The price we pay is that the sample reaches peak pressure through stepwise loading, and not through the single shock typical of most continuum measurements. By using very thin samples, we can reduce the total time to achieve the peak stress, but it still takes several steps to reach this state. Alternatively, experiments on thick samples can be carried out to achieve single-shock loading, but it is not clear how to achieve high spatial and temporal resolution in the investigation of a time-dependent process such as a chemical reaction.

Second, because of stepwise loading (SWL), the peak temperature attained in these experiments, for a given peak pressure, is lower than that attained in single-shock experiments. The peak pressure for the unreacted nitromethane is determined solely by the shock response of the optical windows and is very precisely known. In the absence of reliable temperature measurements under shock loading, both single-shock and SWL experiments rely on calculated temperatures to interpret the nitromethane response. As such, the development of a good equation of state is an important need. The equation of state used for liquid nitromethane in the present work was developed by Duvall [54] and Winey [55] and is based on the universal liquid Hugoniot [56] with constants adjusted to match the data from Lysne and Hardesty [57]. Details regarding the specific heat calculation and the determination of $(\partial P / \partial T)_V$ may be seen elsewhere [55]. Although considerable effort has gone into the equation of state development, it is difficult to put error bars on temperature calculations. Reliable temperature measurements in shock wave experiments continue to be an important need.

4.3.2. Results and Discussion

Selected results from time-resolved absorption measurements [51] and Raman scattering measurements [52,53] on pure and sensitized nitromethane (nitromethane with 0.1 wt.% of ethylenediamine additive) shocked to 14 GPa peak pressures are presented here. The electronic spectroscopy measurements are easier to obtain and have been valuable in establishing the onset of the time-dependent irreversible changes due to chemical decomposition. In the UV-vis absorption measurements, we were only able to examine the n–π^* transition centered at 275 nm; the electronic charge distribution is localized around the NO_2 group. All of the electronic spectroscopy data discussed below are taken from Ref. 51.

Figure 4.2 is a typical plot of absorbance vs. wavelength for shocked pure nitromethane. At times corresponding to shock reverberations (spectra 15–20), the band edge shifts to longer wavelengths. Upon reaching peak pressure and up to 400 ns later (spectrum 28) the band edge maintains its position indicating that the edge shift is a pressure-induced effect. A similar plot for sensitized nitromethane shown in Fig. 4.3 shows that the band edge shifts during shock reverberations (spectra 23 to 28) and continues to shift even after peak pressure is reached in the material. The time-dependent shift of the band edge, at constant pressure, suggests the onset of a chemical reaction. A plot of the band edge shifts for sensitized nitromethane (Fig. 4.4) shows that the shifts are larger for sensisensitized nitromethane and show time dependence even at peak pressures as low as 9.8 GPa. In contrast, the band edge shift in pure nitromethane remains stationary at 14.7 GPa.

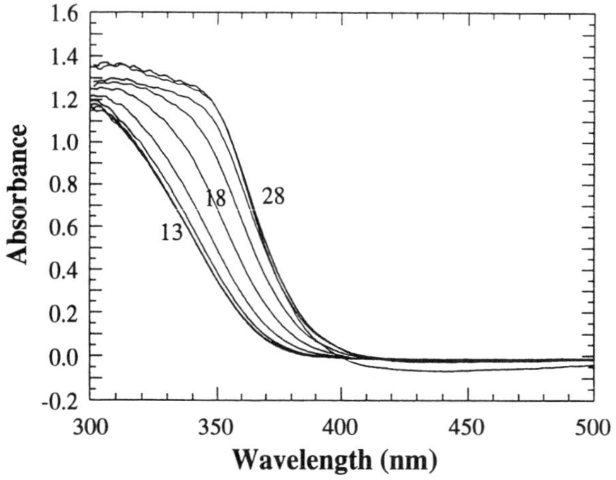

Figure 4.2. Absorption spectra of pure nitromethane while it is being shocked to 14.7 GPa. Successive spectra are separated by 50 ns. Only spectra 13–20, 27, and 28 are shown for clarity. Spectra 1 to 15 are identical. Shock enters the material between spectra 15 and 16, and peak pressure is attained by spectrum 20 [51].

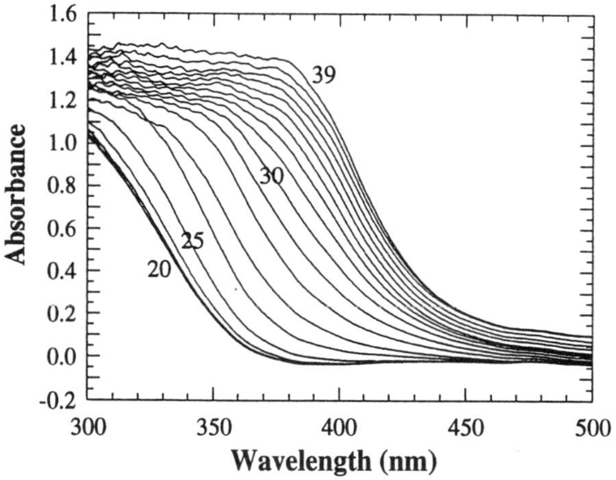

Figure 4.3. Absorption spectra of sensitized nitromethane while it is being shocked to 13.2 GPa. Only spectra 20–39 are shown. Spectra 1 to 23 are identical. The shock wave entered the liquid between spectra 23 and 24, and peak pressure is attained by spectrum 30 [51].

To further establish that the data for sensitized nitromethane corresponded to irreversible changes, unloading experiments were conducted; in these experiments, the shock pressure is unloaded to very small values while maintaining the samples in a state of uniaxial strain. In pure nitromethane, the pressure-induced change in the absorbance during shock loading (spectra 10 to 19, Fig. 4.5a) is reversed upon unloading (spectra 19 to 25, Fig. 5b). In contrast, the absorbance changes observed in sensitized nitromethane during shock loading (spectrum 20 to 26 of Fig. 4.6a) continue (spectra 26 to 31 of Fig. 4.6b) even during unloading. The irreversibility of the absorbance changes in shocked sensitized nitromethane is a clear indication of the onset of an irreversible chemical reaction.

Three points are emphasized with regard to the absorption spectroscopy results. First, pure nitromethane does not show reaction onset in our experiments up to pressures of 14 GPa. This result does not contradict the continuum results that showed reaction at 8.0-GPa pressures, mainly due to the difference in loading conditions (single shock vs. shock reverberation) which result in different peak temperatures. This comparison again underscores the importance of the equation of state to characterize the temperature of the shocked material. Calculated temperatures from the equation of state used in our work yield $T = 770$ K for our nitromethane experiments at 14 GPa, and $T = 1050$ K for nitromethane singly shocked to 8.0 GPa, strongly suggesting that the shock decomposition in pure nitromethane may be a thermally activated process. Recent work by Winey [55]

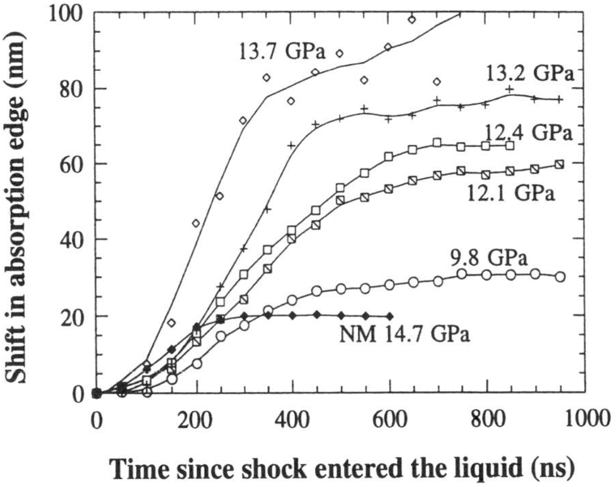

Figure 4.4. Variation with time of the shift in the absorption edge from five reverberation experiments on sensitized nitromethane, and one experiment on pure nitromethane with peak pressure of 14.7 GPa, showing the larger and time-evolving shifts of the absorption edge in sensitized nitromethane. Peak pressure is reached between 200 and 300 ns for these experiments [51].

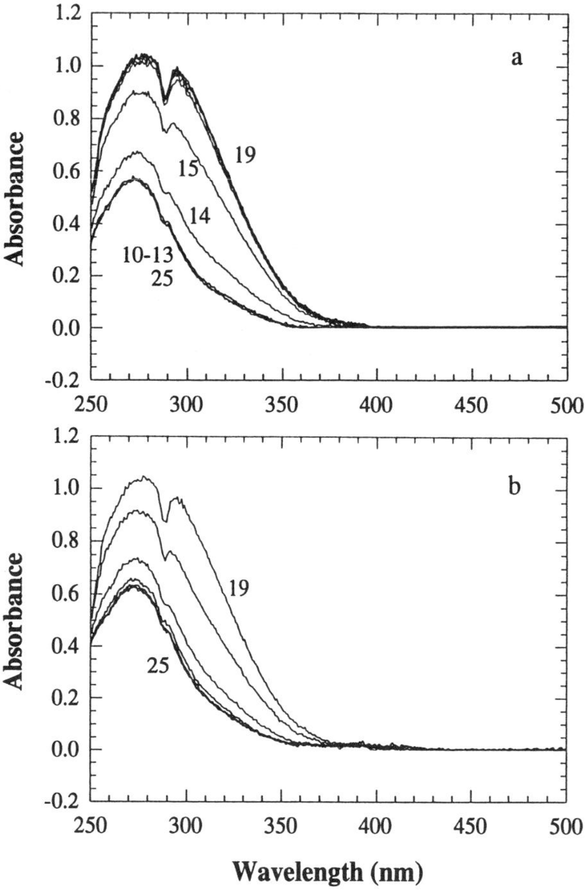

Figure 4.5. Variation in the absorption spectrum of pure nitromethane in an unloading experiment with 14.1 GPa peak pressure: (a) spectra 10–19 (loading phase); (b) spectra 19–25 (unloading phase). In (a), we have included spectrum 25 to demonstrate that the absorption spectrum reverts to its original position and peak absorbance upon pressure release [51].

has suggested that temperatures of 950 K are needed to see the onset of a chemical reaction on a microsecond time scale with our absorption spectroscopy. Second, we clearly observe sensitization in our absorption measurements as a lowering of pressure threshold for initiation of a chemical reaction. This microscopic description may prove useful in providing a definition for explosive sensitivity [50]. Finally, it is noted in our UV-visible results of shocked nitromethane that details of the molecular processes cannot be obtained easily from such data.

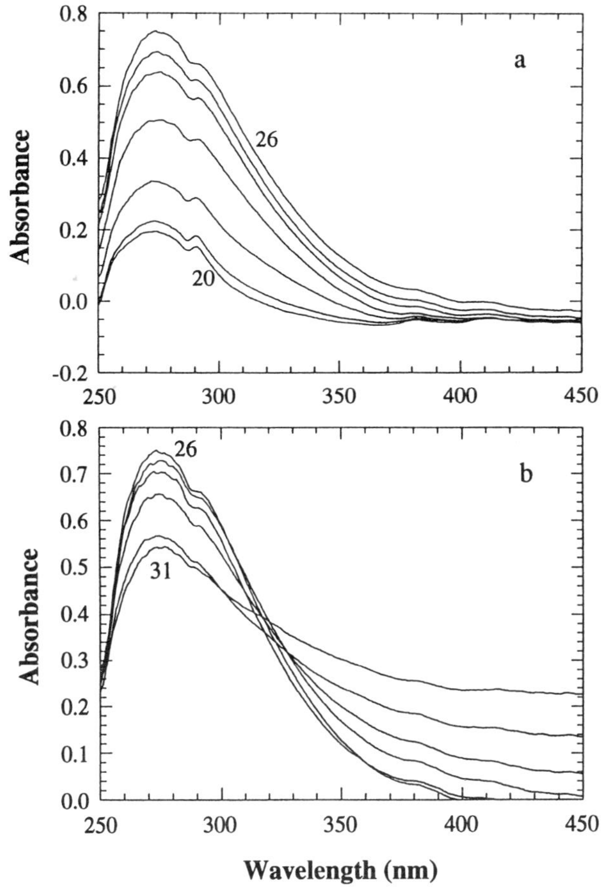

Figure 4.6. Variation in the absorption spectrum of sensitized nitromethane in an unloading experiment with 13.8 GPa peak pressure: (a) spectra 20–26 (loading phase); (b) spectra 26–31 (unloading phase) [51].

To obtain information about the molecular changes, we have invested considerable effort in obtaining good time-resolved Raman measurements in shocked nitromethane. A recent article has presented the first evidence of the ability to continuously monitor molecular changes associated with the onset of a chemical reaction in shocked nitromethane [52]. In the following paragraphs, we summarize the important findings from our Raman measurements.

In shocked nitromethane, we are able to monitor several of the intramolecular modes: CN stretch (917 cm^{-1}), superposed NO_2 stretch/CH_3 band (1400 cm^{-1}/1377 cm^{-1}), and the CH_3 stretch (2968 cm^{-1}). Studies of unreacted shocked nitromethane showed hardening of different modes and a relative enhancement of the NO_2 stretch/ CH_3 bend. A representative example of the changes in the Raman spectrum in pure nitromethane shocked to 14 GPa is shown in Fig. 4.7. Higher-resolution experiments at different pressures showed that the Raman modes exhibit increased widths (the CH_3 stretch broadens most) and large frequency shifts which vary nonlinearly with pressure, as shown in Fig. 4.8. Consistent with our UV-visible measurements, we observed no indication of permanent chemical changes. The broad widths we measured at shock pressures are consistent with an inhomogenous distribution of molecular states, which may include precursor states. Moreover, the curvature of the vibrational shifts with pressure shown in Fig. 4.8 is similar to that observed in hydrogen at pressures just below a phase transition [58]. The Raman measurements therefore probe molecules that may be precursors to the chemical reaction.

The steep hardening of the CH_3 stretch, which initially is the stiffest vibration, is inconsistent with compression of only the intramolecular bonds [59]. This pressure-induced hardening can be understood if other bonds, e.g., intermolecular forces involving the hydrogen, are present. This intermolecular potential is expected to vary from one molecule to another and is partly responsible for the increased width of the CH_3 stretch at shock pressures. Since the convo-

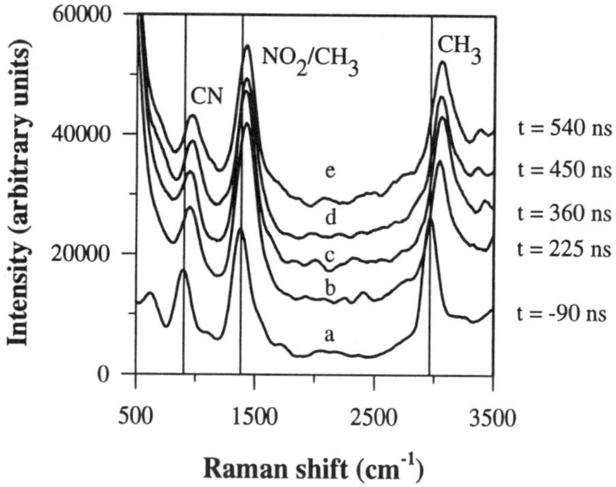

Figure 4.7. Representative Raman spectra of nitromethane shocked to 14 GPa obtained at (a) ambient pressure; (b) during shock-up; (c), (d), and (e) after peak pressure is reached. $t = 0$ corresponds to the time when the shock wave initially enters the sample. Vertical lines mark the frequency positions of the various modes at ambient pressure. For clarity, spectra (b) to (e) are vertically offset successively by 5000 units [52].

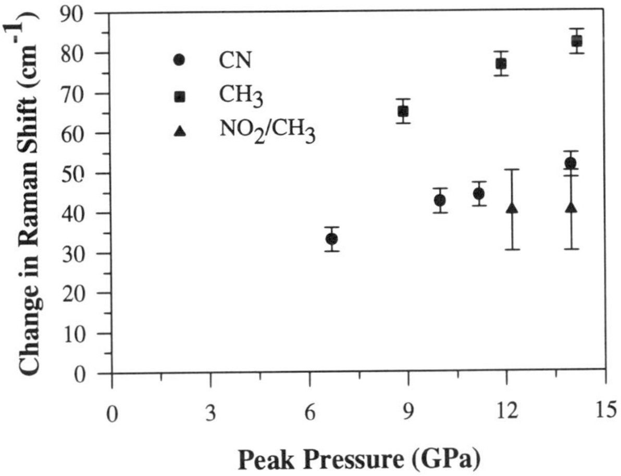

Figure 4.8. Frequency shifts of the nitromethane vibrations from their ambient positions plotted vs. pressure. The CH_3 stretch vibration at 2968 cm^{-1} shows the largest frequency shift. The origin is a datum for each of the three modes listed [53].

luted NO_2 stretch/CH_3 bend vibration did not broaden as much, the CH_3 bending mode does not exhibit this inhomogeneous distribution. It is thus likely that at most one hydrogen per molecule is involved with the intermolecular interaction.

To investigate the molecular changes associated with the chemical decomposition of sensitized nitromethane, a series of Raman experiments was performed. During shock reverberation, the Raman modes are observed to harden as they do in pure nitromethane. Upon attaining peak pressure as low as 12 GPa, slight CN softening is observed, and a time-dependent increase in the spectral background is noted. These become more dramatic at the 14 GPa peak pressure shown in Fig. 4.9, where representative spectra are presented. Initial hardening with shock reverberation is observed; however, CN softening and disappearance, and the time evolution of a spectral background, probably due to evolving intermediates, at constant peak pressure are seen. These results have been confirmed with experiments at higher spectral dispersion. Unlike the pure nitromethane spectra, major changes observed in sensitized nitromethane are irreversible.

The chemical reaction observed does not consume all of the nitromethane molecules at 14 GPa as evidenced by the presence of the NO_2 stretch and CH_3 modes at the end of the time window in Fig. 4.9. This implies that sensitization is achieved through a lowering of the activation energy for initial reactions. This is consistent with models proposed by Cook and Haskins [47,48] and Constantinou [49] that amine addition results in the formation of molecular complexes with lowered activation energies for exothermic reactions. The energy released from the decomposition of the complexes then leads to subsequent reactions involving CN scission, observed as CN mode softening and broadening and finally disappearance as seen in Fig. 4.9. Due to the low concentration of amines, we could not identify in this work the nature of the complexes responsible for the initial reaction: hydrogen bonded [47,48] or charge transfer [49]. We are currently evaluating potential methods to probe these complexes. The observed CN mode softening, however, is not consistent with the formation of aci-ions [43–45] because the double-bonded CN in the aci-ion is expected to have a higher vibrational frequency.

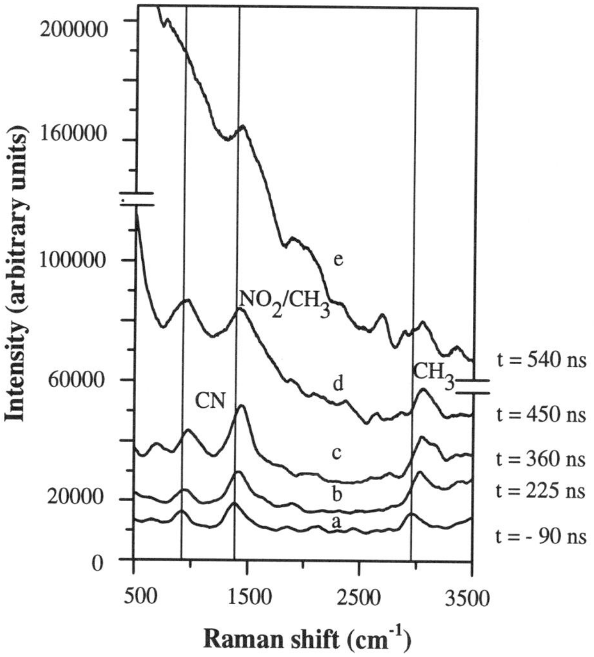

Figure 4.9. Representative Raman spectra of sensitized nitromethane shocked to 14 GPa obtained at times similar to Fig. 4.7. Vertical lines mark the frequency positions of the various modes at ambient pressure. Spectra (b) to (e) are vertically offset successively by 5000 units. The CN vibration softens in spectrum (d) and then disappears in spectrum (e). An increase in the background is observed after peak pressure is reached [52].

Finally, studies of pure nitromethane shocked to higher, reacting pressures [55] have been completed recently by Winey. This work shows that, unlike sensitized nitromethane, CN bond scission is not the first step in the chemical decomposition of pure nitromethane. The presence of the amine is therefore not merely a catalyst for the nitromethane reaction; instead, it changes the molecular processes involved in shock decomposition.

4.4. Summary and Conclusions

Through the liquid nitromethane work summarized in this chapter, we have demonstrated that time-resolved optical spectroscopy can be used to probe the early stages of shock-induced chemical decomposition in condensed explosives. In our spectroscopic measurements on pure nitromethane shocked to 14 GPa (through stepwise loading), we see evidence for increased intermolecular reactions but not chemical reactions. In sensitized nitromethane, onset of chemical reaction occurs for peak pressures as low as 9.8 GPa through the lowering of the activation energy for the initial reaction. From the Raman measurements on sensitized nitromethane, we observe CN scission at the initial stage of the chemical reaction. The continuing work on pure nitromethane [55] suggests that the reaction paths for the pure and sensitized nitromethane are different.

From the liquid nitromethane work, we note several important areas for future investigations. Reliable temperature measurements are an important need because without such data it is difficult to define the threshold conditions for chemical decomposition. We also need to probe the material after the reaction has started to identify intermediate products. The nature of the molecular complex that causes CN scission in sensitized nitromethane merits further study. Work on these problems is currently underway.

A natural extension of our current work is to apply the spectroscopic methods described here to the study of energetic solids. The role of crystal defects and nonhydrostatic deformations on chemical decomposition needs to be carefully explored to understand shock initiation in solid explosives. Scientific issues will likely be best addressed through studies on selected single crystals [60,61].

In closing, studies of shock-induced chemical reactions are at an exciting stage. Theoretical developments coupled with molecular probes have the potential to address scientific questions of long standing. Optimal progress will require sustained, comprehensive efforts on selected materials.

Acknowledgments

It is a pleasure to acknowledge the many valuable contributions of C.P. Constantinou and J.M. Winey to the nitromethane project. Discussions with these individuals have been most helpful. Professor George Duvall is sincerely thanked for helpful discussions and for his work on the nitromethane equation of state. This work was supported by O.N.R. Grant N00014-93-1-0369; Dr. R.S. Miller is thanked for his strong interest in this work.

References

[1] W. Fickett and W.C. Davis, *Detonation*, University of California Press, Los Angeles (1979).

[2] R. Chéret, *Detonation of Condensed Explosives*, Springer-Verlag, New York (1993).

[3] M.C. Cowperthwaite and J.T. Rosenberg, in *Proceedings of the Sixth Symposium on Detonation*, Office of Naval Research, Arlington, VA, p. 786 (1976).

[4] Proceedings of Sixth through Tenth Symposia (International) on Detonation, Office of the Chief of Naval Research, Arlington, VA (1976 through 1993).

[5] E.L. Lee and C.M. Tarver, *Phys. Fluids* **23**, p. 2362 (1980).

[6] G.E. Duvall, K.M. Ogilvie, R. Wilson, P.M. Bellamy, and P.S.P. Wei, *Nature* **296**, p. 846 (1982).

[7] C.S. Yoo and Y.M. Gupta, *J. Phys. Chem.* **94**, p. 2857 (1990).

[8] C.S. Yoo, G.E. Duvall, J. Furrer, and R. Granholm, *J. Phys. Chem.* **93**, p. 3012 (1989).

[9] C.S. Yoo and Y.M. Gupta, *J. Chem. Phys.* **93**, p. 2082 (1990).

[10] P.D. Horn and Y.M. Gupta, *Phys. Rev. B* **39**, p. 973 (1989).

[11] R. Gustavsen and Y.M. Gupta, *J. Appl. Phys.* **69**, p. 918 (1991).

[12] R. Gustavsen and Y.M. Gupta, *J. Chem. Phys.* **95**, p. 451 (1991).

[13] S.C. Schmidt, D.S. Moore, D. Schiferl, and J.W. Shaner, *Phys. Rev. Letters* **50**, p. 661 (1983).

[14] D.S. Moore, S.C. Schmidt, and J.W. Shaner, *Phys. Rev. Lett.* **50**, p. 1819 (1983).

[15] N.C. Holmes, W.J. Nellis, W.B. Graham, and G.E. Walrafen, *Phys. Rev. Lett.* **55,** p. 2433 (1985).

[16] Y.M Gupta, P.D. Horn, and C.S. Yoo, *Appl. Phys. Lett.* **55**, p. 33 (1989).

[17] C.S. Yoo, Y.M. Gupta, and P.D. Horn, *Chem. Phys. Lett.* **159**, p. 178 (1989).

[18] J.M. Boteler and Y.M. Gupta, *Phys. Rev. Lett.* **71**, p. 3497 (1993).

[19] R. Gustavsen and Y.M. Gupta, *J. Appl. Phys.* **75**, p. 2837 (1994).

[20] S.M. Gallivan and Y.M. Gupta, *J. Appl. Phys.* **78**, p. 1557 (1995).

[21] A.M. Renlund, S.A. Sheffield, and W.M. Trott, in *Shock Compression of Condensed Matter—1985* (ed. Y.M. Gupta), Plenum Press, New York, p. 237 (1986).

[22] H. Suzuki, *Electronic Absorption Spectra and Geometry of Organic Molecules: An Application of Molecular Orbital Theory*, Academic Press, New York (1967).

[23] G.M. Barrow, *Introduction to Molecular Spectroscopy*, McGraw-Hill, Singapore (1962).

[24] Y.M. Gupta, *High Press. Res.* **10**, p. 713 (1992).

[25] Y.M. Gupta, in *Shock Compression of Condensed Matter—1991* (ed. S.C. Schmidt, R.D. Dick, J.W. Forbes, and D.G. Tasker), Elsevier, Amsterdam, p. 15 (1992).

[26] G.E. Duvall, *Optical Spectroscopy of Dynamically Compressed Liquids*, Final Technical Report under Contract No. N00014-77C-0232, Office of Naval Research, Arlington, VA (June 30, 1986).

[27] M.M. Sushchinskii, *Raman Spectra of Molecules and Crystals*, Israel Program for Scientific Transactions, Ltd., Jerusalem (1972).

[28] S. Nagakura, *Mol. Phys.* **3**, p. 152 (1960).

[29] G. Malewski, M. Pfeiffer, and P. Reich, *J. Mol. Struct.* **3**, p. 419 (1969).

[30] J.R. Hill, D. S. Moore, S.C. Schmidt, and C.B. Storm, *J. Phys. Chem.* **95**, p. 3039 (1991).

[31] F.D. Verderame, J.A. Lannon, L.E. Harris, W.G. Thomas, and E.A. Lucia, *J. Chem. Phys.* **56**, p. 2638 (1972).

[32] P.J. Miller, S. Block, and G.J. Piermarini, *J. Phys. Chem.* **93**, p. 462 (1989).

[33] D.T. Cromer, R.R. Ryan, and D. Schiferl, *J. Phys. Chem.* **89**, p. 2315 (1985).

[34] G.J. Piermarini, S. Block, and P.J. Miller, *Phys. Chem.* **93**, p. 457 (1989).

[35] J.W. Brasch, *J. Phys. Chem.* **84**, p. 2084 (1980).

[36] R. Engelke and J.B. Bdzil, *Phys. Fluids* **26**, p. 1210 (1983).

[37] S.A. Sheffield, R. Engelke, and R. Alcon, in *Proceedings of the Ninth Symposium (International) on Detonation*, Office of the Chief of Naval Research, Arlington, VA, p. 39 (1989).

[38] D.R. Hardesty, *Combust. Flame* **27**, p. 229 (1976).

[39] M.D. Cook and P.J. Haskins, in *Proceedings of the 19th International Annual Conference of* ICT *on Combustion and Detonation Phenomena*, Fraunhofer-Institute fur Chemische, Technologie, Explosivstoffe, Karlsruhe, Germany, p. 85 (1988).

[40] A. Delpeuch and A. Menil, in *Shock Waves in Condensed Matter* (eds. J.R. Asay, R.A. Graham, and G.K. Straub), Elsevier Science Publishing, New York, p. 309 (1984).

[41] S.C. Schmidt, D.S. Moore, J.W. Shaner, D.L. Shampine, and W.T. Holt, *Physica* **139** & **140B**, p. 587 (1986).

[42] A.M. Renlund and W.M. Trott, *Shock Compression of Condensed Matter—1989* (eds. S.C. Schmidt, J.N. Johnson, and L.W. Davison), Elsevier Science Publishing, New York, p. 875 (1990).

[43] R. Engelke, W.L. Earl, and C.M. Rohlfing, *Int. J. Chem. Kinet.* **18**, p. 1205 (1986).

[44] R. Engelke, W.L. Earl, and C.M. Rohlfing, *J. Chem. Phys.* **84**, p. 142 (1986).

[45] R. Engelke, D. Schiferl, C.B. Storm, and W.L. Earl, *J. Phys. Chem.* **92**, p. 6815 (1988).

[46] R.D. Bardo, *Int. J. Quantum Chem.: Quantum Chem. Symp.*, **20** p. 455 (1986).

[47] M.D. Cook and P.J. Haskins, in *Proceedings of the 9th Symposium (International) Detonation*, Office of the Chief of Naval Research, Arlington, VA, p. 1027 (1989).

[48] M.D. Cook and P.J. Haskins, in *Proceedings of the 10th Symposium (International) Detonation*, Office of the Chief of Naval Research, Arlington, VA, p. 870 (1993).

[49] C.P. Constantinou, *The Nitromethane-Amine Interaction*, Ph.D. dissertation, University of Cambridge (1992).

[50] S. Odiot, M. Blain, E. Vauthier, and S. Fliszar, *J. Mol. Struct. (Theochem.)* **279**, p. 233 (1993).

[51] C.P. Constantinou, J.M. Winey, and Y.M. Gupta, *J. Phys. Chem.* **98**, p. 7767 (1994).

[52] Y.M. Gupta, G.I. Pangilinan, J.M. Winey, and C.P. Constantinou, *Chem. Phys. Lett.* **232**, p. 341 (1995).

[53] G.I. Pangilinan and Y.M. Gupta, *J. Phys. Chem.* **98**, p. 4522 (1994).

[54] G.E. Duvall, *Equation of State of Liquid Nitromethane*, tehnical report, Washington State University—Shock Dynamics Center, unpublished.

[55] J.M. Winey, Ph.D. thesis, Washington State University (1995).

[56] R.W. Woolfolk, M. Cowperthwaite, and R. Shaw, *Thermochim. Acta* **5**, p. 409 (1973).

[57] P.J. Lysne and D.R. Hardesty, *J. Chem. Phys.* **59**, p. 6512 (1973).

[58] S.K. Sharma, H.K. Mao, and P.M. Bell, *Phys. Rev. Lett.* **44**, p. 886 (1980).

[59] B.A. Weinstein and R. Zallen, in *Topics in Applied Physics Vol. 54* (eds. M. Cardona and G. Guntherodt), Springer-Verlag, New York, p. 463 (1984).

[60] J.J. Dick, *J. Phys. Chem.* **97**, p. 6193 (1993).

[61] J.J. Dick, R.N. Mulford, W.J. Spencer, D.R. Petit, E. Garcia, and D.C. Shaw, *J. Appl. Phys.* **70**, p. 3572 (1991).

CHAPTER 5

Effects of Shock Compression on Ceramic Materials

Tsutomu Mashimo

5.1. Introduction

Shock wave propagation in a solid can generate conditions of ultra-high pressure (stress) sufficient to induce changes in the elastic rigidity and the crystal and electronic structures of the material. Hugoniot data are even now the most reliable (in situ and macroscopic) experimental information obtainable from shock compression research on solids. We can directly and precisely determine the pressure (stress)–density relation of condensed matter by measurement of Hugoniot parameters (shock velocity and particle velocity), because these parameters are comparable to ultrasonic data: derivative values of pressure with volume. From these data, the dynamic strength, phase transitions, equation of state (EOS), etc. can be studied. However, these experiments provide little information on microscopic effects because it is very difficult to perform in situ microscopic observations. This is due mainly to the very short duration of the shock process, during which the entropy increases and a high-temperature, compressed state that is heterogeneously deformed appears. However, shock compression research has long occupied an important position in the field of high-pressure science due to the aforementioned features, although its monopoly in generating pressures in the 100 GPa range has recently been lost due to development of diamond-anvil cells.

Shock compression of solids has some features that distinguish it from static compression. The state of plane shock compression is completely uniaxial from a macroscopic point of view, even in the plastic region. The microscopic state seems more complicated but is, in fact, unknown. Uniaxial compression generates an anisotropic stress-strain state even in isotropic solids and, for sufficiently strong shocks, the shear stresses induce an elastoplastic transition (yielding). As a result, a locally heterogeneous strain state arises through the action of slip phenomena. In addition to causing an increase in average temperature, shock compression leads to the formation of hot spots or zones along slip planes and/or at defects and sites of porosity. These hot spots may influence structural phase transition. The anisotropic displacement of atoms under shock compression may affect phase transitions in ways that go beyond the normal thermodynamic effect of the increased pressure and temperature, and slip phenomena also affect the phase transition.

Since shock-induced yielding or phase transition often cannot proceed to completion in the submicrosecond to microsecond time interval available in an experiment, nonequilibrium crystal phenomena appear. At the rise and release of compression, solids deform at a very high strain rate, and show large intrinsic strengths. The high strain rate should also affect phase transitions and chemical reactions. Nonequilibrium electrical phenomena sometimes appear in such conditions and sometimes induce changes in electrical conductivity, electrical field, etc. for piezoelectric-effect materials. Shock wave propagation sometimes generates polarization or depolarization signals, particularly at rise and release, in paraelectric and even non-piezoelectric materials. A brief list of characteristic effects of shock compression on solids is given in Table 5.1.

Table 5.1. Effects of shock compression on solids

Stimulus	Observation
High pressure	Equation of state High-pressure phase transitions Piezoelectric effects
Short pulse duration	Nonequilibrium states (mechanical, thermodynamic, electrical) → Relaxation in transition, charge flow, etc.
High strain rate	High strength, viscosity effect, polarization
Uniaxial compression	Anisotropic elastic state, elastoplastic transition (yielding) → heterogeneous state Anisotropic phase transition
Heterogeneous state	Hot spots and zones, melting zones, luminescence, etc.
High temperature	Melting, chemical reaction

It should be noticed that Hugoniot data represent averaged one-dimensional motions of particles of a continuum, whereas the shock compression states of solids are thought to be nonuniform. In addition, atoms must experience some movement normal to the direction of shock propagation due to the yielding to an isotropic state or the atomic displacement of a phase change. Since Hugoniot data do not include unsteady change, it is important to measure the total shock wave profiles at several points to discuss the transition process.

The Hugoniot parameters of many elemental metals and oxide minerals have been measured over a period of approximately 40 years, beginning with work by scientists of the United States and the former Soviet Union [1–6]. Broader studies of the equation of state of solids have been undertaken for application to high-density physics and earth and planetary science. In the early studies, the measurement methods were usually of the discrete type, and the specimen quality was often poor compared with materials available for more recent studies. Ceramics have remained as unexplored materials in the field of shock compression research. Few investigations of ceramics have been reported, and the specimens used in the early studies were almost always of highly porous sintered material. Equations of state (EOS) and phase transitions of ultra-hard oxides, nitrides, borides, etc. have not been extensively investigated, even by the methods of static compression. The importance of such investigations is increasing, motivated by interest in the development of new materials as well as application to studies of mechanical properties and crystal chemistry of solids, and to earth and planetary science.

Ceramics are hard, but are brittle and have a large crystal anisotropy due to their covalent bonding. Yielding and phase transition behavior of solids under shock compression depends strongly on material characteristics (crystal-chemical properties, thermodynamic properties, microstructure, etc.). Shock yielding properties are significant for cratering problems, shock fragmentation, and dynamic mechanical behavior of solids. Understanding these properties is also critical for studies of high-pressure polymorphism and EOS using shock compression, because the high strength of ceramics significantly influences the interpretation of measured compression curves and slip phenomena may directly influence phase transitions.

When making shock wave measurements on solids, it is important to use both high-quality specimens and high-capability measurement facilities. We have been investigating shock-induced yielding, high-pressure phase transitions, equation of state, etc. of ceramics, minerals, semiconductors, and other materials by means of the inclined-mirror method [7], the manganin-gauge method [8], and the electro-

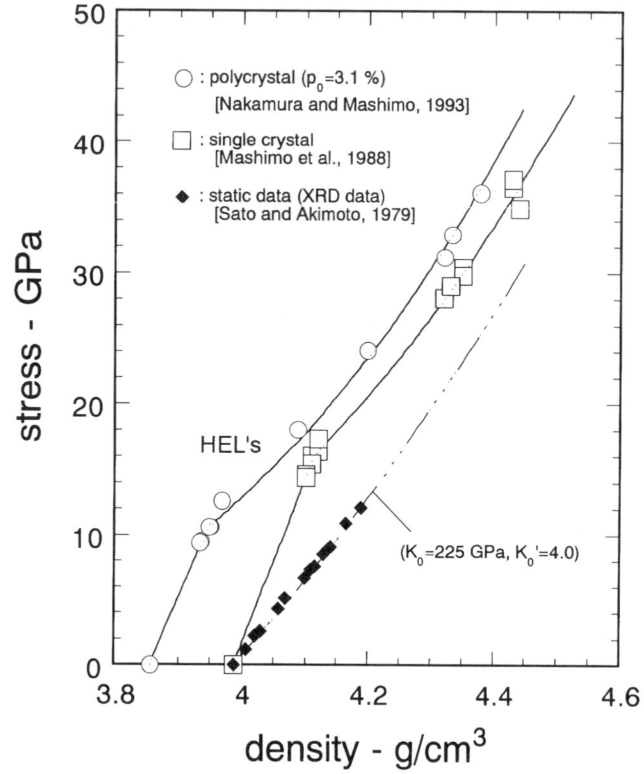

Figure 5.2. Hugoniot-compression curves of Al_2O_3 single crystal [15] and polycrystal [8].

crystal was the larger shear strength and/or the remaining porosity, rather than the temperature increase.

Figure 5.3 shows stress histories in the single crystal and polycrystal, as measured by the manganin-gauge method [19]. Sharp and dispersed two-step structures can be seen at the shock front for the single crystal and the polycrystal, respectively. The dispersed waveform observed for the polycrystal may be due to the porosity and/or the heterogeneous state (the longitudinal wave velocity of the polycrystalline material is the averaged value for crystal grains of differing orientation). The rise shape of the plastic wave in the single crystal is jagged, which may be related to the jagged free-surface profile observed by the inclined-mirror method. At release, a sharp stress drop due to the elastic rarefaction wave from the backing surface is observed for the single crystal, whereas that in the polycrystal is diffused. The stress drops, even for the single crystal, were smaller than

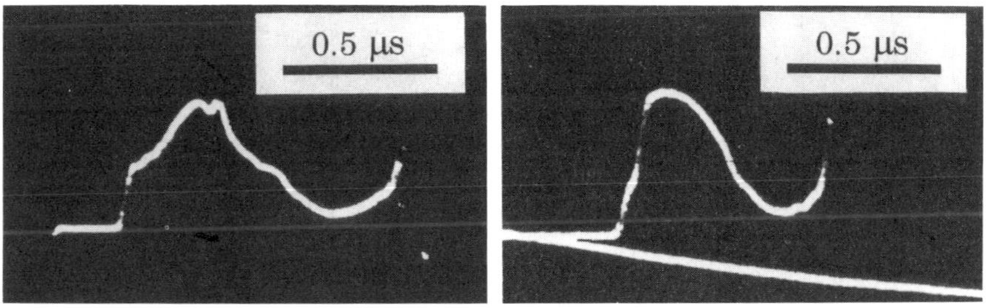

(a) Single crystal (b) Polycrystal

Figure 5.3. Stress histories, measured by the manganin-gauge method, in (a) an Al_2O_3 single crystal impacted by another single crystal at a velocity of 1.21 km/s and (b) an Al_2O_3 polycrystal impacted by tungsten at a velocity of 1.497 km/s [19].

those expected from the Hugoniot offsets. This may also be due to stress relaxation and a smaller yield strength manifest during the expansion process.

5.2.2. Titanium Dioxide (TiO_2)

Rutile-phase TiO_2 is a typical anisotropic crystal which has a tetragonal structure ($a_0 = 0.4594$, $c_0 = 0.2959$ nm) and is expected to exhibit polymorphism analogous to that of AX_2 compounds generally. It is expected that the uniaxial deformation characteristic of plane shock compression directly affects both the elastoplastic and phase transitions. We reported, for the first time, anisotropic Hugoniot data and electrical conductivities of this material for pressures to 40 GPa [20,21]. Syono et al. investigated shock-induced phase transitions and the EOS of high-pressure phases by measurement to pressures over 120 GPa coupled with examination of recovered specimens [22,23]. Here, these studies are reviewed, focusing attention on the effects of uniaxial compression on the phase transition.

Figure 5.4 shows particle-velocity histories obtained from electromagnetic gauges placed at two locations along the <100> and <110> axes of the crystal. These profiles show an obvious difference between the two axes. For the <100> direction, an elastoplastic transition and a diffuse phase transition can be observed. The recorded profile resembles one obtained previously in an experiment on calcite [24]. For the <001> direction, a dispersed elastoplastic transition or an elastoplastic transition followed by a phase transition can be seen in the profile. The final density achieved for a shock propagating in the <100> direction was much greater than that for a shock propagating in the <001> direction.

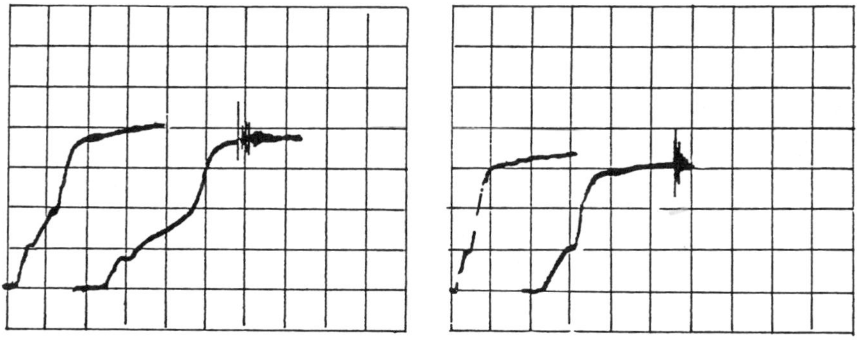

(a) <100> axis (b) <001> axis

Figure 5.4. Particle velocity (U_P) profiles at two points in TiO_2. The vertical scale is 0.281 km/s/div and the horizontal scale is 100 ns/div. In (a), the shock propagated along the <100> axis, the gauge stations were separated by 1.61 mm, and the velocity of the sapphire impactor was 1.86 km/s ($U_{p(peak)}$ = 1.01 km/s). In (b), the shock propagated along the <001> axis, the gauge stations were separated by 1.88 mm, and the velocity of the sapphire impactor was 1.84 km/s ($U_{p(peak)}$ = 0.80 km/s) [20].

Figure 5.5 shows Hugoniot-compression curves for shocks propagating along the <100>, <110>, and <001> axes [20,22], an unknown axis [25], and one of a polycrystal [26]. The phase transition points for the <100>, <110>, and <001> directions are about 12–14, 17, and 34 GPa, respectively. However, the slowly rising profiles observed by the electromagnetic gauge method for propagation in the <001> direction suggest the possible existence of another phase transition point. It would be useful to have a VISAR measurement in the low-pressure region to clarify the yielding relaxation process and phase transitions. A new high-pressure phase, which shows a different pressure–density trend from that reported by McQueen et al. and Alt'shuler et al. [22], was found between 70 and 100 GPa by Syono et al. They reported that this phase may be the fluorite phase, and the final phase may be a denser phase.

Kusaba et al. studied specimens recovered after shock compression at pressures up to 72 GPa [23]. They reported that the yields of α-PbO_2 type TiO_2 for specimens oriented for propagation in the <100> direction were much larger than yields for specimens oriented in the <001> direction. This anisotropic yield was consistent with the anisotropic Hugoniot-compression curves. They assumed that the high-pressure phase was either fluorite or distorted fluorite type and that the phase conversion to the α-PbO_2 type was induced spontaneously in the pressure-release process, on the basis of the high-pressure behavior of MnF_2. In addition, they gave a clear explanation of

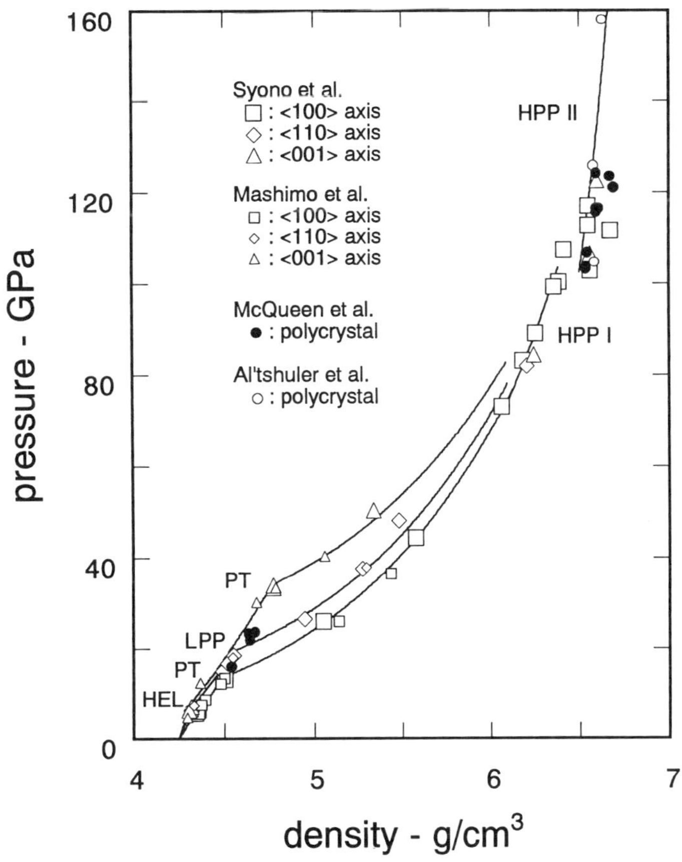

Figure 5.5. Hugoniot-compression curves of TiO_2 single crystals for shocks propagating along the <100>, <110>, and <001> axes, along with a curve for polycrystalline material [20,22,25,26]. The data in Ref. 20, obtained using the electromagnetic-gauge method, were reanalyzed considering gauge movement (transformation from Lagrangian to Eulerian coordinates).

the mechanism of the displacive phase transition, which is assumed to be induced by uniaxial displacement of atoms, on the basis of observations by transmission electron microscopy (TEM).

As a result, it was demonstrated on rutile-phase TiO_2 that the uniaxial shock compression directly affects both the displacive-type phase transition and the elastoplastic transition. Study of this anisotropic phase transition provided many suggestions for improving our understanding of the effects of shock compression on phase transition, as well as material syntheses, crater formation, and other problems.

5.2.3. Zirconia

Zirconia (ZrO_2) exhibits different properties depending on the crystal structure, microstructure, doping material, etc. Zirconia, in its pure form, exists in a stable monoclinic crystal structure. Through doping with other metal oxides such as magnesia (MgO), calcia (CaO), or yittria (Y_2O_3), zirconia can be stabilized in a cubic or tetragonal structure. The stabilized cubic zirconia is used as an ionic conductor having the advantage of high electrical conductivity at high temperature, which arises from vacancies. The stabilized or partially stabilized tetragonal zirconia has an extremely large fracture toughness for ceramics. Under static loading conditions, the high value, which is comparable to values for metals, is due to transformation toughening [27]. The polymorphism of ZrO_2 provides a useful analogy to AX_2 compounds.

It was reported that monoclinic pure zirconia transformed to orthorhombic-I, orthorhombic-II, and unknown phases under static compression [28–30]. However, the compression curve in the high-pressure region was unknown. Hugoniot-compression curves of pure ZrO_2 (monoclinic phase) have been, so far, measured only for sintered polycrystals [1] and twinned crystals [31]. We have recently investigated the shock compression behavior of Y_2O_3- and CaO-doped cubic zirconia single crystals and polycrystals and Y_2O_3-doped tetragonal zirconia polycrystals [32,33]. These zirconias showed very different behaviors despite being of the same basic material.

Y_2O_3-Doped Cubic Zirconia (YCZ)

Streak photographs by the inclined-mirror method for a 9.6-mol% Y_2O_3-doped single crystal (shocked along the <100> axis) and polycrystals of Y_2O_3-doped cubic zirconia (YCZ) are shown in Fig. 5.6 [32]. For the single crystal, kinks due to elastoplastic and phase transitions are clearly visible at points 3 and 4, respectively. However, for the polycrystal, a diffused kink and a continuous diffused kink were observed. This dispersed waveform may be caused by the heterogeneous state and/or porosity of the polycrystal.

The Hugoniot-compression data for shocks propagating along the <100> and <110> axes of the single crystal and a curve for the polycrystalline material are shown in Fig. 5.7 [32]. The HEL stresses for shocks propagating along the <100> and <110> axes were approximately 14 and 25 GPa, respectively, whereas that of the polycrystal was approximately 13 GPa. The difference in HEL stress between the single crystal and the polycrystal may be due to the porosity of the latter (about 1.5%). These anisotropic HELs can be reasonably well

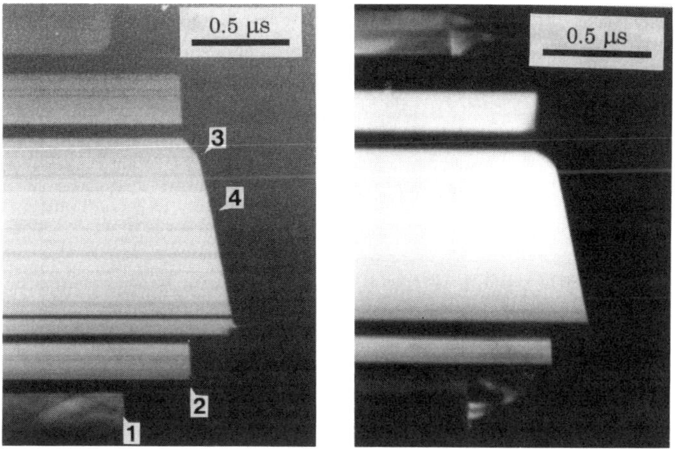

(a) (b)

Figure 5.6. Streak photographs by the inclined-mirror method of (a) an Y_2O_3-doped cubic zirconia single crystal shocked along the <100> axis by a copper impactor having a velocity of 2.993 km/s and (b) a polycrystal impacted by tungsten at a velocity of 2.427 km/s [32].

understood in terms of the primary [111] cleavage plane in the cubic structure. It is assumed that cleavages can most easily grow for shock compression along the <100> direction, since the angles of the <100> axis against the [111] plane are all 35.26°, whereas those of the <110> axis are 90° and 45° (1:1). The yield stresses along both the <100> and the <110> axes were estimated by the von Mises yield condition to have almost the same value of 9–11 GPa, in spite of the large difference in HEL stress. In the plastic region, the single-crystal Hugoniot curves showed relaxation toward an isotropic compression state by the phase transition point, whereas the polycrystal preserved considerable offset from the isotropic state. These results clearly showed the difference in yielding behavior between single crystals and polycrystals.

The YCZ single crystal (9.6-mol% doped) began to transform at approximately 53 GPa under shock compression. This phase transition point was higher than those of either the monoclinic to orthorhombic-II phase transition of >15 GPa (at 600 °C) [28] or another transformation to an unknown phase (tetragonal [29] or orthorhombic-III phases [30]) in the range 35–42 GPa (at room temperature), respectively, under static compression on pure zirconia. An isothermal compression curve analyzed on the basis of the Debye model and the Mie–Grüneisen model and the fitting curve to the Birch–Murnaghan equation are also shown in Fig. 5.7. The volume change be-

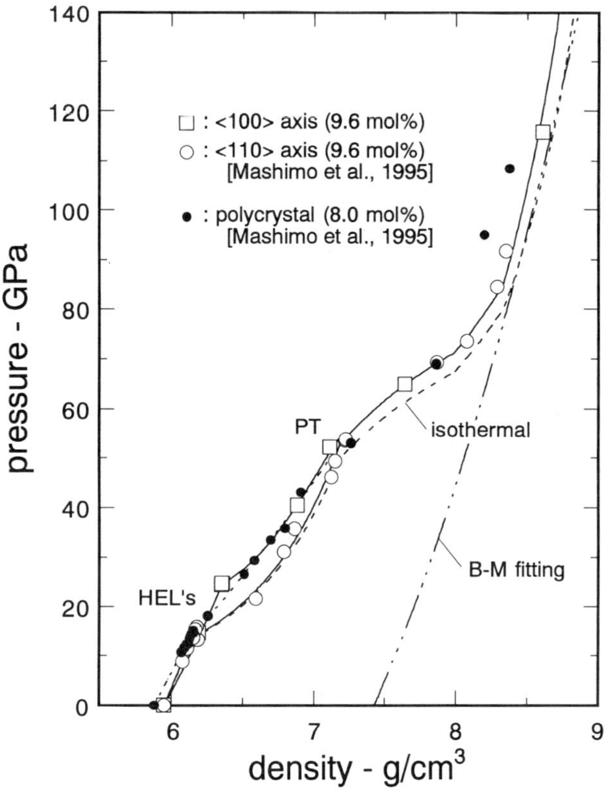

Figure 5.7. Hugoniot-compression curves and the thermal analytical results for single-crystal and polycrystalline Y_2O_3-doped cubic zirconia [32].

tween the cubic phase and the final phase at zero pressure was estimated to be about 20%, which was larger than that expected for the cubic to orthorhombic-II phase transition [32]. Considering the higher transition point (53 GPa) and the larger volume change, it was concluded that the final phase was the unknown phase [29,30]. The bulk moduli of the final phase was estimated to be 510 GPa, which was roughly consistent with that of the high-pressure phase of HfO_2 (K_{B0} ~550 GPa) [34].

Y_2O_3-Doped Teragonal Zirconia (YTZ)

Figure 5.8 shows particle-velocity histories, obtained by the VISAR method, in a 3.0-mol% Y_2O_3-doped tetragonal zirconia (YTZ) specimen in high- and low-pressure regions [35]. Two interesting features were found in these profiles. In the high-pressure region, a three-step shock front was observed for the test at a peak stress of 59 GPa, which agrees with an observation by the inclined-mirror method. In the low-pressure region, an unusually large spall strength (> 1.5 GPa)

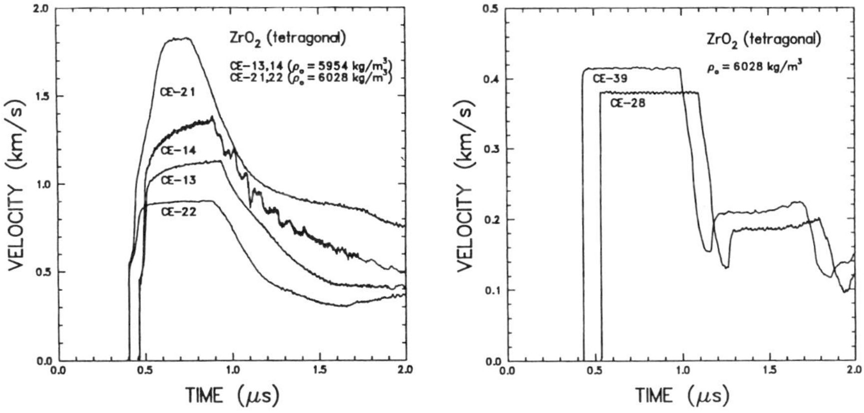

(a) High-pressure region (b) Low-pressure region

Figure 5.8. Particle-velocity histories of the Y_2O_3-doped tetragonal zirconia polycrystal by the VISAR method in (a) the high-pressure region and (b) the low-pressure region [35].

was observed upon release. An instability was also observed (by the manganin-gauge method) upon release from a shock state of about 28 GPa. This may reflect either a phase transition to the monoclinic phase or the spall behavior [33].

Shock velocity (U_s) versus particle velocity (U_p) Hugoniot data for YTZ and 8.0-mol% YCZ polycrystals are shown in Fig. 5.9 [33]. A feature of the YTZ Hugoniot is that the second-shock velocities were much higher than those of the cubic material and, furthermore, were faster than the bulk sound velocity. The first and second transition (T1 and T2) stresses were in the ranges 15–17 and 33–35 GPa, respectively [33,35]. The latter point was higher than that of the tetragonal to orthorhombic-II phase transition point (by 16 GPa at 300–600 °C) under static compression, while the tetragonal to orthorhombic-I phase transition was not observed on the 2-mol% Y_2O_3 doped YTZ material [28]. Two explanations for these transition points were suggested:

i) The first transition point corresponds to the beginning of elastoplastic transition (while the yielding is disturbed by the transformation toughening and, as a result, the second shock velocity is anomalously fast), and the second transition point corresponds to the beginning of a structural phase transition.

ii) The first and second transition points correspond to the beginning of the tetragonal to orthorhombic-II phase transition in the elastic region and the regular elasto-plastic transition with structural phase transition, respectively.

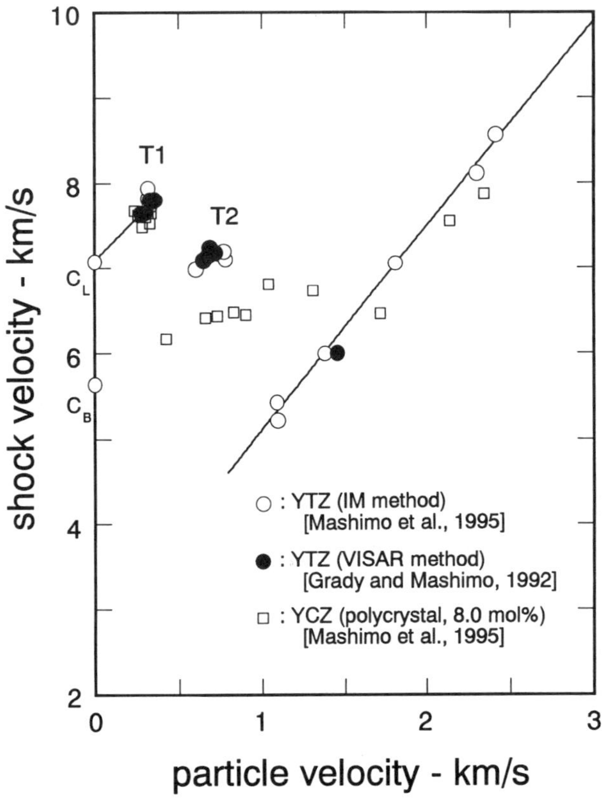

Figure 5.9. Particle velocity–shock velocity Hugoniot results of YTZ polycrystal together with one of the 8.0-mol% YCZ polycrystal [33].

The volume change for the transition from the tetragonal phase to the final phase at zero pressure was estimated to be about 18%. This volume change was larger than that expected for the tetragonal–orthorhombic phase transition, considering that the starting phase (tetragonal structure) was denser than the static compression datum of about 11% on the 2.0-mol% doped YTZ [28]. Considering the larger volume change, differently oriented crystal grains, and the result for the YCZ [32], it was concluded that the final phase was not the orthorhombic-II phase, but an unknown phase [29,30]. The estimated bulk modulus of the final phase, 550 GPa, was also consistent with those of the YCZ and HfO_2.

A visible feature of the YTZ specimens recovered after shock compression above the transition points T1 and T2 was that they remained intact with cracked surfaces, whereas the recovered samples of alumina or cubic zirconia (Y_2O_3 or CaO doped) were crushed

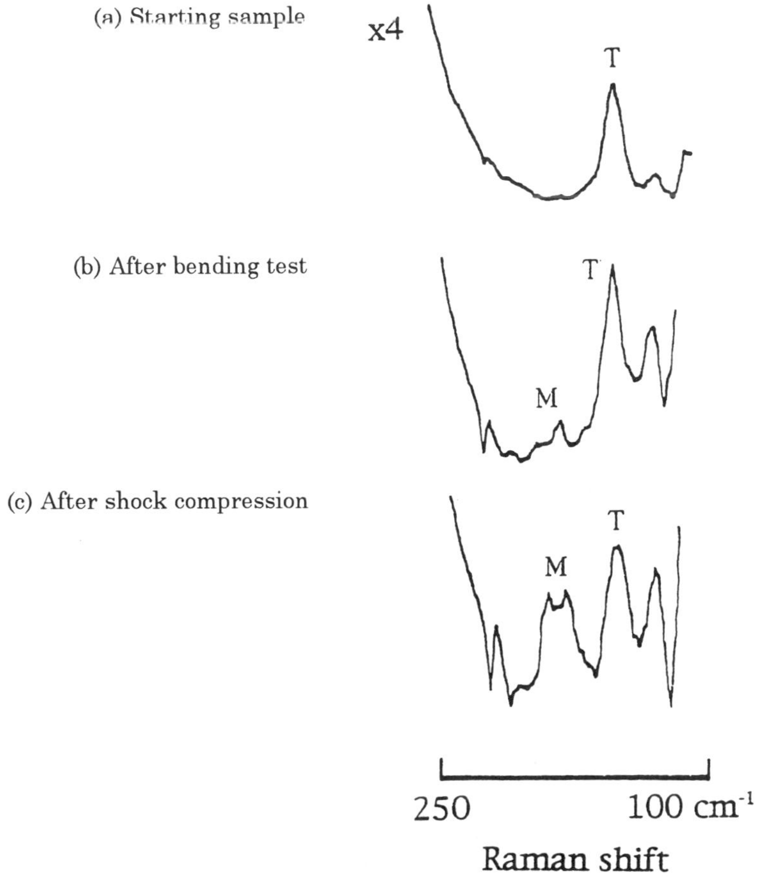

Figure 5.10. Raman spectra of YTZ specimens (a) before and (b) after bending and (c) shock compression at 32 GPa [33]. The monoclinic phase is designated M, and the tetragonal phase is designated T.

and looked like a pressed powder. This indicated that the YTZ maintained a large toughness even under shock compression. The Raman spectra from the fractured surfaces of the YTZ specimens are shown in Fig. 5.10 [33]. Although the Raman band peaks of the monoclinic phase [37] were not observed for the specimen before impact, these band peaks were observed after shock compression (32 GPa). The amplitudes were larger than those after a bending test. For the shock compressions of 11 and 25 GPa in peak stress, the monoclinic-phase bands were also observed. The twined monoclinic phases were also observed by TEM examination [33]. These facts might be related to the anomalous shock compression and release behaviors (fine three-step structure or large spall strength at release), although it was difficult to determine when these monoclinic phases were formed. The possibility of transformation toughening will be discussed in a later section.

5.2.4. Aluminum Nitride (AlN)

Aluminum nitride, which has a wurtzite-type structure at ambient condition, transforms to B1-type structure under static compression [38–40]. However, the transition-point pressures were comparatively scattered. The shock compression behavior of this material is also interesting from the point of view of whether it exhibits an elasto-plastic or elastoisotropic response, because this material has high thermal conductivity despite being a ceramic. It has been pointed out that many brittle materials with low thermal conductivity behave as elastoisotropic solids, whereas many metals behave as elastoplastic solids.

Figure 5.11 shows a typical streak photograph of the pure AlN polycrystal, by the inclined-mirror method [41]. The porosities were less than 0.3%, and the total impurity value was less than 0.1 wt% The grain sizes were about 5–7 μm. Two kinks, corresponding to the elastoplastic and structural phase transition points, respectively, are clearly visible at points 3 and 4 on the image of the inclined mirror.

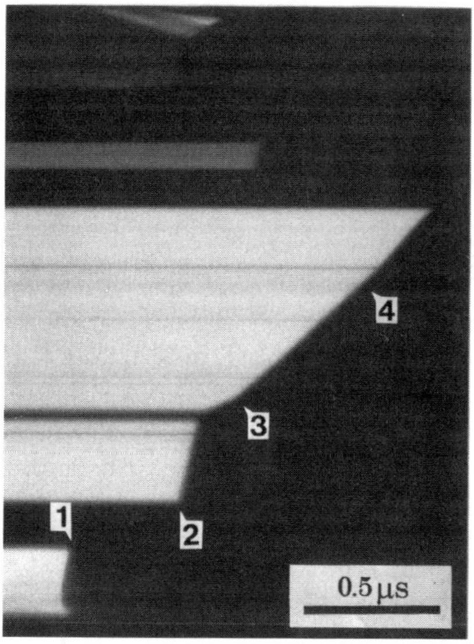

Figure 5.11. Typical streak photograph by the inclined-mirror method of AlN polycrystal impacted by a tungsten impactor having a velocity of 1.395 km/s [41].

Figure 5.12 shows Hugoniot-compression curves of the AlN in the low-pressure region, together with data obtained by the manganin-gauge method [42] and methods of static compression [40]. The HEL stress of the polycrystal was determined to be 9–10 GPa. The Hugoniot data converge to the static compression curve above the HEL. This indicates that this material catastrophically loses its shear strength there. The structural phase transition points of the pure polycrystal were around 20 GPa, whereas the transition point depended on the specimen thickness, etc. The temperature increase and thermal pressure at the structural phase transition point of the polycrystal were calculated to be 44 °C and 0.1 GPa, respectively. The phase transition point was consistent with a recent result obtained by the VISAR method for a polycrystal with almost the same grain size and porosity [43], but was lower than the value of about 22 GPa obtained for a polycrystal with a finer grain size of about 2 μm [44]. These facts indicate that the phase transition point might depend on the grain size.

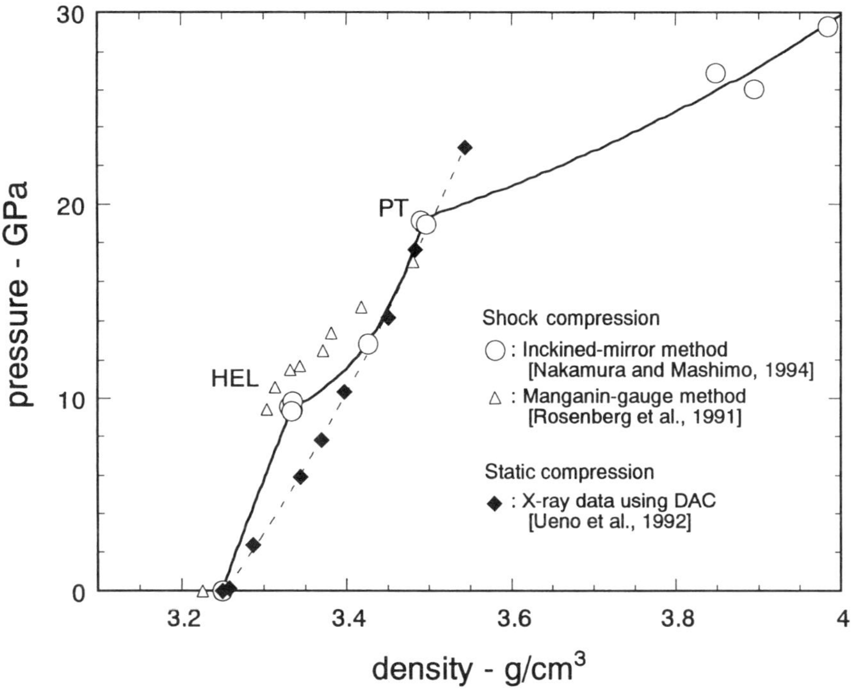

Figure 5.12. Hugoniot-compression curve of an AlN polycrystal in the low-pressure region.

In addition, all the manganin-gauge signals showed a loss of resistance beyond the transition pressure [45]. This indicates that an electric property change including metallic transition may occur in the ceramic and may be related to the change of optical property under static compression [39]. A Hugoniot measurement experiment at a higher pressure has now been made, permitting discussion of the EOS of the high-pressure phase.

5.2.5. Silicon Nitride (Si_3N_4) Ceramics

Silicon nitride (Si_3N_4) ceramics have good mechanical properties such as high strength and high resistance to abrasion, even at high temperature. Many high-performance Si_3N_4 ceramics have been developed recently, including grain-controlled, whisker-doped, and nano-composite materials. However, the dynamic mechanical properties of these new ceramics have not been well investigated. We performed Hugoniot measurements up to 40 GPa on grain-controlled monolithic types (A, B, C) and the whisker-doped type (D) of Si_3N_4 ceramics.

Figure 5.13 shows Hugoniot-compression curves of the monolithic type specimens A, B, and C whose microstructures are different [46]. The specimens A, B and C have pole-like grains, whose grain sizes were about 0.5–1.2, 0.25–0.4, and 0.15–0.3 μm in diameter, respectively. The A, B, C, and D specimens had the porosities of about 4%, 1%, 1%, and 1.5%, respectively. These specimens contained almost the same quantities of Y_2O_3, Al_2O_3, etc. as binders. The HEL stresses of the types A, B, and C were determined to be 10–12.5, 14–16.5, and 17–20 GPa, respectively, although the porosities (except for type A) were almost the same. The difference in HEL stress was attributed to differences in grain dimension. The plastic Hugoniot-compression curves were almost linear, which indicated that Si_3N_4 ceramics behave as quasi-elastoplastic solids, maintaining considerable shear strength in the plastic region.

Figure 5.14 shows SEM micrographs of the starting materials and the relation between the linear density and the HEL stress for the monolithic-type specimens A, B, and C. The HEL stress increases with the increase in linear density, although, except for type A, the porosities are almost same. A larger linear density means a smaller grain dimension and, consequently, an increase in the grain boundary area. The increase in strength of fine-grained polycrystalline material under shock compression can be reasonably understood if we consider that the growth, multiplication, and motion of cracks or dislocations are disturbed by grain boundaries.

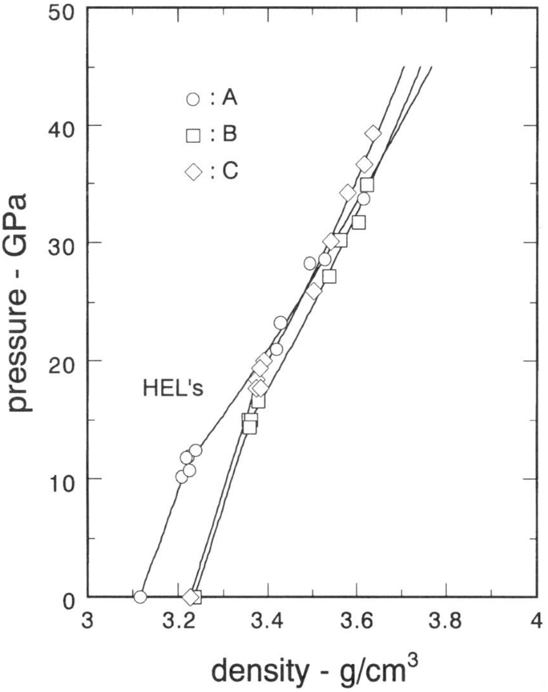

Figure 5.13. Hugoniot-compression curves of the monolithic-type Si_3N_4 ceramics with different microstructures [46].

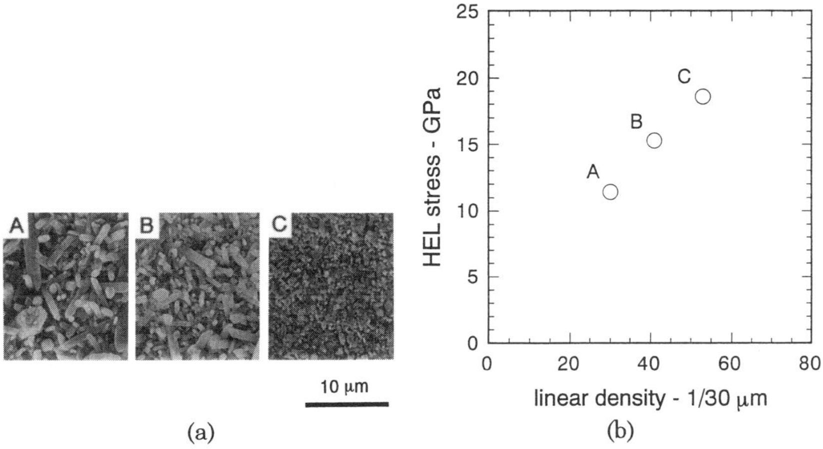

Figure 5.14. (a) SEM photographs of the starting materials and (b) the relation between the linear density and the HEL stress of the monolithic-type Si_3N_4 ceramics [46,47]. Linear density is defined as the number of grains on a line 30 μm long in the SEM photograph.

For the SiC-whisker-doped material, a highly dispersed kink due to elastoplastic transition was observed in the inclined-mirror image [11], whereas transitions of the monolithic type specimens were sharp. The base material of the 20-mass% SiC-whisker-doped specimen (type D) was the type-B material. In this case, the HEL stress was determined to be over 17 GPa, which was larger than that of the monolithic type B. The dispersed kink in the waveform and the high HEL stress may be related to the doping of SiC whiskers.

5.2.6. Boron Carbide (B_4C)

Boron carbide has very high hardness despite its low density and is expected to be used in many fields, such as the space and aeronautics industries. The Hugoniot data, obtained by the inclined-mirror method combined with an explosive system, were scattered in the low-pressure region [13]. The stress histories of the polycrystals with various porosities showed instabilities behind the elastic precursor [48]. Furthermore, Kipp and Grady measured the particle-velocity histories of some hard ceramic materials by the VISAR method and reported the results of B_4C to be noisy [49].

Figure 5.15 shows a streak photograph, by the inclined-mirror method, of pure B_4C [50]. The porosities were less than 0.5%. The

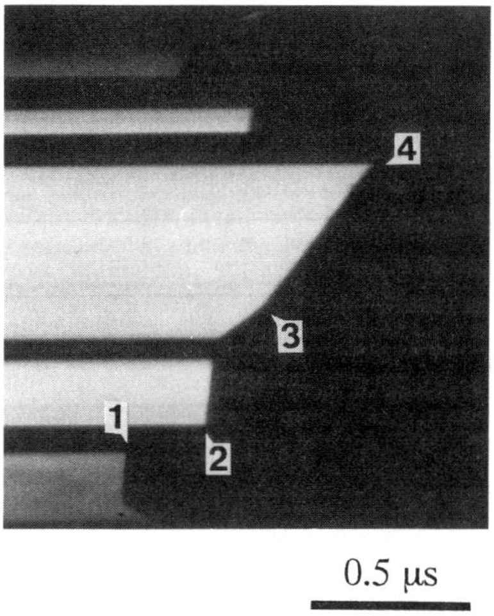

Figure 5.15. Streak photograph, by the inclined-mirror method, of a pure B_4C polycrystal impacted by a tungsten plate at a velocity of 1.406 km/s [50].

main impurities were O and N (1.2 wt%), and the other impurity value was about 0.1 wt%. In the figure, the plastic wave arrived at the specimen rear surface at point 3. The shock velocity was near 15 km/s, much faster than for other ceramics such as the Al_2O_3 polycrystal (approximately 10.8 km/s) [8], the monolithic-type Si_3N_4 ceramic (type C) (11.3 km/s) [38], the pure AlN polycrystal (11.1 km/s) [43], etc. The HEL stress was determined to be higher than 17 GPa, which is very large despite the lower density. A jagged free-surface profile on a millimeter scale can be seen between points 3 and 4, as in the case of Al_2O_3 single crystal. This may also reflect a heterogeneous free-surface profile in the plastic wave front, or spall behavior.

Figure 5.16 shows gapped-flat-mirror streak photographs from an experiment on B_4C and Si_3N_4 ceramics [51]. These images correspond

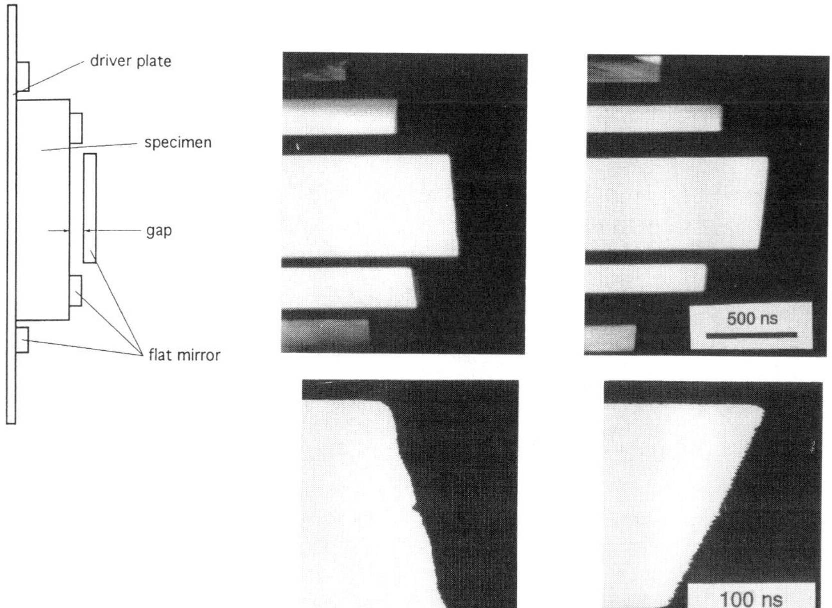

(a) Experimental setup (b) B_4C polycrystal (c) Si_3N_4 polycrystal

Figure 5.16. Streak photographs by the gapped-flat-mirror method of (a) pure B_4C polycrystal when the gap between the free surface and the flat mirror was 0.460 mm and the velocity of the tungsten impact plate was 1.619 km/s and (b) a similar photograph for a Si_3N_4 polycrystal when the gap was 0.430 mm, and the tungsten impactor had a velocity of 1.355 km/s [51]. The bottom photographs present the same results (only the gapped-flat-mirror image) with a magnified time scale (about 2.7 times), obtained by image processing.

to free-surface motion profiles in the plastic region. The gapped-flat-mirror image of B_4C is jagged, as was the inclined-mirror image (Fig. 5.15), and like that of the alumina single crystal, whereas the image for Si_3N_4 is smooth. This fact reflects a heterogeneous deformation of millimeter scale and indicates that a macroscopic slip system (e.g. cracks) plays an important role in elastoplastic transition of this material under shock compression and/or rarefaction processes.

5.2.7. Barium Titanium Oxide ($BaTiO_3$)

Electrical response of ferroelectric material is interesting from the point of view of dielectric transition, as well as of the piezoelectric effect and various applications. The electrical responses to shock compression are thought to be caused by piezoelectricity, ferroelectric–paraelectric transition, and dielectric breakdown due to elastoplastic deformation. The latter phenomenon has been investigated in PZT in the low-pressure region [52–55]. We investigated the electrical response of a typical ferroelectric material, $BaTiO_3$, in the high-pressure region above the HEL to clarify the causes of the electrical response of ferroelectric materials [56].

Figure 5.17 shows assemblies for making parallel and normal mode electrical measurements and Fig. 5.18 shows voltage histories obtained using a parallel mode assembly. In Fig. 5.18, the remnant polarizations are oriented parallel, randomly (nonpoled), and anti-parallel to the shock wave propagation direction, respectively, and the peak stresses were 29 GPa. The signals for parallel and anti-parallel polarizations show increasing voltage with time and wave-forms that are almost reflected images of each other, whereas the signal of the nonpoled specimen shows an almost zero constant value.

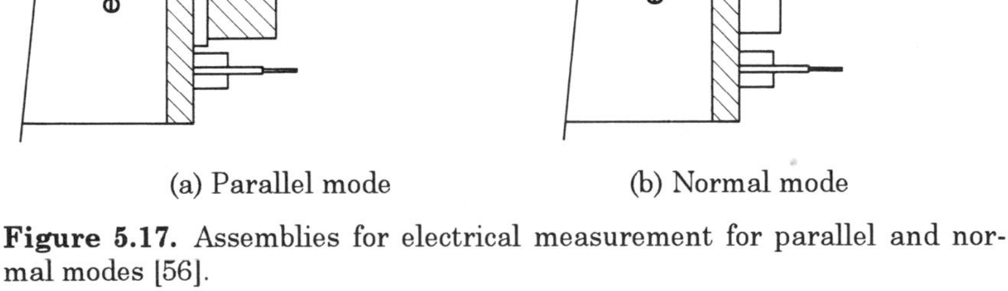

Figure 5.17. Assemblies for electrical measurement for parallel and normal modes [56].

These facts show that the electrical response was due to the remnant polarization rather than shock-induced polarization or a thermoelectric effect. These voltage profiles can be understood by a simple analytical model based on a ferroelectric–paraelectric transition [56]. The dielectric constants under shock compression were roughly consistent with those measured under static conditions for the paraelectric phase.

Figure 5.19 shows current histories obtained using the parallel and normal mode assemblies, respectively, when the peak stress was 12 GPa. In the parallel mode, most of the surface charge flowed out as soon as the elastic shock wave entered the specimen. In the normal mode, an almost-constant current flowed while the elastic shock wave propagated in the specimen and electrode. This indicated that the depolarization continues at the shock front of the elastic wave, and the surface charges at the specimen–electrode interface constantly

(a) Parallel mode

(b) Nonpoled mode

(c) Antiparallel mode

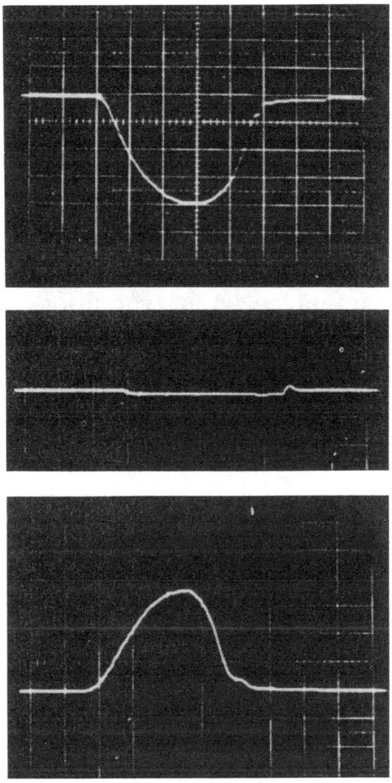

Figure 5.18. Voltage histories for (a) parallel, (b) nonpoled, and (c) antiparallel modes, when the peak stress was 29 GPa [56]. Scales: vertical = 597 V/div, horizontal = 0.2 μs/div.

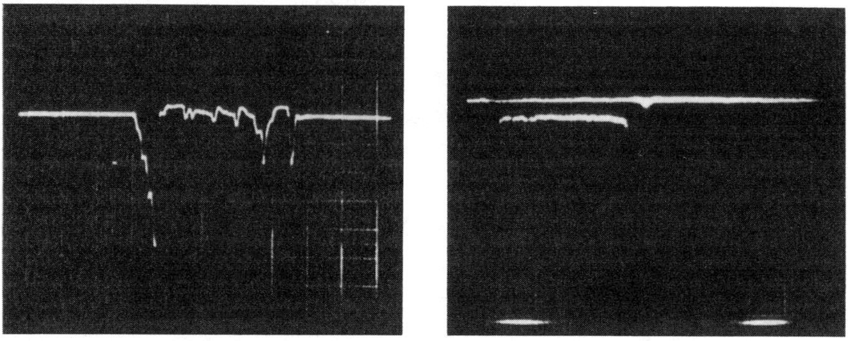

(a) Parallel mode (b) Normal mode

Figure 5.19. Current histories for (a) parallel and (b) normal modes, when the peak stress was 29 GPa [56].

flowed. The total charges were roughly consistent with those due to the remnant polarization before shock compression.

Considering the voltage profiles and amplitudes, it was concluded that the electrical response of $BaTiO_3$ subject to shock compression above the HEL was due to the ferroelectric–paraelectric transition or dielectric breakdown caused by the elastoplastic transition rather than piezoelectricity. However, it was believed that the electrical response was chiefly caused by the depolarization due to the ferroelectric–paraelectric transition, because the voltage was very high, the current for the normal mode stopped just when the elastic wave arrived at the rear surface, and the loss of charge was small [56]. The transition point under static compression was reported to be 2 GPa [57], which was lower than the HEL stress (4.8–6.4 GPa).

5.3. Yielding Mechanism and Correlation with Material Characterization

When the shear stress resulting from a uniaxial strain state of shock compression exceeds a certain value, which is related to the strength, a relaxation to the isotropic stress state occurs and, as a result, a dissipation process, the elastoplastic transition (yielding), is induced. The question to be answered is: What kind of slip mechanisms underlie the elastoplastic transition of solids under shock compression?

Ceramics are hard and have high theoretical strength and large crystal anisotropy, but are brittle. This is generally understood to be caused by the low dislocation density, the small dislocation mobility, the small number of potential slip modes, and the large amount of energy required for nucleation and multiplication of dislocations. These properties are thought to be attributable to the covalent or ionic chemical bonding states of ceramics, which are much different than those of metals. The yielding phenomena of brittle materials are confusing, even under static conditions, and those under shock compression (shock yielding phenomena) are obscure even now [11,80], because it is very difficult to identify the slip mechanisms that are active under shock compression.

At the elastic precursor, solids show large intrinsic strengths (comparable to the theoretical values) due to the very high strain rate. Measured HEL stresses for various kinds of materials are shown in Table 5.2. The HEL stresses of many ceramics are larger than 10 GPa, whereas those of metals, ionic materials, and molecular materials do not exceed a few GPa. Ceramics show a wide variety of HEL stresses and plastic wave front profiles compared with metals, which may be related to the range of mechanisms causing slip in various materials, the heterogeneous state of shock compression, and the effect of material microstructure. This prevents us from directly inferring the EOS from the Hugoniots of high-strength materials in the low-pressure region. In addition, the HEL stress depends on the driving stress, specimen thickness, porosity, and grain size. Recently, many high-performance ceramics, whose crystal structure, microstructure, or composition is controlled, and whose mechanical properties are sometimes improved by formation of process zones at crack tips (process-zone toughening), have been developed. It is an interesting problem whether the process-zone toughening works under shock compression as it does under quasistatic conditions.

5.3.1. Yielding Mechanisms

Yielding of solids can be classified into elastoplastic behavior, in which considerable shear strength is preserved above the HEL, and elastoisotropic behavior, in which shear strength is catastrophically lost, as shown in Fig. 5.20. Real materials often exhibit responses intermediate between these extremes, so the utility of the classifications may be limited to comparison. In addition, a toughening solid whose strength increases under shock compression, as shown by the dotted line in the figure, is expected to exist. The differences among these materials decreases with increasing pressure. Strictly speaking, a perfect elastoisotropic solid does not exist, because under dynamic conditions even a liquid may resist shear deformation due to its viscosity and the high strain rate. The strength of a solid at the HEL or in the plastic region does not always agree with the value at release or with the spall strength. Particularly, many ceramics have large compressive strength but small tensile strength under static conditions.

The differences in yielding behavior are thought to be caused by differences in the slip mechanism. Conceivable slip mechanisms producing shock yielding phenomena in solids include cracks (cleavages and microcracks), dislocations (including twins, stacking faults, etc.), and partial melting zones along slip planes. The main differences among them appear in the bonding strength and the dimension of elastic clusters and in the preserved macroscopic shear strength. It is believed that the reduction of shear strength due to cracks is large, and

Table 5.2. Hugoniot elastic limit data for various materials

Material	Crystal state and porosity (%)	HEL stress (GPa)	Thickness mm	Measurement method[a]	Ref.
Metals					
Be	single	0.14–0.55	3–10	QG, V	58
Al alloy	poly	1.3	3–10	V	59
Fe	poly, 0%	1.1	6–40	OL	60
	poly, 0%	0.9–1.1	3–20	V	61
Cu	single <001>	0.2	5	QG	62
	single <110><111>	0.1	5	QG	62
Ta	poly	<2	2.5–19	QG	63
W	poly, <0.4%	3.9	3–10	V	64
Ionic crystals					
LiF	single <100>	1.5	0.5–5	QG	65
	<100>	0.25	1–3.5	V	66
NaCl	single <100>	0.03	6–13	QG	68
	<110>	0.08	6–13	QG	68
	<111>	0.74	6–13	QG	68
Semiconductors					
Si	single <100>	9.2, 8.4	3–26, 3	IM	69,70
	<110>	5.0, 5.6	3–26, 3	IM	69,70
	<111>	5.4, 5.6	3–26, 3	IM	69,70
Ge	single <100>	5.8	6–13	IM	71
	<110>	4.8	6–13	IM	71
	<111>	4.7, 4.4	6–13, 3–8	IM, QG	71,72
GaP	single <221>	6.7–8.4	3	IM	73

Table 5.2 (cont.). Hugoniot elastic limit data for various materials

Material	Crystal state and porosity (%)	HEL stress (GPa)	Thickness (mm)	Measurement method[a]	Ref.
Ceramics					
BeO	poly, 5.6%	8	3–13	IM	13
MgO	single <001>	3.5–9	2.5–10	IM	74
Al_2O_3	single	12–21	2.5–13	WI	12
	single	14–17	3–6	IM	15
	poly, 0.3%	11	3–13	IM	14
Fe_2O_3	single, 0.2%	13	2–3	IM	75
SiO_2	single Z-cut	12–14	5–25	WI	76
	single X-cut	5–8	5–25	WI	76
	single Y-cut	8–10	5–25	WI	76
Cubic ZrO_2	single <100>	14	2.7–5	IM	31
Y_2O_3 doped	<110>	25	2.7–5	IM	31
Mg_2SiO_4	single <100>	6.5	(2–3)	IM	77
	<010>	12	(2–3)	IM	77
	<001>	9.1	(2–3)	IM	77
$LiNbO_3$	single, a-axis	6.6	2.8	IM	78
	single, c-axis	2.4	2.8	IM	78
$BaTiO_3$	poly, .7–4.4%	4.8–6.4	2–5	IM	58
AlN	poly, 1%	9.4	3–13	MG	42
	poly, <0.3%	9–10	3–5	IM	41
	poly, 1%	~10	2.5–10	V	44
Si_3N_4	poly, 1%	14–16.5	4–6	IM	46
SiC	poly, 4.0%	8	6.4	IM	79
	poly, 1.3%	15	4	V	49
B_4C	poly, 0.8%	14–16	3–10	IM	13
	poly, <0.5%	17–19	4	IM	50

[a] QG: quartz gauge, V: VISAR, OL: optical lever, IM: inclined mirror, WI: wire image, MG: manganin gauge.

the elastic clusters are large compared with those formed during plastic deformation due to dislocations. The loss of strength caused by partial melting should be even larger. If a process zone forms along slip zones active under shock compression, the strength may remain constant or increase.

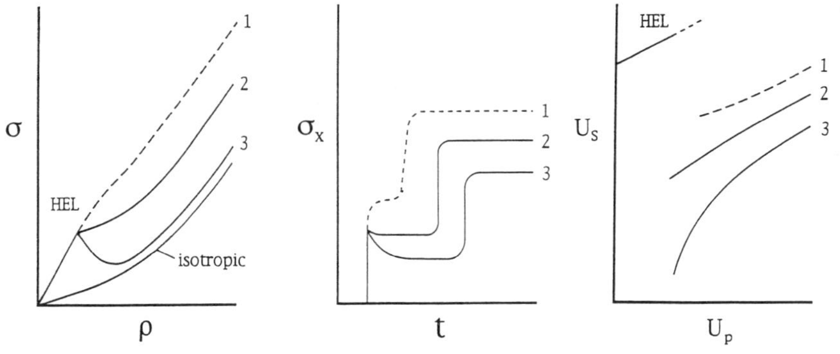

(a) $\sigma-\rho$ (b) σ_x-t (c) $U_s - U_p$

Figure 5.20. Model diagram of (a) the $\sigma-\rho$ Hugoniots, (b) shock-wave profiles, and (c) $U_s - U_p$ Hugoniots for (1) a toughening solid, (2) an elastoplastic solid, (3) an elastoisotropic solid, all at relatively low pressure.

Figure 5.21 shows a model illustration of typical slip mechanisms associated with a crack or melting zone, a dislocation band, a complex slip zone (crack, melting zone, and/or dislocation), and a process zone. Here, it is suggested that the mechanisms of shock yielding phenomena can be classified into the six types given in Table 5.3 [81].

Perfectly plastic deformation should be based on dislocations which uniformly multiply or grow. Many metals behave as elastoplastic solids due to the metallic bonding and high thermal conductivity. However, even metals lose shear strength, because the deformation is not always uniform. In fact, stress relaxation of metals has been observed [64,82]. It was found that even ionic materials behave as elastoplastic solids in the low-stress region [65,66,68], and the traces of dislocations were observed in recovered LiF single crystals [83,84]. For such low-strength materials, the shear strength can be ignored at sufficiently high pressures. In addition, it has been found that ceramics also deform plastically at high temperature.

The brittle-fracture deformation is mainly caused by cracks, including cleavages. The suggested cause of plastic-brittle zone deformation is complex slip zones consisting of dislocations and/or cracks. Cracks or complex slip zones are apt to grow along cleavage planes or planes of maximum shear stress (usually about 45 degrees to the shock propagation direction in continuum material). These types of deformation occur with a wide range of frictional resistance between the elastic clusters, which are surrounded by slip zones. The materials may behave as quasi-elastoplastic solids or quasi-elastoisotropic solids. Many brittle materials may be included within these classes.

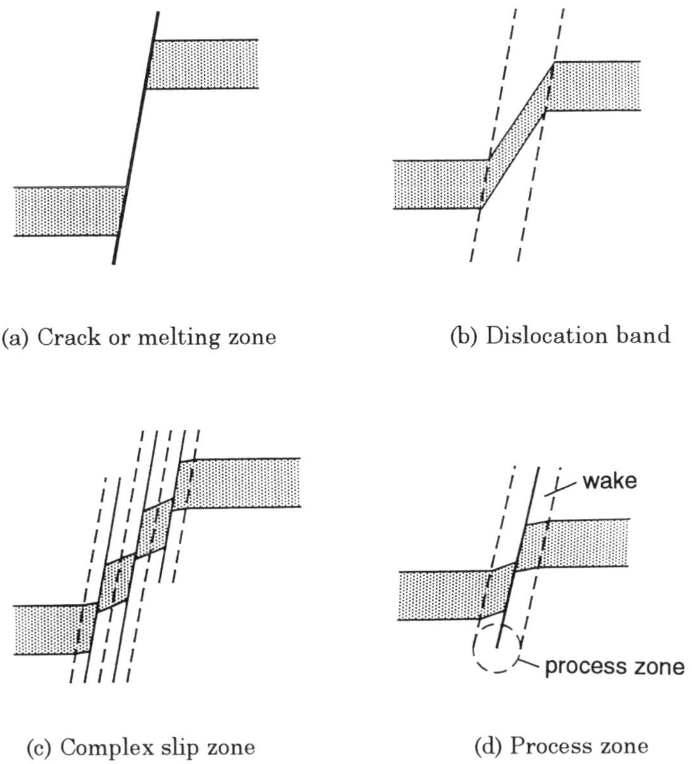

Figure 5.21. Model illustration of the slip systems (a) by crack or melting zone, (b) dislocation band, (c) complex slip zone, and (d) process zone.

Actually, in the low-stress region, cracks may play an important role, and in the high-stress region, the complex slip zones with dislocations and/or melting zones may become significant.

Recovery experiments have led to the observation of the planar slip structures of the zones or cracks (including lamellae structure) in specimens of quartz [85,86], olivine [87,88], periclase [87,89], etc. Above the HEL, many cracks were observed along certain crystal planes, whereas below the HEL, only irregular cracks were observed in recovered specimens of some minerals [90,91]. However, no clear in situ evidence that cracks or melting zones appear under shock compression has been reported, although deformation by dislocations has been recognized by many researchers, as mentioned above. From this point of view, the author believes that the jagged profiles of the free surface of millimeter scale, which were observed on the Al_2O_3 single crystal and the B_4C polycrystal by the gapped-flat-mirror method, were important as supporting evidence for crack formation

Table 5.3. Tentative classification of shock yielding phenomena

Deformation type	Main slip mechanism	Reduction of strength	Yielding type	Material
Plastic	dislocations (incl. twins and stacking faults)	small	elastoplastic	metals
Plastic-brittle zone	cracks and dislocations	small moderate	elastoplastic	metals and brittle materials (high stress region)
Brittle fracture	cracks (incl. cleavages)	moderate large	elastoplastic elastoisotropic	high-density-structure brittle material low-density-structure brittle material (low-stress region)
Partial melting zone	melting zones, etc.	very large	(perfect) elasto-isotropic	low-melting-temp. and low-thermal-conductivity material
Lattice destruction	melting lattice instability? vacancies?	very large	(perfect) elasto-isotropic	low-melting-temp. material low-density-structure material unstable-phase material?
Toughening	process zones	very small or none	toughening	composite material? transformation toughening material?

and growth as a yield mechanism. It is suggested that the elastoplastic transition appears in such materials when more than a critical number of cracks or slip zones form or grow. In fact, we observed a free-surface profile with a few steps, even in the elastic precursor, which indicated the appearance of a small amount of cracking [17,51].

The author assumes that similar jagged profiles of the free surface are produced in other brittle materials but that they are not observed because their scale is below the resolving capability of the instrumentation. The cluster sizes of shock yielding may be determined by the interaction of the shock compression with material characteristics such as strength and crystal state.

For quartz (SiO_2) single crystals, cross-shaped luminescences, having directions parallel to the cleavage planes, were observed under shock compression by in situ observation using an electronic framing camera, although similar luminescences are not observed in fused silica [92]. It has been pointed out that, for some minerals, higher driving stresses produce greater temperature increases, smaller deformation cluster sizes, and lower macroscopic shear strengths. As a result, the state might approach one of isotropic stress [80,87]. This is consistent with the fact that above a certain stress, luminescence of quartz was not observed [92].

Among the five deformation types, without the lattice destruction type, the partial-melting-zone mechanism is most effective in contributing to the loss of shear strength. In this mechanism, the slip occurs in the partial-melting zone along cracks or slip zones [80,93]. However, the author assumed that elastoisotropic deformation is caused not only by the partial-melting mechanism but by other microscopic changes causing lattice destruction. The lattice destruction mechanism can result from instabilities due to melting, structure transition, amorphization, or a high concentration of vacancies. The partial-melting deformation mechanism may occur in low-melting temperature and low-thermal-conductivity solids, and the lattice destruction mechanism may occur in solids with low-melting, low-packing-density structures, solids with high vacancy concentration, or solids in an unstable phase. It has been pointed out that silica, some silicates, and some semiconductors having low-packing-density structure behave as elastoisotropic solids [69–72,76–78]. Amorphous phase parts were observed along slip zones, where lamellar structures formed, in recovered specimens of silica, olivine, feldspar, etc. [94–96]. They may be related to amorphization or melting along slip planes, where hot spots and/or zones appear. It was also assumed that these amorphous phases were the reversion products of high-density, short-range-order solids [97]. The CaO-doped cubic zirconia also behaved as an elastoisotropic solid [98]. The author assumed that this was related to the high concentration of vacancies due to the doping of CaO. However, it is very difficult to strictly confirm the slip mechanisms. Probably, several slip mechanisms operate concurrently in many cases.

5.3.2. Correlation with Material Characterization

Crystal State and Thermal Properties

It has been pointed out [80,99] that thermal conductivity strongly affects the yielding behavior of brittle materials, with the uniformity of yielding increasing in materials of higher conductivity. The author suggested that the chemical-crystal properties should be ranked as major factors determining the shock yielding behavior, rather than the thermophysical properties [15,100]. However, the practical correlations between these properties and yielding have not been well understood. Conceivable characterization factors include chemical-crystal properties, crystal state (single crystal, polycrystal, or amorphous), thermal conductivity, microstructure, porosity, phase transition, etc.

Some experimental results showed differences in Hugoniot or stress history even for the same compound. The crystal state can be identified as an influential factor in shock yielding. Figure 5.22 shows a model diagram of stress–density (σ–ρ) Hugoniots, shock wave profiles, and shock velocity–particle velocity (U_s–U_p) Hugoniots of single crystal, polycrystal, and amorphous materials, respectively, at relatively low stress. Single crystals show high HEL stress compared with polycrystals. In addition, the yielding behavior of single crystals depends on the shock propagation direction relative to the orientation of cleavage planes. This dependence was observed in the Hugoniot results for cubic zirconia (Sec. 5.2.3), TiO_2 (Sec. 5.2.2), and LiF single crystals [65–67]. Behind the HEL, single crystals show a larger stress relaxation than polycrystals. The author believes that, although the reduction in strength is conspicuous, single crystals of many high-strength ceramics retain some shear strength behind the elastic shock front. For polycrystals, the shape of the rising portion of the waveform is disperse because the phase velocity is an average of values for each axis direction. The porosity may decrease their lower HEL stress. The elastic stress drop at stress release (expansion process) in single crystals is larger than that for polycrystals, as shown in Fig. 5.3 for Al_2O_3. This may be due to the large elastic regions that remain in the shock compressed single crystal. These tendencies were observed in Al_2O_3 and zirconia (YCZ).

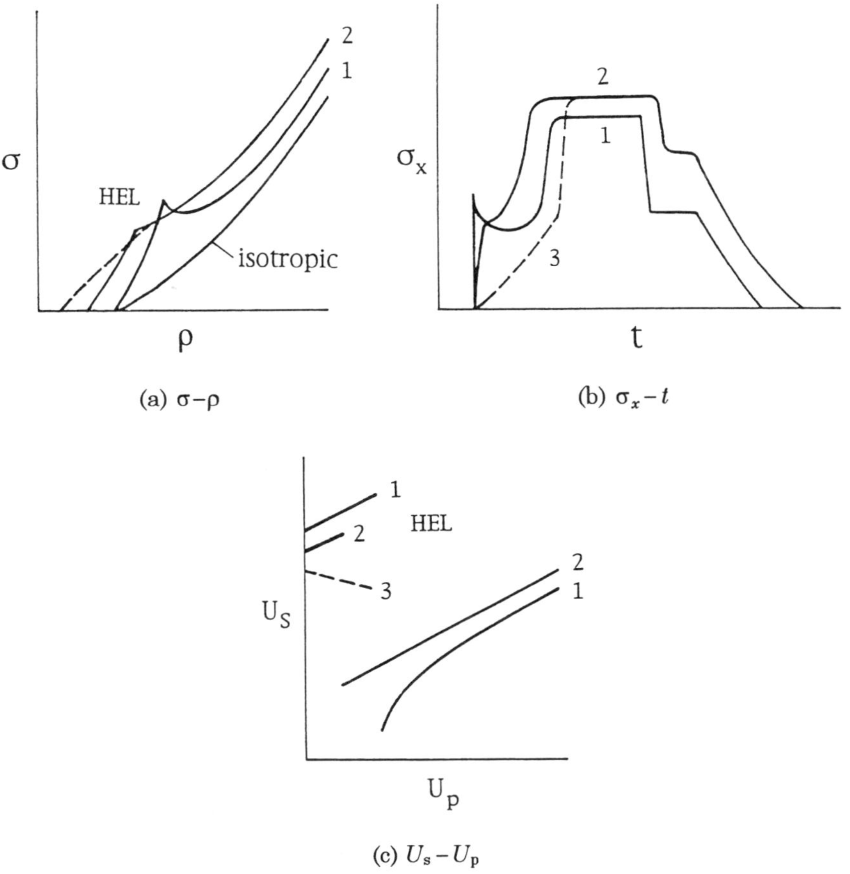

Figure 5.22. Model diagram of (a) the $\sigma-\rho$ Hugoniots, (b) shock-wave profiles, and (c) $U_s - U_p$ Hugoniots for (1) single crystal, (2) polycrystal, and (3) amorphous materials, all at relatively low pressure.

Figure 5.23 shows a model illustration of deformation images in single-crystal and polycrystalline brittle materials under shock compression. It is assumed that cleavages or cracks can easily form or propagate along cleavage planes or planes of maximum shear stress in single crystals and, as a result, the elastic cluster sizes may be large. However, it is assumed that in polycrystals, the growth, multiplication, and motion of cracks or dislocations are disturbed by grain boundaries, stacking faults, impurities, and interactions among these defects. Many microcracks or dislocations appear or multiply at these locations. The elastic cluster size becomes small and the uniformity of deformation is increased. As a result, the material may become hard, and the stress behind the elastic shock front increases, as shown in Fig. 5.22. Supporting this explanation is the fact that jagged free-surface profiles of millimeter scale in the plastic region were observed in Al_2O_3 single crystals, whereas waveforms in polycrystals were smooth.

Crystalline and amorphous materials may also show differences in shock yielding behavior, even of the same compound. In the rising part of the waveform, some amorphous materials show a dispersed elastic shock front, due to the negative first pressure derivative of elastic moduli, as shown in Fig. 5.23 [76,101]. It was pointed out that fused silica and some polymers preserved a certain shear strength [102,103], although quartz undergoes immediate transformation to an isotropic state behind the elastic shock front [76]. It is assumed that, in amorphous material, dislocations or cracks cannot easily move or grow, because the amorphous structure has no long-range lattice ordering and no cleavage plane. Recently, an unique shock wave structure with low velocity (failure wave) was found in some glasses [104,105]. The origin of this wave is an interesting problem in relation to crack behavior.

Although it has been suggested that heterogeneous shock yielding of brittle materials is caused mainly by high local temperatures (attributable to their low thermal conductivity), it was concluded that thermal conductivity is not a major factor determining the shock yielding behavior. This conclusion was drawn because it was confirmed that AlN behaves as an elastoisotropic solid [43,44], although its heat conductivities are as large as those of metals [106,107]. It is assumed that deformation by cracks is not significantly affected by the

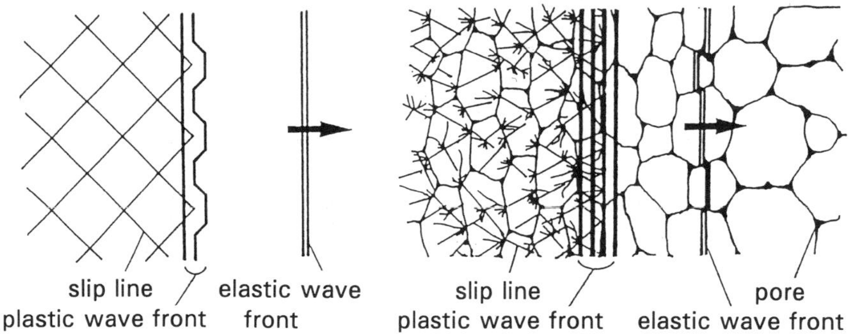

(a) Single crystal (b) Polycrystal

Figure 5.23. Model illustration of the deformation images in (a) single crystal and (b) polycrystal of brittle material under shock compression.

thermal uniformity, but that low thermal conductivity enhances the partial-melting phenomenon. It was pointed out that tungsten (W) and tantalum (Ta) also lose shear strength behind the elastic precursor, despite being metals, but preserve a certain shear strength in the plastic region [63,64]. Tungsten is thought to be a critical material, since it has high hardness and brittleness compared with the other metals, but also a high thermal conductivity. On the other hand, diamond (C) and dense boron nitride (BN) have the highest hardness and the highest thermal conductivity. It would be interesting to investigate the shock yielding properties of these materials.

Effects of Microstructural Control and Phase Transition (Process-Zone Toughening?)

It was pointed out that formation of process zones increases the stored deformation energy at crack tips and, as a result, causes the nonlinear deformation and high toughness of certain ceramics under static conditions. It is expected that if process zones are formed at the tips of cracks or slip zones under shock compression, the rise profile of the shock wave will become nonlinear and the HEL stress or Hugoniot offset will increase, as shown in Figs. 5.21a and 5.21b (type 1). In fact, it was found that fine-grained Si_3N_4 ceramics had a higher HEL stress than the normal material [38]. It may be that fine-grained polycrystalline materials can preserve larger shear strength because the growth or motion of cracks or dislocations is impeded as they encounter the many grain boundaries present. In addition, we confirmed that the SiC-whisker-doped Si_3N_4 ceramics showed high HEL stresses [11]. This is also caused by the effects of doping of high-strength whiskers, which impede the growth or motion of cracks or dislocations. It is assumed that microcracks play an important role in these materials.

The Y_2O_3-doped tetragonal zirconia [33,35] showed anomalous shock wave behavior compared with the cubic zirconia, in spite of being of the same basic compound. Examples are the fine three-step structure at rise and the large spall strength at release. Specimens of YTZ recovered following shock compression were intact with cracks at the surface, although those of alumina and cubic zirconia were severely crushed and looked like a compacted powder. In addition, it was found that a certain amount of monoclinic phase was formed in shock compression experiments. These facts supported the conclusion that transformation toughening occurred under shock compression, particularly at release.

Available experimental information is not sufficient to permit discussion of the effects of microstructural control on shock phenomena.

It is desirable that other high-performance ceramics such as the nano-composite and fiber-doped ceramics be examined in a study of toughening mechanisms. Process-zone toughening may be examined by measuring precisely the shock wave profile, by observing process-zone wakes (trace), or by measuring the residual strain energy of recovered specimens.

5.4. Effects of Shock Compression on Shock-Induced Phase Transition

Shock wave propagation in solids can induce first- or second-order phase transitions [99,108,109]. Figure 5.24 shows a model diagram of the pressure-density Hugoniot of a solid. We can get information about the phase transition point and the EOS of the high-pressure phase by the Hugoniot measurement. However, the phase transition process under shock compression is sometimes different from that under static compression, due to the short pulse duration, uniaxial compression, heterogeneous state, and other differences between the two experiments. They disturb in situ microscopic observations. In addition, nonequilibrium electrical states can be produced by electric or magnetic transitions. Table 5.4 shows the conceivable peculiar effects of shock compression on phase transitions.

5.4.1. Short Pulse Duration

The Hugoniot-compression curves with phase transition always show a wide range mixed phase region, as in Fig. 5.24, even for displacive transitions such as occur in carbon [110] and rutile-phase TiO_2, or structural phase transitions with electronic transition such as occur in lithium niobate ($LiNbO_3$) [111] and hematite (Fe_2O_3) [75]. If the transition proceeds to completion on a nanosecond time scale, the mixed phase region should not appear, for example, in pure electronic transitions [111]. The author assumes that its presence may be caused by a heterogeneous stress–strain–temperature state produced during the compression process and by the fact that the Hugoniot data are mainly determined by the state just behind the shock front. A partial phase transition may occur in material in such a heterogeneous state and may proceed behind the shock front. As a result, the high-pressure phase and low-pressure phase regions exist together even for non-diffusion-type transitions.

It is assumed that a second-order phase transition occurring under shock compression is completed almost instantaneously. This type phase transition can produce a nonequilibrium electromagnetic state and can induce electrical charge flows or eddy currents. High

Table 5.4. Conceivable effects of shock compression on phase transition in solids

Stimulus	Effect
High pressure	Structural phase transition Electronic phase transition Dielectric and magnetic phenomena
Short pulse duration	Nonequilibrium transition phenomena (mixed phase state, relaxation, etc.)
Uniaxial compression	Anisotropic phase transition (in displacive transition) Stress-induced phase transition Amorphization?
Heterogeneous state	Mixed phase state, lamellar structure, melting zones, etc.
Shear deformation	Nucleation, activation effect
High temperature	Melting, activation effect

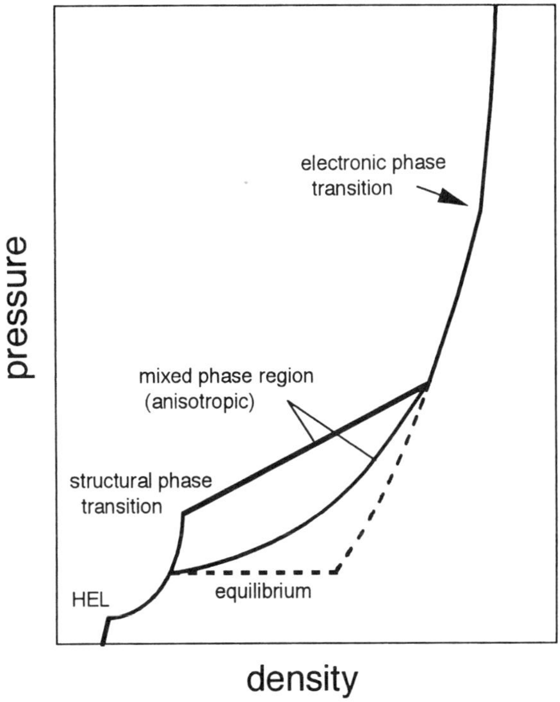

Figure 5.24. Model diagram of the pressure–density Hugoniot of a solid, in which elastoplastic transition, structural phase transition, and electronic phase transition occur.

voltages or large currents have been measured on ferroelectric materials [52–56], and eddy currents were measured on ferromagnetic or ferrimagnetic materials [99,113,114].

5.4.2. Uniaxial Compression

The uniaxial strain produced by plane shock compression sometimes affects shock-induced phase transitions, particularly those of the displacive type. The anisotropic Hugoniot-compression curves in the mixed phase region, which were observed on ilumenire ($FeTiO_3$) [115], rutile (TiO_2) [20,22,23], carbon [116], etc., are thought to be evidence of direct effects of uniaxial compression. Particularly, it was well understood microscopically that the uniaxial displacement of atoms directly induces the displacive phase transition on rutile. Similar effects may be observed on other anisotropic crystals. It is desirable that Hugoniot measurements on the single crystals, even of previously studied materials, be performed to examine the effects of uniaxial compression.

Even for nucleation-type phase transitions, anisotropic effects may be observed through anisotropy in the yielding behavior. Uniaxial compression induces slip phenomena along cleavage or maximum shear stress planes, where hot spots or hot zones appear. It is assumed that nuclei can easily form or grow there and, as a result, new phases or melting zones can appear. In fact, lamellar structures were observed in recovered specimens of some silicates, etc. and may related to shear band phenomena and phase transition.

Recently, the amorphization of H_2O [117], GeO_2 [118], $AlPO_4$ [119], etc. by static compression has attracted considerable attention. It is also well known that amorphous phases can be obtained by shock compression of quartz [120,121] and feldspars [122–125]. Amorphous phases were also found in recovered specimens of GeO_2 [126], C [127], etc. The process of formation of the amorphous phases is almost unknown. Some of them may be directly related to the shear deformation of shock compression, in addition to phase transition or melting.

5.4.3. Effect of Crystal State and Microstructure

Some effects of crystal state and microstructure on phase transition were identified in our studies. The cubic and tetragonal zirconias showed very different phase transition behavior despite being of the same basic material [32,33]. In addition, cubic zirconia (YCZ) polycrystals showed a diffuse transition point, whereas we observed sharp transitions for single crystals [32]. This may be caused by the existence of grain boundaries, varying grain orientation, and/or porosity of polycrystals. Porosity may affect phase transition through a heterogeneous temperature increase. The shock compression behavior and the phase transitions of powders are more confusing. For AlN [43,45,46], different transition points were observed for polycrystals having different grain sizes. The grain size effects were also observed under static pressure [128]. These results are important in discussing the phase transition mechanism and also show us the importance of using good quality specimens in shock compression experiments.

5.5. Concluding Remarks

The purpose of this chapter is to show typical effects of shock compression on the shock yielding behavior and phase transitions of ceramics and to discuss the mechanisms and the correlations with their material characterization. A tentative classification of shock yielding phenomena of solids was suggested, and the mechanisms and correlations with material characterization, including transformation and microstructure, were discussed on the basis of the experimental results. Particularly, it was suggested that the jagged free-surface profiles observed during shock compression of Al_2O_3 single crystals and B_4C polycrystals were supporting evidence for crack formation and growth as a yielding mechanism. The possibility of process-zone toughening under shock compression was also discussed. The typical effects of short pulse duration, uniaxial compression, etc. on shock-induced phase transitions were summarized.

However, even now, available Hugoniot data and observational results are insufficient to permit complete discussion of shock yielding and the phase transition phenomena in solids. It is expected that shock wave measurement experiments on both newly developed materials and high-quality specimens of materials that have already been investigated will be performed. Particularly, the yielding behavior, phase transitions, and EOS of hard ceramics such as carbides, borides, nitrides and diamond are expected to be investigated in detail. Amorphous materials and some kinds of intermetallic compounds and polymers are also expected to be investigated to clarify yielding properties.

Taking notice of measurement methods, it is desirable that in situ microscopic observation be routinely performed to directly discuss transition mechanisms and to identify high-pressure phases. It is also desirable to use the VISAR method for the investigation of ceramics. Measurements of strain rate and relaxation are significant for the study of shock yielding phenomena and phase transitions.

Few such investigations have been performed on brittle materials, although investigation of metals is well advanced [129,130]. In addition, the author expects that this method makes measurements on small specimens possible and breaks new measurement ground on metastable phase materials such as high-pressure phases, and amorphous phase and superlattice compounds. In addition, special dynamic compression methods such as combined compression–shear shock waves [16,131–133] and quasi-isentropic plane waves [134] can provide interesting information about the equation of state, phase transition, and yielding properties. Particularly, the measurement of shear waves is expected to give direct information on yielding.

At the same time, detailed microscopic observation of recovered specimens is also expected to be conducted, using recently advanced technologies or facilities, to microscopically clarify the mechanism of yielding and phase transition, since it is very difficult to perform in situ microscopic measurements under shock compression. Furthermore, theoretical studies are also expected to be advanced. Analytical theories of shock yielding phenomena of brittle materials have been suggested [135,136], and simulation of phase transitions by molecular-dynamic analysis has started [137,138]. However, shock yielding and phase transition of brittle materials are very confusing because they are affected by many complicated factors (slip mechanisms, nonuniformity, thermal and chemical factors, melting, transition mechanisms, etc.), and, particularly, the heterogeneous stress–strain–temperature state. It is thought to be very difficult to simulate strictly shock-wave profiles or compression curves of brittle materials with full consideration of these factors. The author believes that shock compression research can progress by the good interactions and combinations of the measurement and recovery experiments, and also the theoretical approaches.

References

[1] S.P. Marsh (ed.), *LASL Shock Hugoniot Data*, University of California Press, Berkeley (1980).

[2] M. van Thiel (ed.), *Compendium of Shock Wave Data, Vol. 1*, technical report UCRL-50108 (Vol. 1), Lawrence Radiation Laboratory, Livermore, California (1966).

[3] R.G. McQueen, S.P. Marsh, J.W. Taylor, J.N. Fritz, and W.J. Carter, in *High-Velocity Impact Phenomena* (ed. R. Kinslow), Academic Press, New York, p. 244 (1970).

[4] L.V. Al'tshuler, *Sov. Phys. Usp.* **8**, p. 52 (1965).

[5] T.J. Ahrens and V.G. Gregson, Jr., *J. Geophys. Res.* **69**, p. 4839 (1964).

[6] R.G. McQueen, S.P. Marsh, and J.N. Fritz, *J. Geophys. Res.* **72**, p. 4999 (1967).

[7] T. Mashimo, A. Nakamura, and S. Hamada, in *SPIE-1801*, p. 170 (1993).

[8] A. Nakamura and T. Mashimo, *Jpn. J. Appl. Phys.* **32**, p. 4785 (1993).

[9] T. Mashimo and K. Nagayama, *Jpn. J. Appl. Phys.* **25**, Suppl. 25-1, pp. 103–105 (1986).

[10] T. Mashimo, S. Ozaki, and K. Nagayama, *Rev. Sci. Instrum.* **55**, p. 226 (1984)

[11] T. Mashimo, in *Shock Wave in Materials Science* (ed. A. Sawaoka), Springer-Verlag, Tokyo, pp.113–144 (1993).

[12] R.A. Graham and W.P. Brooks, *Phys. Chem. Solids* **32**, p. 2311 (1971).

[13] W.H. Gust and E.B. Royce, *J. Appl. Phys.* **42**, p. 276 (1971).

[14] T.J. Ahrens, W.H. Gust, and E.B. Royce, *J. Appl. Phys.* **39**, p. 4610 (1968).

[15] T. Mashimo, Y. Hanaoka, and K. Nagayama, *J. Appl. Phys.* **63**, pp. 327–336 (1988).

[16] T. Mashimo, in *Shock Waves in Condensed Matter—1987* (ed. S.C. Schmidt and N.C. Holmes), North-Holland, Amsterdam, p. 285 (1988).

[17] T. Mashimo, unpublished data.

[18] T. Sato and S. Akimoto, *J. Appl. Phys.* **50**, p. 5285 (1979).

[19] T. Mashimo, unpublished data.

[20] T. Mashimo, K. Nagayama, and A. Sawaoka, *J. Appl. Phys.* **54**, p. 5043 (1983).

[21] T. Mashimo, K. Nagayama, and A. Sawaoka, in *Proc. 8th AIRAPT High Pressure Conf.*, p. 239 (1982).

[22] Y. Syono, K. Kusaba, M. Kikuchi, and K. Fukuoka, in *High-Pressure Research in Mineral Physics* (ed. H. Manghanani and Y. Syono), p. 385 (1987).

[23] K. Kusaba, M. Kikuchi, K. Fukuoka, and Y. Syono, *Phys. Chem. Miner.* **154**, p. 238 (1988).

[24] D.E. Grady, R.E. Hollenbach, and K.W. Schuler, *J. Geophys. Res.* **83**, p. 2839 (1978).

[25] R.G. McQueen, J.C. Jamieson, and S.P. Marsh, *Science* **155**, p. 140 (1960).

[26] L.V. Al'tshuler, M.A. Podurets, G.V. Simakov, and R.F. Trunin, *Sov. Phys. Solid State* **15**, p. 969 (1973).

[27] R.C. Garvie, R.N. Hannink, and R.T. Pasoe, *Nature* **256**, p. 713 (1975).

[28] O. Ohtaka, S. Kume, and E. Ito, *J. Am. Ceram. Soc.* **71**, p. C-448 (1988).

[29] H. Arashi, T. Yagi, S. Akimoto, and Y. Kudoh, *Phys. Rev.* **B41**, p. 4309 (1990).

[30] J.M. Leger, R.F. Tomaszewski, A. Atouf, and A.S. Pereira, *Phys. Rev.* **B47**, p. 14075 (1993).

[31] T. Mashimo, K. Nagayama, and A. Sawaoka, *Phys. Chem. Miner.* **9**, p. 237 (1983).

[32] T. Mashimo, A. Nakamura, K. Kodama, K. Kusaba, K. Fukuoka, and Y. Syono, *J. Appl. Phys.* **77**, p. 5060 (1995).

[33] T. Mashimo, A. Nakamura, M. Nishida, S. Matsuzaki, K. Kusaba, K. Fukuoka, and Y. Syono, *J. Appl. Phys.* **77**, p. 5069. (1995).

[34] J.M. Leger, A. Atouf, P.E. Tomaszewski, and A.S.Pereira, *Phys. Rev.* **B48**, p. 93 (1993).

[35] D.E. Grady and T. Mashimo, *J. Appl. Phys.* **71**, p. 4868 (1992).

[36] T. Mashimo, *J. Appl. Phys.* **63**, p. 4747 (1988).

[37] T. Ogata, M. Kihara, K. Nakamura, and K. Kobayashi, *J. Ceram. Soc. Jpn.* **96**, p. 310–316 (1988).

[38] H. Vollstadt, E. Ito, M. Akaishi, S. Akimoto, and O. Fukunaga, *Proc. Jpn. Acad.* **66**, Ser B, p. 7 (1990).

[39] I. Gorczyca, N.E. Christensen, P. Perlin, I. Grzegory, J. Jun, and M. Bockowski, *Solid State Commun.* **79**, p. 1033 (1991).

[40] M. Ueno, A. Onodera, O. Shimomura, and K. Takemura, *Phys. Rev.* **B45**, p. 10123 (1992).

[41] A. Nakamura and T. Mashimo, in *High-Pressure Science and Technology–1993* (ed. S.C. Schmidt, J.W. Shaner, G.A. Samara, and M. Ross), American Institute of Physics, New York, p. 303 (1995).

[42] Z. Rosenberg, N.S. Brar, and S.J. Bless, *J. Appl. Phys.* **70**, p. 167 (1991).

[43] D.E. Grady, Private communication.

[44] M.E. Kipp and D.E. Grady, in *Proc. EURO DYMAT 94,* in press (1995).

[45] T. Mashimo, unpublished data.

[46] A. Yamakawa, T. Nishioka, M. Miyake, K. Wakamori, A. Nakamura, and T. Mashimo, *J. Ceram. Soc. Jpn. Int. Edition* **101**, p. 1322 (1993).

[47] T. Mashimo, A. Nakamura, A. Yamakawa, T. Nishioka, and M. Miyake, in *Dynamic Plasticity and Structural Behavior*, p. 547 (1995).

[48] N.S. Brar, Z. Rosenberg, and S.J. Bless, *J. Appl. Phys.* **69**, p. 7890 (1991).

[49] M.E. Kipp and D.E. Grady, in *Shock Compression of Condensed Matter—1989* (ed. S.C. Schmidt, J.N. Johnson, and L.W. Davison), North-Holland, Amsterdam, p. 377 (1990).

[50] T. Mashimo, M. Uchino, and A. Nakamura, in *Proc. 20th Internat. Conf. High-Pressure Photograph & Photonics,* SPIE Vol. 2513, SPIE Press, Bellingham, WA, p. 792 (1995).

[51] T. Mashimo and M. Uchino, *J. Appl Phys.* **81**, p. 7064 (1997).

[52] F.W. Neilson, *Bull. Am. Phys. Soc.* **2**, p. 302 (1957).

[53] C.E. Reynolds and G.E. Seay, *J. Appl. Phys.* **32**, p. 1401 (1961).

[54] W.J. Halpin, *J Appl. Phys.* **37**, p. 153 (1966).

[55] P.C. Lysne, *J. Appl. Phys.* **48**, p. 1024 (1977).

[56] T. Mashimo, K. Toda, K. Nagayama, T. Goto, and Y. Syono, *J. Appl. Phys.* **59**, p. 748 (1986).

[57] S. Minomura, M. Tanaka, B. Okai, and H. Nagasaki, *Jpn. J. Appl. Phys.* **28**, Suppl., p. 404 (1970).

[58] L.E. Pope and J.N. Johnson, *J. Appl. Phys.* **46**, p. 720 (1975).

[59] L. Davison, A.L. Stevens, and M.E. Kipp, *J. Mech. Phys. Solids* **25**, p. 11 (1974).

[60] S. Minshall, *J. Appl. Phys.* **26**, p. 463 (1955).

[61] L.M. Barker and R.E. Hollenbach, *J. Appl. Phys.* **45**, p. 4872 (1974).

[62] O.E. Jones and R.A. Graham, in *Accurate Characterization of the High-Pressure Environment* (ed. E.C. Lloyd), U. S. National Bureau of Standards, Washington, DC, p. 229 (1971).

[63] P.P. Gillis, K.G. Hoge, and R.J. Wasley, *J. Appl. Phys.* **41**, p. 2145 (1970).

[64] J.R. Asay, L.C. Chhabildas, and D.P. Dandekar, *J. Appl. Phys.* **51**, p. 4774 (1980).

[65] Y.M. Gupta, G.E. Duvall, and G.R. Fowles, *J. Appl. Phys.* **46**, p. 532 (1975).

[66] J.R. Asay, D.L. Hicks, and D.B. Holdridge, *J. Appl. Phys.* **46**, p. 4316 (1975).

[68] W.J. Murri and G.D. Anderson, *J. Appl. Phys.* **41**, p. 3521 (1970).

[69] W.H. Gust and E.B. Royce, *J. Appl. Phys.* **42**, p. 1897 (1971).

[70] T. Goto, T. Sato, and Y. Syono, *Jpn. J. Appl. Phys.* **21**, p. L369 (1982).

[71] R.A. Graham, O.E. Jones, and J.R. Holland, *J. Phys. Chem. Solids* **27**, p. 1519 (1960).

[72] W.H. Gust and E.B. Royce, *J. Appl. Phys.* **43**, p. 4439 (1972).

[73] T. Goto and Y. Syono, in *Shock Waves in Condensed Matter—1981* (ed. W.J. Nellis, L. Seaman, and R.A. Graham), American Institute of Physics, New York, p. 320 (1982).

[74] T.J. Ahrens, *J. Appl. Phys.* **37**, p. 2532 (1966).

[75] T. Goto, J. Sato, and Y. Syono, in *High Pressure Research: Application in Geophysics* (ed. M.H. Manghnani and S. Akimoto), p. 595 (1982).

[76] J. Wackerle, *J. Appl. Phys.* **33**, p. 922 (1962).

[77] Y. Syono and T. Goto, in *High Pressure Research: Application in Geophysics* (ed. M.H. Manghnani and S. Akimoto), p. 563 (1982).

[78] T. Goto and Y. Syono, *J. Appl. Phys.* **58**, p. 2548 (1985).

[79] W.H. Gust, A.C. Holt, and E.B. Royce, *J. Appl. Phys.* **44**, p. 550 (1973).

[80] D.E. Grady, *J. Geophys. Res.* **85**, p. 913 (1980).

[81] T. Mashimo, in *High-Pressure Science and Technology—1993* (ed. J.R. Asay, R.A. Graham, and G.K. Straub), American Institute of Physics, New York, p. 757 (1995).

[82] J.W. Taylor and M.H. Rice, *J. Appl. Phys.* **34**, p. 364 (1963).

[83] J.E. Flinn, G.E. Duvall, G.R. Fowles, and R.F. Tinder, *J. Appl. Phys.* **46**, p. 3752 (1975).

[84] P. Kumar and R.J. Clifton, *J. Appl. Phys.* **50**, p. 4747 (1979).

[85] D. Stöffler, *Fortschr. Miner.* **49**, p. 50 (1972).

[86] A.V. Ananin, O.N. Breusov, A.N. Dremin, S.V. Pershin, and V.F. Tatsii, *Combust. Expl. Shock Waves* **10**, p. 426 (1974).

[87] W.F. Müller and U. Hornemann, *Earth Planet Sci. Lett.* **7**, p. 251 (1969).

[88] W.U. Reimold and D. Stöffler, in *Proc. 9th. Lunar Planet Sci. Conf.* p. 2805 (1978).

[89] M.J. Klein, *Phil. Mag.* **12**, p. 735 (1965).

[90] J.F. Bauer, in *Proc. 10th Lunar Planet Sci. Conf.*, p. 2573 (1979).

[91] H. Mori, *J. Jpn. Crystallogr. Soc.* **27**, p. 179 (1985).

[92] P.J. Brannon, C.H. Konrad, R.W. Morris, E.D. Jones, and J.R. Asay, technical report SAND82-2469, Sandia National Laboratories, Albuquerque, New Mexico (1983).

[93] A.J. Granz, *Phys. Chem. Miner.* **16**, p. 221 (1988).

[94] W. Engelhardt and D. Stöffler, in *Shock Metamorphism of Natural Minerals* (ed. B. French and N. Short), Mono Press, Baltimore, p. 159 (1968).

[95] R. Jeanloz, T.J. Ahrens, J.S. Lally, G.L. Nord. Jr., J. M. Christie, and A.H. Heuer, *Science* **197**, p. 457 (1972).

[96] A.G. Bogdanov, S.A. Popov, and V.S. Rundenko, *Acad. Sci. USSR Proc. Chem. Sect.* **201**, p. 1011 (1971).

[97] P.S. DeCarli and D.J. Milton, *Science* **147**, p. 144 (1965).

[98] T. Mashimo, M. Kodama, and K. Nagayama, *Adv. Ceram.* **24**, p. 329 (1988).

[99] L. Davison and R.A. Graham, *Phys. Rept.* **55**, p. 255 (1979).

[100] T. Mashimo, in *Shock Waves in Condensed Matter—1987* (ed. S.C. Schmidt and N.C. Holmes), North-Holland, Amsterdam, p. 285 (1988).

[101] H. Sugiura, K. Kondo, and A. Sawaoka, *J. Appl. Phys.* **52**, p. 3375 (1981).

[102] S.J. Bless, N.S. Brar, and A. Rozenberg, in *Shock Waves in Condensed Matter—1987* (ed. S.C. Schmidt and N.C. Holmes), North-Holland, Amsterdam, p. 309 (1988).

[103] P.F. Chartagnac, *J. Appl. Phys.* **53**, p. 948 (1982).

[104] G.I. Kanel, S.V. Razorenov, and V.E. Fortov, in *Shock Compression of Condensed Matter—1991* (ed. S.C. Schmidt, R.D. Dick, J.W. Forbes, and D.J. Tasker), North-Holland, Amsterdam, p. 451 (1992).

[105] N.S. Brar, S.J. Bless, and Z. Rosenberg, *Appl. Phys. Lett.* **59**, p. 3396 (1991).

[106] Sumitomo Electric Industries Co. Ltd., Private communication.

[107] A. Horiguchi, F. Ueno, and A. Tsuge, *Toshiba Rev.* **44**, p. 616 (1986).

[108] G.E. Duvall and R.A. Graham, *Rev. Mod. Phys.* **49**, p. 523 (1977).

[109] Y. Syono, in *High Pressure Explosive Processing of Ceramics* (ed. R.A. Graham and A.B. Sawaoka), Tera Tech, Switzerland, pp. 479–400 (1987).

[110] W.H. Gust and D.A. Young, in *High Pressure Science & Technology* (ed. K.D. Timmerhaus and M.S. Barber), Plenum, p. 944 (1979).

[111] T. Goto and Y. Syono, *J. Appl. Phys.* **58**, p. 2548 (1985).

[112] L.V. Al'tshuler and A.A. Bakanova, *Sov. Phys. Usp.* **11**, p. 678 (1969).

[113] G.W. Anderson and F.W. Neilson, *Bull. Am. Phys. Soc.* **2**, p. 302 (1957).

[114] D.E. Grady, G.E. Duvall, and E.B. Royce, *J. Appl. Phys.* **43**, p. 1948 (1972).

[115] D.A. King and T.J. Ahrens, *J. Geophys. Res.* **81**, p. 931 (1976).

[116] T. Sekine, Private communication.

[117] O. Mishima, L.D.Calvert, and E. Whalley, *Nature* **310**, p. 393 (1984).

[118] M. Maden, P. Gilletm, C. Jullien, and G.D. Price, *Phys. Chem. Miner.* **18**, p. 7 (1991).

[119] M.B. Kruger and R. Jeanloz, *Science* **249**, p. 647 (1990).

[120] P.S. DeCarli and J.C. Jamieson, *J. Chem. Phys.* **31**, p. 1675 (1959).

[121] T. Mashimo, K. Nishii, T. Soma, and A. Sawaoka, *Phys. Chem. Miner.* **5**, p. 367 (1980).

[122] D. Stöffler and U. Hornemann, *Meteorite* **7**, p. 371 (1972).

[123] R.V. Gibbons and T.J. Ahrens, *Phys. Chem. Miner.* **1**, p. 95 (1977).

[124] M. Kimura, T. Goto, and Y. Syono, *Contr. Miner. Petrol.* **61**, p. 299 (1977).

[125] M. Okuno, F. Marumo, and Y. Syono, *Miner. J.* **12**, p. 197 (1985).

[126] N. Suresh, G. Satish, G.C. Gupta, S.K.S. Sangeeta, and S.C. Sabharwal, *J. Appl. Phys.* **76**, p. 1530 (1994).

[127] D.J. Erskine and W.J. Nellis, *Nature* **349**, p. 317 (1991).

[128] S. Kawasaki, T. Yamanaka, S. Kume, and T. Ashida, *Solid State Commun.* **76**, p. 527 (1990).

[129] L.C. Chhabildas and J.R. Asay, *J. Appl. Phys.* **50**, p. 2749 (1979).

[130] J.W. Swegle and D.E. Grady, *J. Appl. Phys.* **58**, p. 692 (1985).

[131] A.S. Abou-Sayed, R. J. Clifton, and L. Hermann, *Exp. Mech.* **6**, p. 127 (1976).

[132] Y.M. Gupta, *Appl. Phys. Lett.* **29**, p. 694 (1976).

[133] L.C. Chhabildas and J.W. Swegle, *J. Appl. Phys.* **51**, p. 4799 (1980).

[134] L.M. Barker and D.D. Scott, technical report SAND84-0432, Sandia National Laboratories, Albuquerque, New Mexico (1984).

[135] J. Stainberg, *J. Appl. Phys.* **65**, p. 3417 (1988).

[136] F.L. Addessio and J.N. Johnson, *J. Appl. Phys.* **67**, p. 3275 (1990).

[137] D.H. Robertson, D.W. Brenner, and C.T. White, *Phys. Rev. Lett.* **25**, p. 3132 (1991).

[138] N. J. Wagner, B.L. Holian, and A.F. Voter, *Phys. Rev.* **A45**, p. 8457 (1992).

CHAPTER 6

Response of High-Strength Ceramics to Plane and Spherical Shock Waves

J. Cagnoux and J.-Y. Tranchet

6.1. Introduction

The word "ceramics" comes from the Greek "χεραμος" meaning either potter's clay or pottery, depending on the author. Nowadays, this term also includes materials produced from chemically defined constituents (e.g., oxides, fluorides, borides, carbides, and nitrides) which are usually prepared components rather than natural raw materials. This chapter addresses these last materials and, more precisely, those among them with very high mechanical properties.

Ceramics are generally composed of one or more metallic elements combined with a nonmetallic one. Most of them have ionic bonds; others have covalent bonds. These two types of bond give ceramics their stability and high rigidity. Because of their low level of symmetry, crystal lattices of ceramics possess an insufficient number of slip systems to allow easy deformability in standard conditions. Covalent crystals are also highly resistant to dislocation movement. The difficulty of plastically deforming ceramics in standard conditions explains their brittleness and low tensile strength.

The first shock compression studies addressing these materials were conducted in the seventies. The interpretation of the shock wave behavior of ceramics was based on the brittle nature of their failure observed in standard conditions. The shock behavior of ceramics has been reinterpreted in the last decade to provide a basis for the use of these materials in armor elements. For that reason, only ceramics presenting a real potential in ballistic applications were addressed in the published studies. These ceramics are principally materials such as Al_2O_3, ZrO_2, B_4C, TiB_2, AlN, Si_3N_4, and SiC that possess high mechanical properties. At the present time, it seems that brittleness alone cannot explain many of the large variety of behaviors ceramics exhibit under high dynamic pressure.

6.2. Elements of Experimental Strategy

> If a constitutive equation is to be useful in the solution of practical problems, it can, at best, apply only to *limited ranges* of strain, strain rate, temperature etc.
> (W. Prager [1])

We believe that this opinion, stated in 1967, is still timely. Preparation of models of behavior and optimization of the microstructure of materials can only be conducted with regard to a particular application. Consequently, a sequential analysis of the loading applied to the ceramics has to be performed for every application [2]. This can be done by numerical simulation using a simplified model. Each element of the loading sequence is then studied in the laboratory with the help of appropriate experiments. Each experiment becomes integrated into the global strategy of the sequential analysis, taking into account, as initial conditions, the residual state of the material after the preceding experiment and preparing for the immediately following one. In many problems, uniaxial deformation by a plane shock wave obtained by plate-impact experiments is a conventional and preferred way to obtain experimental data. In the high dynamic pressure range, there are few laboratory experiments allowing access to triaxial deformation data. Among those that are available, we note, particularly, the loading by a longitudinal divergent spherical wave initiated by a pyrotechnic device. The relevance of this experiment is due to its noteworthy complementarity with plate impact (see Sec. 6.4.) and to the similarity of the loading obtained with that arising in many applications (particularly ballistic impacts).

6.3. Uniaxial Deformation by a Plane Shock Wave

Uniaxial deformation by plane shock waves is produced in plate-impact experiments. In past years, this type of experiment allowed collection of data on the micromechanical and mesomechanical responses of ceramics after analysis of samples carefully recovered without further damage (soft-recovered). It also yielded phenomenological data through analysis of various recorded macromechanical parameters. The objective of this section is to discuss the response of ceramics as observed at these different scales.

6.3.1. Exploitation of Soft-Recovered Samples

The first publications devoted to plane shock wave propagation in ceramics [3–5] mentioned cracking as essential process of deformation in these materials, even though it had not been observed in an experimental program specifically addressed to the issue. Nowadays, the increasing accuracy of metrology (for instance VISAR diagnostics) allows interpretation of the nature of the irreversibilities by analyzing recorded time-resolved signals. Nevertheless, these explanations remain hypotheses unsupported by direct observations. For this reason, since the beginning of the eighties, many teams have been trying to analyze soft-recovered samples after plate-impact experiments especially designed with this aim in view. Most of the configurations are derived from the star-on-star configuration developed by Kumar and Clifton [6]. However, these configurations are, in general, limited to low impact velocities (about 10–300 m/s) [7]. For higher velocities, capsules designed to protect the samples are increasingly used [8]. Table 6.1 draws up a nonexhaustive inventory of of soft-recovery experiments of ceramics, showing the great diversity of the implemented techniques. The analyses performed on soft-recovered ceramics reveal three classes of irreversibility: cracking, microplasticity, and phase change [9–18]

Table 6.1. Soft-recovery experiments on ceramics

Soft-recovery configuration	Material	Stress range (GPa)	References
Star impactor on circular target/free flight	Al_2O_3	0–9	19
	BeO	0–10	19
Circular impactor on circular target/capsule	Al_2O_3	2–7	11
	Al_2O_3	0–4.6	12
Circular impactor on circular target/free flight	TiB_2	3–14	14
	TiB_2	0–7	20
	Si_3N_4	2–15	21
Circular impactor on star target/free flight	Al_2O_3	13–20	22,23
	SiC	10–17	22,23
	B_4C	9–18	23
	TiB_2	3–20	22,23
	ZrO_2	2–15	23
Star impactor on square target/free flight	Al_2O_3	0–1.5	13,24
Circular impactor on rectangular target/capsule	Al_2O_3	5 – 23	18
	Al_2O_3	20	15,25
Star impactor on star target/free flight	Al_2O_3	0–12	10,17,26,27

Microplastic phenomena and phase changes can be attributed without risk to the loading phase induced by the plane shock wave because of the high confinement necessary for their occurrence.

In contrast, the validity of observations concerning cracking phenomena is contingent on the ability of the soft-recovery technique to trap lateral stress release waves and capture the sample without a second prejudicial shock or flexure. Observation of cracks in a soft-recovered sample does not imply that they were produced by uniaxial deformation due to the plane shock wave.

6.3.2. Nucleation of Cracks

In the middle stress range (from a few GPa to about 10 GPa), microplastic activity concerns only a few grains [10–14]. For increasing Hugoniot stress, cracking occurs when the elastoplastic response of grains and grain boundaries is not able to accommodate imposed deformations. In Al_2O_3, intergranular cracks are mainly observed [10,13,17,19,26,27]. One should notice that these cracks may have been nucleated either by grain fracture (intra) or grain boundary fracture (inter). The predominance of intergranular cracks is also observed in TiB_2 [14,23], SiC [23], and Si_3N_4 [21]. In opposition, in Y_2O_3-doped tetragonal ZrO_2 and B_4C, only transgranular cracks are observed [23]. For every sample observed, cracks always occur with random orientation. Because the largest number of observations have been made on alumina, the discussion will rely on the behavior of this material.

The first cause of nucleation may be crushing of either an intragranular pore or one located at a triple point, in accordance with a mechanism more particularly described by Brace and Bombolakis [28]. Louro and Meyers [12] have described this mechanism as an important source of cracks produced by compression of aluminas.

A second cause of crack nucleation may be related to the sintering-induced residual stresses that result from crystal anisotropy. Cagnoux and Cosculluela [29] have explained the difference between thresholds of cracking in A16-2 and AL23 aluminas by referring to the analysis of Evans and Fu, which shows that, in polycrystalline alumina, sintering residual stresses are significant for grain sizes larger than 20 μm [30]. Raiser et al. [13] have observed many intergranular cracks at triple points in Vistal alumina. They have emphasized the importance of these triple points that concentrate residual stresses and have suggested this mechanism as a factor in the nucleation of intergranular cracking [24]. On the other hand,

Cosculluela [17] has proposed local anisotropy of elastic deformations induced by the shock wave as a mechanism nucleating the first intergranular cracks to appear in Al_2O_3.

In the general framework of ceramic compression, the place of microplasticity in fracture mechanisms is mentioned by Rice [31]. Nucleation mechanisms based on plasticity have recently been discussed by Lankford [32]. Intergranular cracks are often observed to occur along alumina grains containing an important network of dislocations [12,13,17]. Winkler and Stilp [14] have shown dislocation pileup at cracked grain boundaries in TiB_2. Figure 6.1 shows an intergranular crack next to a grain containing twins in AL23 alumina [17]. Nevertheless, these observations do not prove that crack nucleation occurs as a result of twinning or pileup of dislocations at grain boundaries. Another scenario may be that a fracture is produced by relaxation of external stresses on the plastically deformed structure. With this scenario, we can wonder if some intergranular cracks are nucleated during unloading.

At the present state of the art, the greatest care must be taken before establishing a hierarchy among the different nucleating mechanisms that are possible in the middle stress range (from a few GPa to about 10 GPa). However, some common conclusions can be drawn from the various studies conducted up to the present time. The first conclusion is that reduction of grain size in pure or quasi-pure aluminas (>99.5% Al_2O_3) leads to an increase in the cracking threshold

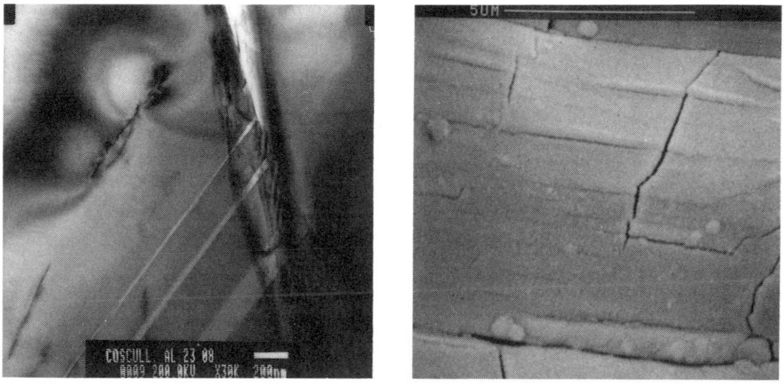

(a) (b)

Figure 6.1. (a) Transmission electron micrograph of Al23 alumina (5.6 GPa) showing twins and intergranular cracking. (b) Scanning electron micrograph of AL23 alumina (20 GPa) showing emergence of twins or deformation bands and intergranular cracking.

[17,24,27,29]. This is consistent with the last two nucleation mechanisms mentioned, because small grain sizes allow reduction of sinter-induced residual stresses as well as of twinning possibilities and stresses at the front of dislocation pileups. Moreover, in association with a homogeneous population of grain size, heterogeneities of local elastic deformations due to loading can be reduced. Second, a low average thickness of the secondary phase at the grain boundaries (in particular, a glassy phase) results in an increase in the threshold for intergranular cracking [17,24,29]. Indeed, a very low threshold of intergranular cracking is observed in aluminas in which a high proportion of glassy phase is present [11,26,29].

For the high-stress range (>10–15 GPa), most of the grain presents plastic activity [15,16,17,18,22]. Clear cases of nucleation due to microplasticity have been mentioned. Wang and Mikkola [18] have observed, on a single crystal of alumina, crystallographic cracks and randomly oriented cracks probably due to branching of the first ones. Similar processes occur in polycrystals. Figure 6.1 shows intragranular cracks oriented with regard to emergence of slip bands or twins as well as secondary branched cracks in AL23 alumina. This double network constitutes the main part of the intragranular cracking in this material.

In brief, several nucleating mechanisms occur in the middle stress range, making use of the variety of compositions and microstructures. The mechanisms described can operate individually or be coupled. In the high stress range, crack nucleation by twinning or slip band formation dominates the process of comminution of aluminas: it completes the pullout of grains by intergranular cracking and leads to intragranular cracking.

6.3.3. Hugoniot Elastic Limit (HEL) and Flow in Compression

Measurements

The HEL appears as a singularity of plane wave propagation in the material, which allows its measurement. Table 6.2. draws up a non-exhaustive list of HEL measurements performed for various ceramics. For the same material, comparison of HEL values obtained by different investigators requires care because of the different accuracies of implemented metrologies and experimental conditions.

In the case of materials having a high shock impedance, the VISAR currently affords the best time resolution combined with the best accuracy for measuring an HEL. This metrology makes it possible to classify histories of particle velocity recorded for different ceramics into three classes (Fig. 6.2).

Factors Influencing the HEL

Among factors intrinsic to the material, chemical composition, and deformation mode exert primary influence over the HEL. When a material contains significant porosity, the nominal value of the HEL is considerably modified [10,45]. Moreover, in Al_2O_3, an important variation of HEL values appears for high-density material (typically

Table 6.2. Some HEL measurements of high-strength ceramics

Material	Type of measurement	References	HEL (GPa)[a]	
Al_2O_3	Inclined-mirror method	3,33	16.1 ± 0.6	$\rho_0 = 3.92$ [3]
	Piezoresistive gauge	33–38	12.4	$\rho_0 = 3.96$ [38]
	VISAR	5,10,17, 19,26,27, 37,39 40	11.9	$\rho_0 = 3.974$ [40]
SiC	Inclined-mirror method	4	8.0 ± 3.0	$\rho_0 = 3.09$
	VISAR	22 23,41, 42	15–16	$\rho_0 = 3.177$ [42]
TiB_2	Inclined-mirror method	4	8.6 ± 3.0	$\rho_0 = 4.51$
	Piezoresistive gauge	43,44	7.1 - 7.9	$\rho_0 = 4.45$ [41]
	VISAR	22, 3,41	HEL① 4.7-5.2	$\rho_0 = 4.452$ [41]
			HEL② 16-18.6	$\rho_0 = 4.40$ [22]
B_4C	Inclined-mirror method	3	16.2 ± 9	$\rho_0 = 2.50$
	Piezoresistive gauge	45	19.4 ± 3	$\rho_0 = 2.52$
	VISAR	23,41,42	18–20	$\rho_0 = 2.506$ [42]
ZrO_2	Inclined-mirror method	46	30	$\rho_0 = 5.954$
100%	Piezoresistive gauge	38	12.2	$\rho_0 = 6.03$
tetrago-	VISAR	23,47	HEL① 16.9	$\rho_0 = 6.028$ [47]
nal			HEL② 30–31	$\rho_0 = 6.028$ [47]
AlN	Piezoresistive gauge	48	9.4 ± 0.2	$\rho_0 = 3.226$
	VISAR	49	HEL① 10	$\rho_0 = 3.262$
			HEL② 22	$\rho_0 = 3.262$
Si_3N_4	VISAR	21	12.1	$\rho_0 = 3.15$

[a] Reported HEL are the highest values measured among the mentioned references.

above 3.90 g/cm^3). In the first analysis, the influence of grain size can be implicated. Following this direction, Cagnoux and Cosculluela [29] have ordered the HELs of quasi-pure aluminas (>99.5% Al_2O_3) prepared by the same process, according to a Hall–Petch law. Elsewhere, Mashimo [50] has measured HELs of Si_3N_4, finding them to be inversely proportional to grain size. In fact, parameters to be taken into account include not only grain size but also grain distribution, the nature of intergranular phase and its proportion, porosity location (in grains, triple points), pore geometry, etc. To our knowledge, there are no definitive data establishing the place of those parameters that are not easy to investigate individually without modifying others.

Concerning factors extrinsic to the material, some authors have observed a precursor decay with increasing propagation distance and decreasing shock level in alumina [3,34,35,40], whereas others have not [27,39]. A precursor decay has also been observed in Si_3N_4 [21]. Recently, Grady, using a wave-profile normalization technique, has eliminated the role of the propagation distance in several ceramics and has observed no effect of shock amplitude on SiC and B_4C [42]. Concerning the loading rate, Cagnoux and Longy [39] have not observed any effect on alumina for strain rates between 5.4×10^4 s^{-1} and 5.8×10^5 s^{-1}.

Physical Interpretations and Phenomenological Aspects

The essentially brittle nature of failure of ceramics in standard conditions has led many authors to explain the HEL as a threshold of cataclastic failure [5,20,36,51,52]. Today, exploitation of VISAR particle velocity histories and analysis of soft-recovered samples call the generality of these conclusions into question.

Analysis of shocked aluminas demonstrates plastic flow beginning at stresses lower than the HEL and rising with shock level. Quasi-pure aluminas have been soft-recovered by Longy [26] and found to be free of cracks for stresses up to 1.5 × HEL. Microplasticity of grains provided the only observed source of inelastic deformation. On the other hand, for every alumina, the particle velocity histories always correspond to the profile type presented in Fig 6.2. On this basis, Longy and Cagnoux [10,27] have been able to assert that the HEL corresponds to a threshold of coupling between microplastic activity and macromechanical behavior, as in metals, keeping in mind that a quantitative study is necessary to validate this hypothesis. It is proposed that the ramp-wave structure following the HEL is due to work hardening. This proposition is supported by the high resistance to dislocation movement in these materials and by the existence

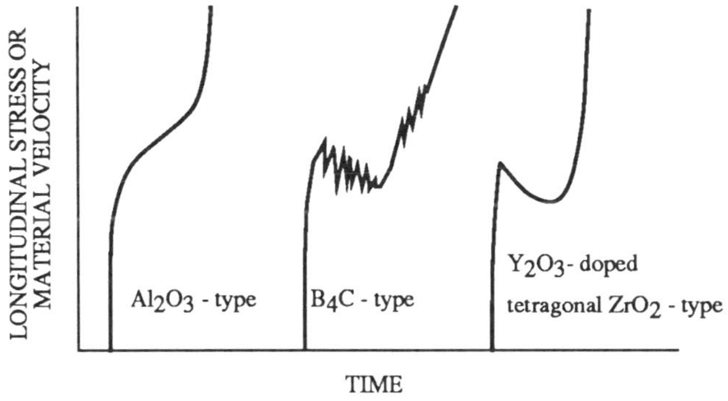

Figure 6.2. Different types of elastic precursor recorded in high-strength ceramics.

of many barriers corresponding to grain boundaries. From a phenomenological point of view, a loss of shear strength is observed above the HEL (Fig. 6.3), as shown by measurement with piezoresistive gauges [53,54] and by measured differences between shock and hydrostatic compression [3,23,33]. The above-mentioned work hardening appears in Fig. 6.3 as an overshoot of the shear strength reached at the HEL. Above approximately $1.5 \times$HEL, this shear strength becomes quasi-constant [3,23,33]. At 20 GPa, it should be noticed that Howitt and Kelsey [15] have shown the existence of a sufficient number of slip planes to satisfy the von Mises flow criterion. Behavior of aluminas will be described as being quasi-elastoplastic [50].

The influence of damage on elastoplastic behavior has not been explicitly shown even if, from a micromechanical point of view, the competition observed in compression between accommodations due to microplasticity and to cracking leads, de facto, to a coupling.

Even if the laboratory analyses performed on shocked samples of other ceramics do not suceed as well as those performed on aluminas, some phenomenological considerations allow flow interpretation. Thus, the ramped structure of the wave profile following the HEL of SiC, Si_3N_4, AlN (HEL①), and TiB_2 (HEL②) and the ability of these materials to support a shear stress greater than that measured at the HEL [55] (Fig. 6.3) lead to the same conclusion as was drawn for Al_2O_3.

Cases of double-wave profiles cannot be interpreted in this way. Thus, because of the erratic form of the elastic precursor recorded for B_4C (Fig. 6.2), Grady has supposed that phenomena of a heterogene-

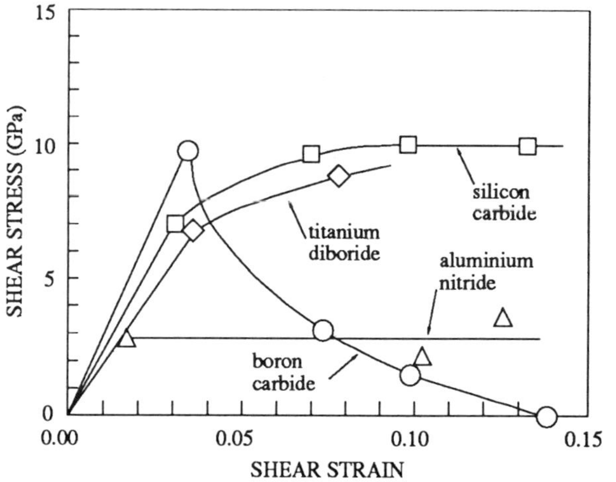

(a)

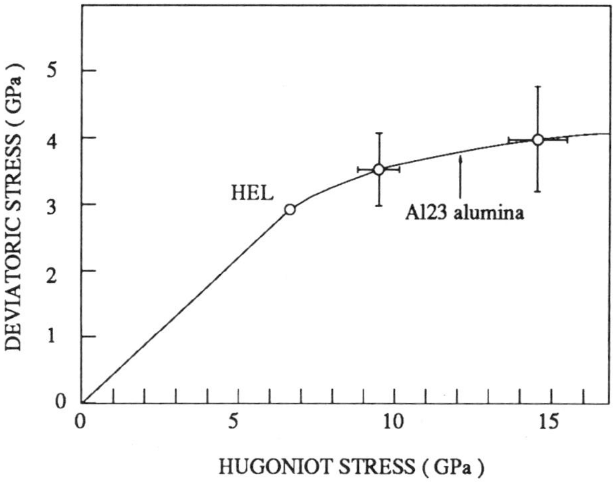

(b)

Figure 6.3. (a) Shear stress vs. shear strain for different high-strength ceramics [55]. (b) Deviatoric stress vs. Hugoniot stress for AL23 alumina [23,53].

ous nature cause the flow, considering the possibility of a structural densification related to a phase-change-like volume collapse [42]. Only analysis of soft-recovered samples can explain the micromechanical and mesomechanical nature of flow in B_4C. From a phenomenological point of view, a total loss of shear strength is observed for a Hugoniot stress corresponding to approximately $2 \times$ HEL [3,23, 42]. This behavior is described as quasi-elastohydrodynamic [50].

In some ceramics, the HEL is related to a phase change. Conventionally, the HEL② of AlN is discussed in this sense [49,57]. For Y_2O_3-doped tetragonal ZrO_2, which exhibits a profile type shown in Fig. 6.2, x-ray analysis of a sample soft-recovered at $0.95 \times$ HEL shows that the material is constituted almost entirely of the initial tetragonal phase and that microplastic activity had taken place [23]. Despite the lack of analysis of samples soft-recovered above the HEL, recorded wave profiles exhibit a double-wave structure that is probably due to a phase transition [23]. The behavior is described as quasi-elastohydrodynamic. Grady has also discussed the interpretation of HELs observed in zirconia [47,56]. Thus, a cubic-phase ZrO_2 exhibits an HEL at 5–6 GPa associated with a profile type similar to that of aluminas [41,47]. It appears that zirconias exhibit different responses depending on their initial phase, grain size, and porosity, due to complex coupling between microplastic activity and phase changes [47].

6.3.4. Release Behavior and Tension

Study of release wave features is an interesting way to refine an understanding of material behavior. Measurement of residual spall strength also provides an important diagnostic. Nevertheless, the behavior of ceramics in dynamic tension will not be specifically discussed in this chapter because of the important influence of the imposed loading (notably strain rate) on the measured spall strength. In every case, the VISAR is the metrological means preferred for its resolution and accuracy. It is important that release wave experiments be configured to avoid any artifact coming from other materials than the ceramic under investigation. That is why symmetrical plate impact is a commonly used configuration. Direct impact of a ceramic impactor on a window material is also an interesting possibility [39].

Kipp and Grady [58] have conducted a study of release waves in B_4C and Y_2O_3-doped tetragonal ZrO_2 from a Hugoniot stress lower than the HEL. These authors have shown the complex structure of elastic restitution, especially for ZrO_2, in a stress range where damage is absent according to the measured spall strength.

Coupling between damage and elastoplastic behavior of aluminas can be discussed by means of experiments performed above the HEL. Figure 6.4 presents particle velocity histories measured during symmetrical plate impacts on two aluminas. Profile no. 1305, recorded for A16-2 alumina, shows a release wave exhibiting low dispersion, confirming an undamaged soft-recovered material. In contrast, profile no. 138, recorded for AL23 alumina, shows a dispersive release

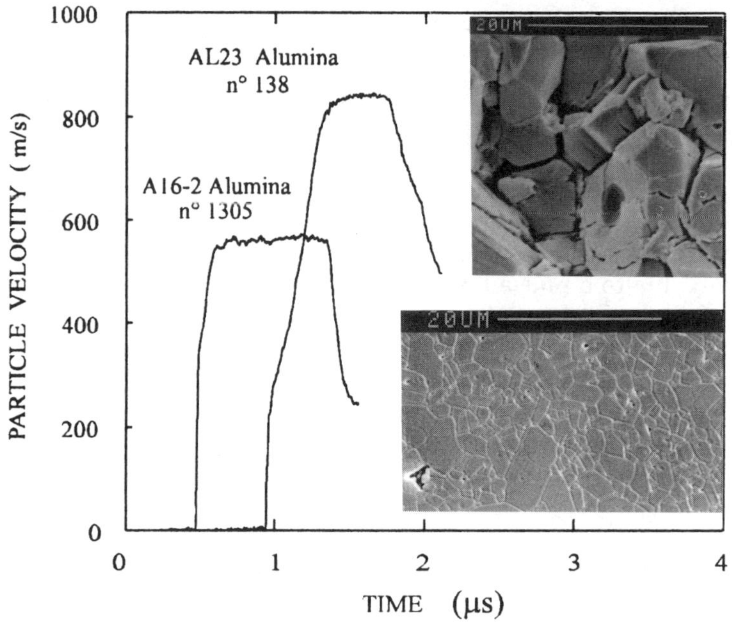

Figure 6.4. Particle velocity histories for two aluminas recorded in symmetrical plate-impact experiments and scanning electron micrographs of the corresponding soft-recovered samples.

wave corresponding to a very damaged material. Furthermore, analysis of a series of particle velocity profiles recorded on AL23 alumina gives the stress-strain paths illustrated in Fig. 6.5. At the beginning of unloading, the constrained modulus has a higher value than at standard conditions, then decreases to a lower one. This indicates that coupling, during unloading, between damage and elastoplastic behavior, is only activated below a critical state of confinement. Two explanations are proposed: i) cracks nucleated during loading are only activated during unloading below a critical pressure threshold; ii) cracks nucleate during unloading according to the scenario mentioned in Sec. 6.3.2.

In Fig. 6.5, the elastoplastic nature of the behavior of AL23 alumina becomes visible in release from a Hugoniot stress of 20 GPa. The coupling between damage and threshold of plastic flow on unloading has not been quantified. A clear case of elastoplastic restitution in release is also observed on SiC [41,42]. The nondispersive feature of the release wave from a maximum stress of 2 × HEL proves the absence of damage in this last material for this stress range. This deduction is consistent with the ability of some SiC ceramics to support a tension of 0.7 GPa after impact at 1.5 × HEL [59]. TiB_2 and cubic-phase ZrO_2 also exhibit an elastoplastic restitution [41]. Never-

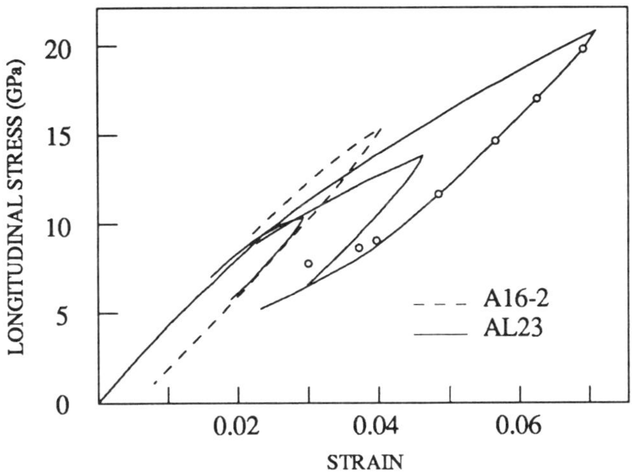

Figure 6.5. Longitudinal stress vs. strain for two aluminas. Solid curves are the results of centered-simple-wave analysis of particle velocity histories. Open circles are the results of simple-wave analysis of two particle velocity histories recorded for two sample thicknesses on AL23 at 20 GPa.

theless, for these two materials, release wave dispersion indicates a coupling between damage and behavior.

In brief, materials propagating a double-wave profile such as aluminas at their HEL level (Fig. 6.3) manifest, in release, a quasi-elastoplastic behavior, as observed in compression. This reversibility of the phenomenological behavior confirms the microplastic origin of the flow.

For B_4C, analysis of particle velocity histories leads to a release path quasi-superposed on the hydrostatic curve confirming the quasi-elastohydrodynamic feature of this material behavior [42].

6.3.5. Compaction Observed on the Hugoniot

Two principal sources of compaction are identifiable in the ceramics that have been studied: pore closing by microplasticity and readjustment of fragmented matter, and crystal lattice contraction by phase change.

A case of pore closure by microplasticity has been established in A16-2 and A16-17 aluminas [10,27]. This process, initiated for stresses lower than HEL, is the only cause of compaction above the HEL as long as cracking does not appear. In other ceramics, this process is more difficult to separate since soft-recovered samples are often found to be cracked.

Cracks appear in general for supplying elastoplastic accommodations. Referring to Sec. 6.3.1, intergranular cracking appears first in many ceramics. If these are quasi dense, pulled out grains do not appear to have readjusted their position in soft-recovered materials so that this cracking should not stimulate a reduction of porosity. In contrast, for very porous materials, grain readjustment can occur and this process plays a role in compaction.

In AL23 alumina, grains are fragmented by intragranular cracking above about 12 GPa. At 20 GPa however, the fragments produced do not seem to be readjusted (see photograph in Fig. 6.4). This observation is consistent with the fact that, at this stress level, the Hugoniot of AL23 alumina is still above that of Lucalox alumina which has a density closed to the theoretical value. The two Hugoniots converge at about 25–30 GPa. We should notice that collapse in this stress range is common to many aluminas [3].

Phase changes of some materials also provide matter a way to occupy the minimum volume. Such phenomena have been demonstrated for AlN and Y_2O_3-doped tetragonal ZrO_2. In ZrO_2, it is not easy to determine the phase change order of intervention, with regard to the above mentioned processes (see Sec. 6.3.3). This order results from competition and coupling among all these phenomena, taking into account every structural and microstructural parameter.

6.4. Triaxial Deformation by a Divergent Spherical Wave

A sample can be subjected to triaxial deformation when a pyrotechnic device initiates propagation of a divergent spherical wave. The advantage presented by these experiments in the study of phenomena associated with impacts or explosions was very soon recognized [60–62]. They allow both characterization of the stresses applied to sample and of its residual state. Because the wave that is generated is longitudinal, it is possible to perform unidimensional numerical simulations.

6.4.1. Exploitation of Experiments

The loading induced by a spherical wave leads to a nonuniform distribution of stress and strain in samples, due to the divergent nature of the propagation. Furthermore, the loading induces triaxial deformation and stress states at every point.

The profiles recorded at a point include information from areas subjected to complex and inhomogeneous loading, making their exploitation for phenomenological analysis of behavior more difficult than is the case for plane waves.

Thus, the main result is the evaluation of radial stress attenuation as determined by measurements performed at different abscissae in the sample. Radial stress is obtained by direct measurement using manganin–constantan gauges. It can also be inferred from particle velocity measurements with IDL, VISAR, or electromagnetic gauges. The Rankine–Hugoniot relations are used to determine the maximum radial stresses [17].

Technically speaking, soft recovery of the sample after divergent spherical wave experiments is easier than for plate-impact experiments, and most of the sample can be recovered "in place." The major difficulties consist in avoiding the effect of lateral release waves that significantly modify the sample state after the spherical wave propagation, and in associating an appropriately characterized mechanical loading with the final state obtained at every point of material. It only remains to identify the local phenomenological connection between the loading history and the final state of the material.

To our knowledge, only aluminas have been studied by spherical wave experiments. The inventory of these experiments is presented in Table 6.3.

6.4.2. Results and Physical Interpretations

Near the explosive, i.e., for applied stresses greater than the HEL ($> 2 \times$ HEL), an intense microplastic activity is observed in alumina grains [17,29], as in plate-impact experiments. Identically, cracking of AL23 alumina is intergranular and intragranular, cracking of T299 alumina being only intragranular. No orientation of cracks is privileged. The dependence of densities of microplastic activity and cracking with the Hugoniot stress identified in plate-impact experiments leads to the observed decrease of these densities with the wave propagation distance in divergent spherical wave experiments.

From a micromechanical point of view, sample observations do not suggest new mechanisms of crack nucleation. However, a large number of cracks has been observed in areas exhibiting an intense microplastic activity. The analysis of AD 85 and AD 995 alumina samples performed by McGinn et al., [65] shows that intragranular cracks are nucleated following intense plastic activity of grains in the form of twinning or formation of dislocation bands. These authors subscribe to the thesis of nucleation of intragranular and transgranular cracks by intense microplastic activity.

Table 6.3. Divergent spherical wave propagation experiments performed on aluminas

Configuration of experiment	Measurement type	Material	References
Unconfined hemi-spherical explosive, unconfined sample	IDL	AL23 alumina	17,29
	Manganin-constantan gauges	T299 alumina[a]	17,29
Unconfined hemi-spherical explosive, confined sample	Soft recovery	AL23 alumina	17,29,63
	Soft recovery	T299 alumina[a]	17,29,63
Spherical explosive, confined sample	Electromagnetic gauges and Soft recovery	AD85 alumina	64
		AD995 alumina	64

[a] Alumina of composition and microstructure identical to that of A16-2.

The ability of T299 alumina to compact without cracking for stress above the HEL was confirmed during a special experiment carried out by Cagnoux [63]. For this experiment, a steel plate was interposed between the explosive and the alumina. The part of the material close to the steel plate was soft-recovered in a compacted and crack-free state, probably protected during the release phase by the boundary conditions at the steel–alumina interface. When the same test was performed on AL23 alumina, it led to a completely cracked sample [63]. This difference of behavior is analogous to that obtained in plate-impact experiments.

For high radial stresses ($\sigma_r > \text{HEL}$), the states of soft-recovered materials seem to be qualitatively equivalent after propagation of a plane wave and a divergent wave. A quantitative study has to be completed to evaluate the influence of the pulse duration. For lower radial stresses ($\sigma_r < \text{HEL}$), radially oriented transgranular cracks are observed in AL23 and T299 alumina samples [17,29,63]. This feature is attributed to the release phase of the divergent spherical wave. During this phase, hoop stresses, more quickly released than radial stress, lead to unconfined stress states, even tension states, that may produce the observed cracking either by local fractures similar to axial splitting of grains or by tensile fracture of grains.

In conclusion, the loading provided by a divergent spherical wave includes a compression phase similar to that obtained by a plane shock wave, followed by a typical release phase. There is a great complementarity between the two types of experiment. Analyses of soft-recovered samples after plate impact allow a differentiation of compression phase effects from those associated with the release phase in spherical geometry. The main interest of pyrotechnic loading experiments lies in the large volume of the soft-recovered material, allowing an easier analysis of phenomena, and, at the same time, providing an important loading gradient.

6.4.3. Phenomenological Analysis by a Hybrid Method

The analysis of soft-recovered samples discloses a chronology of the various phenomena occurring during divergent spherical wave propagation. Nevertheless, it does not allow an evaluation of coupling between the phenomena observed on microscopic scale and the macromechanical response. This analysis can be conducted with a hybrid method proposed by Collombet and Tranchet [66]: It consists in an iterative coupling between identification of parameters of a constitutive model and comparison of numerical simulation to experiment. In this case, the first step of this method consists in comparing numerical simulations performed using an elastoviscoplastic model developed for analyzing plane wave response [53], with particle velocity histories obtained for AL23 alumina [17]. This comparison shows that the rising portions of particle velocity histories are properly reproduced by numerical calculations; this point reinforces the supposed equivalence between the loading of these two types of experiment and the hypothesis of a quasi-elastoplastic behavior expressed in Sec. 6.3.2.

In contrast, the model imperfectly reproduces the decreasing phases of profiles. In this context, the second step of the method consists in developing a mesomechanical model of damage, taking into account the effect of isotropic or anisotropic damage on the macromechanical response during the release phase [67].

After a few iterations, numerical simulations show improvement in reproducing the experimental profiles (Fig. 6.6). These calculations show that the intense cracking, which has a weak influence during the loading phase, produces a high level of coupling during the release phase, even for high confinement [67].

6.5. Conclusions, Prospects, and Recommendations

During the past decade, understanding of the behavior of ceramics under high dynamic pressure has been growing with the numerous

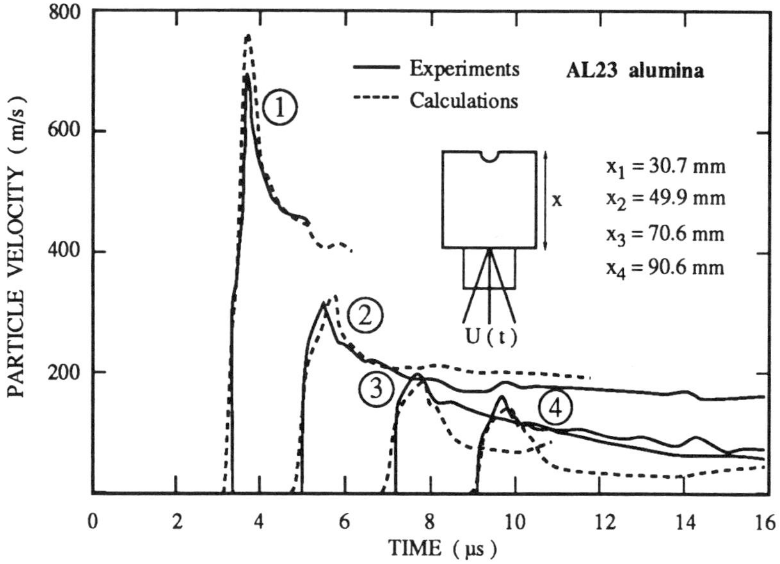

Figure 6.6. Comparisons between IDL experiments obtained for samples with different thicknesses and unidimensional numerical calculations performed with the elastoviscoplastic model for the loading phase [53] and the mesomechanical model of damage for the release phase [67].

experimental data resulting from analyses of soft-recovered samples after shock, and measurements of macromechanical properties. For a ceramic characterized by the chemical nature of its principal component, numerous microstructural parameters participate in the determination of its deformation modes: grain boundary nature, proportion of secondary phase, grain size and crystallographic nature, grain distribution, porosity, etc. In fact, the material behavior is the sum of micromechanical and mesomechanical processes activated in a coupled and complementary way. Among them, microplasticity of grains and grain boundaries seems to be common to many ceramics. Always on the scale of the grains, phase changes are observed in some varieties of ceramics. In soft-recovered samples, cracking appears as a deformation mode common to all ceramics. Nevertheless, conditions of nucleation and activation of these cracks have to be discussed with great care, taking into account the uncertainties in the effectiveness of soft-recovery techniques. Notably, in uniaxial deformation, it is not excluded, in our opinion, that a number of cracks are initiated during the release phase.

Despite the existence of many possible micromechanical deformation processes, two major classes of macromechanical behavior can be identified. In the first, a quasi-elastoplastic behavior has been identified for ceramics such as Al_2O_3, SiC, TiB_2, Si_3N_4, and AlN. For Al_2O_3, phenomenological analysis is based on use of damage models for representing effects of cracking. For the other ceramics mentioned above, phenomenological analysis has to be strengthened. On the other hand, B_4C and some varieties of zirconia exhibit a quasi-hydrodynamic behavior. No clear data allow discussion of the role of coupling between this type of behavior and cracking damage in this case. Despite its unwieldiness, the transmission electron microscope could give answers to open questions.

Even with the advances of the last decade, many gaps remain to be filled. Combining soft-recovery experiments with macromechanical measurements must remain the experimental strategy. When conducting soft-recovery experiments employing plate impact, use of star-shaped targets and impactors is the preferred method for trapping lateral releases. However, these configurations are, in general, limited to low impact velocities (about 10–300 m/s). For higher velocities, the use of capsules designed to protect the sample from secondary shocks is certainly to be promoted. In the field of material soft-recovery, the divergent spherical wave propagation produced in pyrotechnic experiments offers advantages, in particular the possibility of keeping the sample in place and producing a large range of stresses in a single experiment. In contrast, this type of experiment does not allow exploration of the effects of the pulse duration. In any case, the complementarity of plane shock wave and divergent spherical wave experiments has been demonstrated.

For measurement of macromechanical properties, a suitable metrology has to be carefully chosen. In the case of ceramics, materials of high shock impedance, the utility of the VISAR has been pointed out many times in this chapter. The Lagrangian analysis of measured time-resolved profiles is the preferred means of characterizing the stress-strain response of ceramics. These metrology and analysis techniques are generally well understood for plane waves. In contrast, a sustained effort is required to adapt these methods to divergent spherical waves.

In terms of material behavior, the understanding and control of microcracking of shock-loaded ceramics remains a challenge. The structural and microstructural characteristics of materials to be tested in the laboratory have to be determined with the greatest care to ensure that the scientific community converges on the same interpretations. One should advise the different authors to be as explicit as possible in describing these characteristics in their publications. On the other hand, studies aiming at analyzing the place of every intrinsic parameter independently from the others must be conducted. Studies directed toward identification of deformation mechanisms should be performed on samples especially prepared with this aim in view, with commercial ceramics being reserved for validation.

Manufacturing control of ceramics allows exploration of new directions and validation of interpretation. In this, the MTU JS-I alumina prepared by Staehler et al. [40] exhibits remarkable properties due to the size, geometry, and distribution of its grains. Future studies will also have to take into account the opportunities afforded by the behavior of particular compositions of grain boundaries.

All these prospects will only be realizable if the interest shown in the shock wave behavior of ceramics during the past decade continues. In addition to the role of conjunctural and economic parameters, this will depend on the evaluation of the ability of existing commercial ceramics to meet identified requirements, in view of the competition confronting every material.

References

[1] W. Prager, closing comments by session chairmen, in *Mechanical Behaviour of Materials under Dynamic Loads* (ed. U.S. Lindhom), Springer-Verlag, New York, p. 403 (1968).

[2] P. Chartagnac, in *Shock Compression of Condensed Matter—1989* (ed. S.C. Schmidt, J.N. Johnson, and L.W. Davison), Elsevier Science Publishers, Amsterdam, pp. 923–930 (1990).

[3] W.H. Gust and E.B. Royce, *J. Appl. Phys.* **42**, pp. 276–295 (1971).

[4] W.H. Gust, A.C. Holt, and E.B. Royce, *J. Appl. Phys.* **44**, pp. 550–560 (1973).

[5] D.E. Munson and R.J. Lawrence, *J. Appl. Phys.* **50**, pp. 6272–6282 (1979).

[6] P. Kumar and R.J. Clifton, *J. Appl. Phys.* **48**, pp. 4850–4852 (1977).

[7] F. Longy and J. Cagnoux, in *Shock Compression of Condensed Matter—1989* (ed. S.C. Schmidt, J.N. Johnson, and L.W. Davison), Elsevier Science Publishers, Amsterdam, pp. 441–444 (1990).

[8] G.T. Gray III, in *High-Pressure Shock Compression of Solids* (ed. J.R. Asay and M. Shahinpoor), Springer-Verlag, New York, pp. 187–215 (1993).

[9] E.K. Beauchamp, R.A. Graham, and M.J. Carr, *Mater. Res. Soc. Sym. Proc.* **24**, p. 281 (1984).

[10] F. Longy and J. Cagnoux, in *Proceedings of the International Conference on Impact Loading and Dynamic Behavior of Materials* (ed. C.Y. Chiem, H.D. Kunze, and L.W. Meyer), DGM Informationsgeselleschaft mbH, Oberursels, Germany, pp. 1001–1008 (1988).

[11] Y. Yeshurun, D.G. Brandon, A. Venkert, and Z. Rosenberg, *J. Phys. Colloq. C3,* **49**, pp. 11–18 (1988).

[12] L.H.L. Louro and M.A. Meyers, *J. Mater. Sci.* **24**, pp. 2516–2532 (1989).

[13] G. Raiser, R.J. Clifton, and M. Ortiz, *Mech. of Mater.* **10**, pp. 43–58 (1989).

[14] W.D. Winkler and A.J. Stilp, in *Shock Compression of Condensed Matter—1991* (ed. S.C. Schimdt, R.D. Dick, J.W. Forbes, and D.G. Tasker), Elsevier Science Publishers, Amsterdam, pp. 475–478 (1992).

[15] D.G. Howitt and P.V. Kelsey, in *Proceedings of the International Conference on Impact Loading and Dynamic Behavior of Materials* (ed. C.Y. Chiem, H.D. Kunze, and L.W. Meyer), DGM Informations-gesellschaft mbH, Oberursels, Germany, pp. 249–256 (1988).

[16] D.M. Vanderwalker and W.J. Croft, *J. Mater. Res.* **3**, pp. 761–763 (1988).

[17] A. Cosculluela, Ph.D. dissertation, University of Bordeaux (1992).

[18] Y. Wang and D.E. Mikkola, in *Shock Wave and High Strain Rate Phenomena in Materials* (ed. M.A. Meyers, L.E. Murr, and K.P. Staudhammer), Marcel Dekker, Inc., New York, pp. 1031–1040 (1992).

[19] D. Yaziv, Ph.D dissertation, University of Dayton (1985).

[20] L. Ewart and D.P. Dandekar, in *High-Pressure Science and Technology—1993* (ed. S.C. Schmidt, J.W. Shaner, G.A. Samara, and M. Ross), American Institute of Physics, New York, pp.1201–1204 (1994).

[21] H. Nahme, V. Hohler, and A. Stilp, in *High-Pressure Science and Technology—1993* (ed. S.C. Schmidt, J.W. Shaner, G.A. Samara, and M. Ross), American Institute of Physics, New York, pp. 765–768 (1994).

[22] J. Cagnoux, *J. Phys. IV, Colloq. C8,* **4**, pp. 257–261 (1994).

[23] J. Cagnoux, Unpublished data.

[24] G.F. Raiser, J.L. Wise, R.J. Clifton, D.E. Grady, and D.E. Cox, *J. Appl. Phys.* **75**, pp. 3862–3869 (1994).

[25] J.A. Brusso, D.E. Mikkola, J.E.Flinn, and P.V. Kelsey, *Scripta Metallurg.* **22**, pp. 47–52 (1988).

[26] F. Longy, Ph.D dissertation, University of Limoges (1987).

[27] F. Longy and J. Cagnoux, *J. Am. Phys. Soc.* **72**, pp. 971–979 (1989).

[28] W.F. Brace and E.F. Bombolakis, *J. Geophys. Res.* **68**, pp. 3709–3713 (1963).

[29] J. Cagnoux and A. Cosculluela, in *Proceedings of Dynamic Failure of Materials* (ed. H.P. Rossmanith and A.J. Rosakis), Elsevier Applied Science, Amsterdam, pp.73–84 (1991).

[30] A.G. Evans and Y. Fu, in *Fracture in Ceramic Materials* (ed. A.G. Evans), Noyes Publications, USA (1984).

[31] R.W. Rice, *Mater. Sci. Res.* **5**, pp. 195–227 (1971).

[32] J. Lankford, *J. Hard Mater.* **2**, pp. 55–57 (1991).

[33] T. Mashimo, Y. Hanaoka, and K. Nagayama, *J. Appl. Phys.* **63**, pp. 327–336 (1988).

[34] D. Yaziv, Y. Yeshurun, Y. Partom, and Z. Rosenberg, in *Shock Waves in Condensed Matter—1987* (ed. S.C. Schmidt and N.C. Holmes), Elsevier Science Publishers, Amsterdam, pp. 297–300 (1988).

[35] N.K. Bourne, Z. Rosenberg, J.E. Field, and I.G. Crouch, in *Shock Compression of Condensed Matter—1991* (ed. S.C. Schimdt, R.D. Dick, J.W. Forbes, and D.G. Tasker), Elsevier Science Publishers, Amsterdam, pp. 269–273 (1992).

[36] Z. Rosenberg, Y. Yeshurun, and D.E. Brandon, *J. Phys. Colloq. C5,* **46**, pp. 331–341 (1985).

[37] D.P. Dandekar and P. Bartokowski, in *High-Pressure Science and Technology—1993* (ed. S.C. Schmidt, J.W. Shaner, G.A. Samara, and M. Ross), American Institute of Physics, New York, pp. 733–736 (1994).

[38] H. Song, S.J. Bless, N.S. Brar, C.H. Simha, and S.D. Jang, in *High-Pressure Science and Technology—1993* (ed. S.C. Schmidt, J.W. Shaner, G.A. Samara, and M. Ross), American Institute of Physics, New York, pp. 737–740 (1994).

[39] J.Cagnoux and F. Longy, in *Shock Waves in Condensed Matter—1987* (ed. S.C. Schimdt and N.C. Holmes), Elsevier Science Publishers, Amsterdam (1988), pp. 293–296.

[40] J.M. Staehler, W.W. Predebon, and B.J. Pletka, in *High-Pressure Science and Technology—1993* (ed. S.C. Schmidt, J.W. Shaner, G.A. Samara, and M. Ross), American Institute of Physics, New York, pp. 745–748 (1994).

[41] M.E. Kipp and D.E. Grady, in *Shock Compression of Condensed Matter—1989* (ed. S.C. Schmidt, J.N. Johnson, and L.W. Davison), Elsevier Science Publishers, Amsterdam, pp. 337–380 (1990).

[42] D.E. Grady, in *Shock Compression of Condensed Matter—1991* (ed. S.C. Schimdt, R.D. Dick, J.W. Forbes, and D.G. Tasker), Elsevier Science Publishers, Amsterdam, pp. 385–391 (1992).

[43] D.P. Dandekar, in *High-Pressure Science and Technology—1993* (ed. S.C. Schmidt, J.W. Shaner, G.A. Samara, and M. Ross), American Institute of Physics, New York, pp. 729–732 (1994).

[44] D. Yaziv and N.S. Brar, *J. Phys. Colloq. C3,* **49**, pp. 683–687 (1988).

[45] N.S. Brar, Z. Rosenberg, and S.J. Bless, *J. Appl. Phys.* **69**, pp. 7890–7891 (1991).

[46] T. Mashimo, *J. Appl. Phys.* **63**, pp. 4747–4750 (1988).

[47] D.E. Grady and T. Mashimo, *J. Appl. Phys.* **71**, pp. 4868–4874 (1992).

[48] Z. Rosenberg, N.S. Brar, and S.J. Bless, *J. Appl. Phys.* **70**, pp. 167–171 (1991).

[49] M.E. Kipp and D.E. Grady, *J. Phys. IV, Colloq. C8,* **4**, pp. 249–256 (1994).

[50] T. Mashimo, in *High-Pressure Science and Technology—1993* (ed. S.C. Schmidt, J.W. Shaner, G.A. Samara, and M. Ross), American Institute of Physics, New York, pp. 757–760 (1994).

[51] D.J. Steinberg, *J. Phys. IV, Colloq. C3,* **1**, pp. 837–844 (1991).

[52] F.L. Adessio and J.N. Johnson, *J. Appl. Phys.* **67**, pp. 3275–3286 (1990).

[53] J.Y. Tranchet, *J. Phys. IV, Colloq. C8,* **4**, pp. 298–294 (1994).

[54] Z. Rosenberg, D. Yaziv, Y. Yeshurun, and S.J. Bless, in *Proceedings of the International Conference on Impact Loading and Dynamic Behavior of Materials* (ed. C.Y. Chiem, H.D. Kunze, and L.W. Meyer), DGM Informationsgeselleschaft mbH, Oberursels, Germany, pp. 393–398 (1988).

[55] D.E. Grady, in *Shock Compression of Condensed Matter—1991* (ed. S.C. Schimdt, R.D. Dick, J.W. Forbes, and D.G. Tasker), Elsevier Science Publishers, Amsterdam, pp. 455–458 (1992).

[56] D.E. Grady, in *Proceedings of the XIII International AIRAPT Conference*, Bangalore, pp 641–650 (1991).

[57] A. Nakamura and T. Mashimo, in *High-Pressure Science and Technology—1993* (ed. S.C. Schmidt, J.W. Shaner, G.A. Samara, and M. Ross), American Institute of Physics, New York, pp. 303–305 (1994).

[58] M.E. Kipp and D.E. Grady, in *Shock Compression of Condensed Matter—1991* (ed. S.C. Schimdt, R.D. Dick, J.W. Forbes, and D.G. Tasker), Elsevier Science Publishers, Amsterdam, pp. 459–462 (1992).

[59] W.D. Winkler and A.J. Stilp, in *Shock Compression of Condensed Matter—1991* (ed. S.C. Schimdt, R.D. Dick, J.W. Forbes, and D.G. Tasker), Elsevier Science Publishers, Amsterdam, pp. 475–478 (1992).

[60] D.B. Larson, *Int. J. Rock Mech. Mining Sci. Geomech. Abstr.* **19**(4), pp. 157–166 (1982).

[61] J.C. Cizek and A.L. Florence, *Laboratory Investigation of Containment of Underground Explosions*, Technical report, SRI International, DNA -TR -84-11 (1983).

[62] J. Cagnoux, Ph.D. dissertation, Universitié de Poitiers (1985).

[63] J. Cagnoux, in *Shock Compression of Condensed Matter—1989* (ed. S.C. Schmidt, J.N. Johnson, and L.W. Davison), Elsevier Science Publishers, Amsterdam, pp. 445–448 (1990).

[64] R.W. Klopp, D.A. Shockey, L. Seaman, D.R. Curran, J.T. McGinn, and T. de Resseguier, *Mechanical Testing of Ceramics and Ceramic Composites*, American Society of Mechanical Engineers, New York, pp. 41–60 (1994).

[65] J.T. McGinn, R.W. Klopp, and D.A. Shockey, in *Proceedings of Materials Research Society*, Symposium, Boston (1994).

[66] F. Collombet and J.Y. Tranchet, *J. Phys. IV, Colloq. C8,* **4**, pp. 641–646 (1994).

[67] J.Y. Tranchet, Ph.D dissertation (1994).

CHAPTER 7

Initiation and Propagation of Detonation in Condensed-Phase High Explosives

Ray Engelke and Stephen A. Sheffield

7.1. Introduction

When the term *detonation* is mentioned, most people picture a violent, uncontrolled, chaotic event such as that shown in Fig. 7.1. We will present a picture of a detonation as an orderly event that is governed and rigorously describable in terms of the conservation of mass, momentum, and energy and certain material-specific properties of the explosive. When a detonation is viewed experimentally, on the proper time and space scales, the observed wave phenomena are orderly and, when variables are well controlled in the experiment, have some simplicity. The photograph in Fig. 7.2 illustrates this; i.e., an orderly wave progressing along a cylindrical explosive charge.

Figure 7.1. Photograph of a large piece of condensed-phase high explosive detonating at a firing bunker at Los Alamos National Laboratory.

Figure 7.2. Photographs of a detonation wave progressing along a cylindrical explosive charge confined by a thin copper tube 36 mm in diameter (outside) with a 2-mm-thick wall. The detonation front is moving from left to right at a speed near 8 km/s, with the reaction products expanding the copper tube behind the front. The bottom picture was taken 10 μs after the top one. Bright spots on the expanding copper tube are thought to be due to flaws in the copper metal. These pictures are part of a set taken with a high-speed rotating-mirror framing camera. (Photograph courtesy of John Vorthman, Los Alamos National Laboratory.)

In this chapter, we discuss the initiation and propagation of detonation in condensed-phase high explosives. This will be done from both a theoretical and an experimental standpoint. A reasonably well-developed analytical theory of steady one-dimensional planar detonation exists. This theory is based on the Euler equations of chemically reactive compressible flow. Initiation theory is not as well developed, due primarily to a lack of knowledge of how the chemical energy stored in an explosive is released into the reacting flow as a function of the thermodynamic state variables. Our present knowledge of initiation of detonation is rooted in numerical studies that model experimental measurements using assumed energy-release rate forms. This is an active research area.

Most of the processes we shall describe below are thought to occur in both condensed phases and gases. In fact, much of the early understanding of detonation phenomena came from gas-phase studies because of the larger length scales and lower pressures in these detonations. The lower pressures exhibited in gas detonations allow repeated use of the same apparatus (e.g., a shock tube). In contrast, the surrounding apparatus is destroyed in condensed-phase explosive experiments. Furthermore, the chemical kinetics that occur in gas detonations are understood more completely than is condensed-phase exoenergetic chemistry; this is due, primarily, to it being easier to experimentally interrogate gas-phase materials. Because of this, most of our understanding of multidimensional detonation and initiation processes in condensed-phase explosives is phenomenological and is based on experimental studies of one-dimensional (1D) detonation waves and their interaction with the explosive's confinement (i.e., the surrounding inert materials).

Explosives exist in an energetically metastable state that can be controllably destabilized. In mixture explosives (e.g., mining explosives), the energy storage is usually via the proximity of fuel and oxidizer particles. In high-performance explosives, the fuel and oxidizer are usually present in the same molecule (molecular explosives) but isolated by chemical bonds (e.g., TNT [2,4,6-trinitrotoluene]). Detonation proceeds by a process in which the passage of a shock wave triggers the chemical energy release and, in turn, some of this energy sustains the shock wave motion. Pressures produced in high-quality detonating high explosives are typically in the 30 GPa (approximately 300,000 atmosphere) range. Such high detonation pressures mean that the explosive and anything (e.g., steel) in contact with it must be treated as *compressible* materials in theoretical treatments of the phenomena.

In this chapter, we begin by giving a short history of the theory of detonation and of development of explosive materials (Sec. 7.2); only a few important facts are covered to give an idea of the long and interesting development of this area. This is followed by Sec. 7.3 which deals with planar steady detonation theory, including the conservation relations, simple shock wave theory, and the two simple models that treat planar steady detonation. Section 7.4 is a discussion of the general reactive fluid flow equations, heat-release functions, equations of state, and various qualitative features of reactive flow. Theoretical solutions relating to complex systems are only briefly mentioned and most of our discussion of such systems is experimentally based. Experimental aspects of initiation and steady two-dimensional detonation are discussed in Secs. 7.5 and 7.6, respectively. This discussion includes the observed qualitatively different behavior of homogeneous and heterogeneous explosives. Section 7.7 is a compendium of chemical, detonation, and sensitivity data about certain selected high-explosive materials. In Sec. 7.8, we give a discussion of how some of the measurements of processes occurring in initiating and detonating materials are made. Section 7.9 is a short summary of this chapter and Sec. 7.10 is a glossary of some terms and acronyms used in the detonation physics community.

7.2. Brief History of Condensed-Phase Explosive Technology

Over the last several hundred years, explosive technology has progressed from the use of rapid-burning materials (e.g., gun powders) to sophisticated molecular explosives (e.g., very safe, but still high-energy molecules, for example, triamino-trinitro benzene [TATB]) and explosive mixtures tailored to particular applications (e.g., ammonium nitrate/fuel oil [ANFO] based emulsion explosives used in blasting). Most of the major materials and theoretical advances have been made in the last 150 years. One should note that although our understanding of the physics of initiating and detonating materials is substantial, our knowledge of the underlying microscopic chemistry that drives the detonation processes in condensed-phase materials is scant. In the next subsection, we briefly discuss the history of explosive materials development. This is followed by a short discussion (Sec. 7.2.2) of the historical development of the theoretical concepts relating to detonation and the accompanying reactive flow.

7.2.1. Explosive Materials Development

Explosive technology began when the oxidizer potassium nitrate (KNO_3, saltpeter) was discovered, probably in China or India. Fireworks were reported in China in the twelfth century. In England, Roger Bacon made the first pure KNO_3 and mixed it with charcoal and sulfur to make a form of black powder in about 1242. A formula for black powder (in the weight ratio 75.0/15.62/9.38 wt% KNO_3/charcoal/sulfur) was published in Brussels in 1560. This material began to be used in mines in the 1600s.

Nitroglycerin (NG), the first molecular explosive, was discovered in 1846 by Sobrero (a professor at the University of Turin in Italy). It was promisingly powerful but dangerously unpredictable. Early use of NG was accompanied by severe accidents; by 1875, Alfred Nobel had learned that by absorbing NG in *kieselguhr* (a porous siliceous earth) or by making a gelatin with a nitrated cotton material (nitrocellulose), the safety was enhanced. These materials were called *dynamites* and gained almost immediate success in blasting operations.

TNT was discovered in the late 1800s and became the main explosive used in World War I. It is still widely used today. TNT is relatively insensitive, can be easily melted (melting point 79°C), mixed with other materials (such as aluminum), and then cast to a shape. More powerful molecular explosives, such as HMX (cyclotetramethylene tetranitramine), are now of considerable importance for weapons applications. Hundreds of molecular explosives are known, but only relatively few have wide application.

After a disaster in 1947 in which two ships loaded with fertilizer-grade ammonium nitrate (AN) blew up in the Texas City, TX, harbor, it became apparent that AN was an explosive. Shortly after, ANFO (a mixture of 94/6 wt% AN and fuel oil) was found to be a commercially valuable explosive. Through its cheapness and ease of preparation, this material rapidly became a very important blasting agent and remains so today. A blasting agents industry has developed, based on mixtures of nitrates, fuels, and other materials, in which explosives (slurries, emulsions, etc.) are tailor-made for particular applications. These materials account for the bulk of explosives used today.

7.2.2. Physics of Detonation Developments

The first experimental observations (ca. 1880) of detonation waves (in gases) were made by French workers (Berthelot and Vielle [1,2] and Mallard and Le Chatelier [3]). Later, Chapman [4] and Jouguet [5], independently, gave a theoretical treatment of detonation in gases (ca. 1900). In the Chapman–Jouguet (CJ) treatment, a detonation is idealized as a planar mathematical discontinuity that propagates steadily. In the CJ picture, the passage of this discontinuity causes complete release of the stored chemical energy; i.e., there is no spatially resolved chemical reaction zone. The fluid mechanics required to devise the CJ picture is primarily the mass, momentum, and energy conservation-law jump conditions across a shock wave (the Rankine–Hugoniot conditions). An important result of the CJ treatment is that, in any given energetic material, there is a *minimum* velocity at which a steady one-dimensional detonation wave can propagate.

Later (ca. 1940), Zeldovich [6], von Neumann [7], and Doering [8] (ZND), independently, refined the CJ model by i) assuming that the conservation conditions of mass, momentum, and energy for invicid flow apply and ii) relaxing the assumption of an *instantaneous* heat release triggered by the shock. They modeled the heat release as a single forward exothermic rate process. Within these assumptions, one finds that steady planar solutions of the flow and shock jump equations exist. The ZND detonation consists of a steady reaction zone with an attached following flow. The state at the end of a steady reaction zone is the CJ state and it is independent of the rate form. The fluid-mechanical state within the ZND reaction zone is *dependent* on the form of the exothermic rate.

Quite early on, it was experimentally recognized that detonations exist which show behavior significantly more complex than that described by either the CJ or ZND models. For instance, real detonation waves have, for example, i) time-dependent behavior before they achieve steadiness (see Sec. 7.5), ii) dependence on the lateral size of the explosive charge, even when the detonation is traveling steadily (see Sec. 7.6), and iii) inherent longitudinal and transverse instabilities (see Fickett and Davis [9] and Dremin et al. [10]). All of these complications have been studied theoretically and experimentally.

There are a number of books which include discussions of detonation theory in more detail than is possible in this chapter. These generally approach the subject from the viewpoint and interests of the particular author(s). Since this subject involves fluid flow, shock dynamics, chemical reaction, thermodynamics, materials science, equations of state of the materials involved, and high-pressure and high-time-resolution diagnostics, the discussions can be quite different. Several examples of these works can be found in Refs. 11–16. It should be pointed out that we have not intended to be complete in the references cited in this chapter, but rather have tried to present a short coherent picture of the subject as we see it.

7.3. Planar Steady Detonation Theory

The CJ theory of planar steady detonation is the simplest case; therefore, it will be considered first. This will be done by introducing the hydrodynamic jump conditions across a shock followed by a discussion of the Rayleigh line and the Hugoniot curve (a more complete description of shock processes may be found in Ref. 17). This is followed by a discussion of a Chapman–Jouguet detonation (in which the reaction zone is not considered) which leads into a discussion relating to detonations in which the reaction zone is resolved. All this applies to one-dimensional planar conditions.

7.3.1. Hydrodynamic Jump Conditions Across a Shock Wave

The mass, momentum, and energy conservation conditions that relate flow quantities across a planar shock wave (or any steady flow or jump discontinuity as shown in Fig. 7.3) can be expressed in the following forms, respectively:

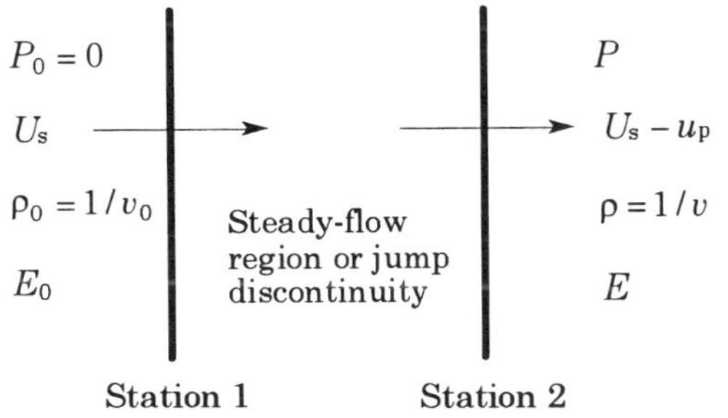

Figure 7.3. Stations and quantities used in obtaining the conservation conditions in steady flow or across discontinuities. The reference frame is that of the materials flowing into the steady flow region or jump discontinuity.

$$\rho_0 U_s = \rho(U_s - u_p), \tag{7.1}$$

$$P = -\rho(U_s - u_p)^2 + \rho_0 U_s^2, \tag{7.2}$$

and

$$P(U_s - u_p) = -\rho(U_s - u_p)\left[E + \frac{1}{2}(U_s - u_p)^2\right] + \rho_0 U_s\left[E_0 + \frac{1}{2}U_s^2\right]. \tag{7.3}$$

In these equations, the subscript 0 corresponds to the value of a variable before passage of the shock wave, whereas those variables without a subscript 0 refer to values immediately after passage. Here, ρ is the mass density, u_p is the mass velocity or particle velocity, U_s is the wave velocity or shock velocity, E is the internal energy per unit mass, and P is the pressure. We have assumed that P_0 is negligible relative to the pressure behind the shock wave. For detonating condensed explosives, this is a very good approximation, since the ratio of ambient pressure to detonation pressure is typically of the order of 10^{-5}. Note also that if the process causing the flow is a wave moving at speed U_s in the laboratory frame, then the Galilean frame used to obtain Eqs. 7.1–7.3 is the one in which the wave is *stationary*; i.e., we have transformed from the laboratory frame to one in which the wave is motionless and matter streams through the plane defined by the wave. For a derivation of these relationships, see Thompson [18].

7.3.2. Rayleigh Line

A valuable relationship can be obtained by eliminating u_p between Eqs. 7.1 and 7.2. One finds that

$$P = K(v_0 - v) \quad \text{and} \quad K \equiv \rho_0^2 U_s^2, \tag{7.4}$$

where $v \equiv 1/\rho$ is the volume per unit mass (specific volume) and K is a constant for a steady wave. Flow across jump discontinuities with initial state $(P_0 = 0, v_0)$ *must* have P, v values that satisfy Eq. 7.4, to conserve mass and momentum. This line in the P,v plane is called the Rayleigh line [19] (see Fig. 7.4); it will prove useful in defining what detonation processes are possible.

7.3.3. Hugoniot Curve

If both U_s and u_p are eliminated from Eq. 7.3 by use of Eqs. 7.1 and 7.2, one finds that

$$(E - E_0) - \frac{1}{2}P(v_0 - v) = 0. \qquad (7.5)$$

Thus, for flow across shocks that start from the initial point $(P_0 = 0, v_0, E_0)$, the final state must lie on the curve defined by Eq. 7.5 in order to conserve mass, momentum, and energy. Given the dependence of E on P and v for a particular material, Eq. 7.5 becomes a relationship between P and v. If the process connecting the initial and final states is a single shock process from ambient conditions, this P, v relationship is called the principal Hugoniot curve of the material [20]. The Hugoniot curve defines all possible final P, v

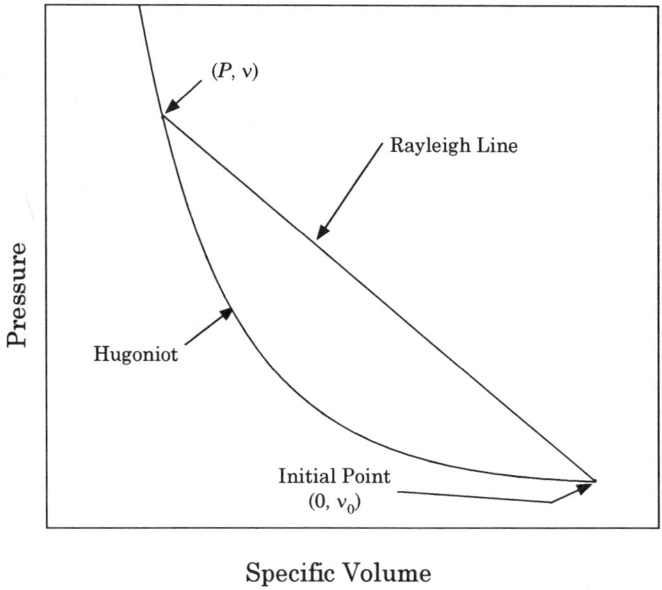

Figure 7.4. Plot of Hugoniot and Rayleigh line in pressure vs. specific volume plane.

shock states reachable from the initial state by a *single* shock process. The curve labeled "Hugoniot" in Fig. 7.4 is a locus of such points.

7.3.4. Chapman–Jouguet Theory

We can now consider the Chapman–Jouguet (CJ) theory of detonation [4,5]. In order to make the discussion concrete, we will specialize to a material in which the internal energy per unit mass is given by

$$E(P, v, \lambda) = \frac{Pv}{\gamma - 1} - q\lambda \quad , \tag{7.6}$$

where γ is a constant characteristic of the material, q is the chemical energy stored in the material per unit mass, and λ is a variable that measures the proportion of the chemical energy that has been released into the flow; $\lambda = 0$ and 1 correspond to no and complete chemical reaction, respectively. The $Pv/(\gamma - 1)$ part of Eq. 7.6 is called a *γ-law equation of state* (EOS). For condensed-phase explosives, a γ value of approximately 3 gives a reasonable description of many observed detonation phenomena. It is important to note that a γ-law EOS is not necessarily a perfect gas EOS. This can be seen by noting that for a perfect gas, γ is the ratio of specific heats, $C_p / C_v = 1 + (2/n)$, where n is the number of degrees of freedom available to the particles composing the gas. Thus, since $3 \le n < \infty$, γ is constrained by $5/3 \ge \gamma > 1$ for a perfect gas. The values of $\gamma \approx 3$ useful in condensed-phase detonation allow a gas to be stiffer than is possible for a perfect gas.

The possible 1D steady planar detonation states must simultaneously satisfy Eqs. 7.4 and 7.5 in order to satisfy the conservation laws. We suppose that i) $E = E(P, v, \lambda)$ is defined by Eq. 7.6, ii) the initial state is $(P, v, \lambda) = (0, v_0, 0)$, and iii) the final state is fully reacted, i.e., $(P, v, \lambda) = (P, v, 1)$, then the Hugoniot condition between the initial and final state is

$$\frac{Pv}{\gamma - 1} - q - \frac{1}{2}P(v_0 - v) = 0. \tag{7.7}$$

Note that except for the v_0 term, Eq. 7.7 depends only on the fully reacted material (i.e., $\lambda = 1$) properties. Elimination of P from Eq. 7.7 by use of the Rayleigh line condition (Eq. 7.4) yields a quadratic equation for the specific volume of the possible detonation states, i.e.,

$$v^2 - Bv + C = 0, \tag{7.8}$$

where $B \equiv 2\gamma v_0 / (\gamma + 1)$ and $C \equiv (\gamma - 1)[1 + 2q / D^2] v_0^2 / (\gamma + 1)$. The value of $B^2 - 4C$ determines the number of real roots of Eq. 7.8. At this point, the value of D, the detonation wave velocity is a free parameter in C related to the slope of the Rayleigh line. When $D=\sqrt{[2q(\gamma^2-1)]}$, $B^2-4C=0$ and Eq. 7.8 has one real root. For D larger (smaller) than this value, Eq. 7.8 has two (zero) real roots.

Thus, the conservation conditions have shown that for

$$D<\sqrt{2q(\gamma^2-1)},$$

no steady 1D planar detonation wave is possible. When

$$D=\sqrt{2q(\gamma^2-1)},$$

there is a unique value of the detonation wave velocity—this is the Chapman–Jouguet (CJ) value (D_{CJ}). It will be shown below that for a wave traveling at D_{CJ}, the flow at the CJ point (see Fig. 7.5) is *sonic* relative to the following flow. Consequently, such a wave is immune from hydrodynamic disturbances from the rear (other than shocks).

Next consider the case where $D>D_{CJ}$; in this situation, there are two solutions. The solution with $P>P_{CJ}$ (see Fig. 7.5) corresponds to a detonation wave that moves subsonically relative to its following flow and therefore can be attenuated by rarefactions from the rear. This solution is rejected as being uninteresting for this reason. The "weak" solution with $P<P_{CJ}$ travels supersonically relative to the following flow. Such detonation waves have been observed in gas detonations (see Ref. 21). The theory of weak detonations is discussed in Sec. 7.4.4.3.

The value of the specific volume at the CJ state is given by $B/2$ of Eq. 7.8; i.e., $v_{CJ}=\gamma v_0/(\gamma+1)$. Substitution of v_{CJ} into the Rayleigh line equation yields the CJ pressure

$$P_{CJ}=\frac{\rho_0 D_{CJ}^2}{\gamma+1}.$$

For $\gamma=3$, one sees that $v_{CJ}/v_0=3/4$; i.e., the volume per unit mass is reduced by 25 % at the CJ state relative to the uncompressed state. Typical measured values of ρ_0 and D_{CJ} for condensed-phase explosives are 1.80 g/cm^3 and 8.0 km/s. Use of these values (and $\gamma=3$) in the P_{CJ} relation yields a CJ pressure of 29 GPa (290 kbar); as a comparison, the yield strength of common steels is in the neighborhood of 0.5 GPa. For these detonation parameters, one *megawatt* of power is generated by an area on the detonation front of approximately 1.9 cm^2.

Using results of the last paragraph, one can prove that the flow at the CJ point is sonic. A few other facts are needed for the proof. First, note that if U_s is eliminated between Eqs. 7.1 and 7.2, one finds that

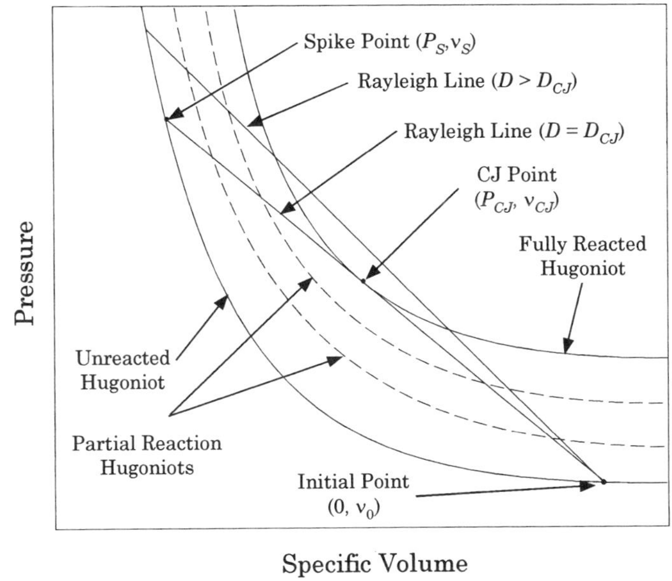

Figure 7.5. Pressure vs. specific-volume plane quantities useful in the discussion of the CJ and ZND models of detonation.

$$u_{\mathrm{p}} = \sqrt{P(v_0 - v)}.$$

Use of P_{CJ} and v_{CJ} in this equation shows that $u_{\mathrm{CJ}} = D/(\gamma+1)$. Next, note that the sound speed in a fluid is given by

$$c = \sqrt{(\partial P/\partial \rho)_S}$$

(the subscript S denotes constant entropy) and that for a γ-law EOS the isentropes are given by Pv^{γ} = constant. Therefore,

$$c_{\mathrm{CJ}} = \sqrt{\gamma\, P_{\mathrm{CJ}}\, v_{\mathrm{CJ}}} = \frac{\gamma D}{\gamma+1}.$$

So that $u_{\mathrm{CJ}} + c_{\mathrm{CJ}} = D$; i.e., a sound signal at the CJ plane travels exactly at the detonation wave speed.

We have proved the sonic character of the flow at the CJ point for a constant-γ EOS. It is important to note that this sonic character at the CJ point is true for any reasonably behaved EOS (see Ref. 9).

7.3.5. Zeldovich–von Neumann–Doering Theory

Zeldovich [6], von Neumann [7], and Doering [8] (ZND) advanced the theory of steady 1D detonation beyond the CJ theory by finding a solution of the steady 1D detonation flow with a *resolved* chemical reaction zone.

Before proceeding to a discussion of the ZND theory, we note that Eqs. 7.1–7.3 are also valid for *any* position behind a shock wave provided the flow from the shock to that position is invicid, 1D, and *steady*. This statement is true because, in such cases, the same balancing of flux differences (e.g., momentum flux) and production terms (e.g., pressure) can be made whether the position of interest is immediately behind the shock or further removed. Therefore, we can use Eqs. 7.1–7.3 and the results derived therefrom to study the ZND reaction-zone structure.

ZND assumed that the detonation process consists of a shock wave that takes the unreacted material from its initial state $(P,v,\lambda)=(0,v_0,0)$ to a *spike point* $(P_s, v_s, 0)$ on the unreacted Hugoniot (see Fig. 7.5). Since the chemistry is now resolved, there is a continuum of partial reaction Hugoniots. For the γ-law form (Eq. 7.6), these are given by

$$\frac{Pv}{\gamma-1} - \lambda q - \frac{1}{2}P(v_0 - v) = 0, \qquad (7.9)$$

where the value of the variable λ determines which partial reaction Hugoniot is being considered. In traversing the ZND reaction zone, one proceeds down the Rayleigh line from the spike point $(P_s, v_s, 0)$ to the CJ point $(P_{CJ}, v_{CJ}, 1)$—crossing the full set of partial reaction Hugoniots in the process (see Fig. 7.5). The variable q in Eq. 7.9 is the difference in enthalpy between reactants and products and is called the *heat of detonation* (see Sec. 7.7.2). Within the ZND model, the chemical energy release in Eq. 7.9 is assumed to be governed by a single chemical progress variable and increased reaction always corresponds to increased exothermicity.

A (P, v, λ) state in the reaction zone corresponding to a particular λ is found by solving a generalized form of Eq. 7.8, in which the quantity C has been replaced by $C(\lambda)$, where

$$C(\lambda) = \frac{\gamma-1}{\gamma+1}\left[1 + \frac{2q\lambda}{D^2}\right]v_0^2. \qquad (7.10)$$

The detonation wave velocity is determined from the common value of the Rayleigh line and the *fully reacted* Hugoniot, i.e., again

$$D = \sqrt{2q(\gamma^2 - 1)},$$

as in the CJ theory. The specific volume as a function of λ, as obtained from Eqs. 7.8 and 7.10, is (at the CJ detonation velocity)

$$v(\lambda)=\frac{\gamma v_0}{\gamma+1}\left[1-\sqrt{1-\left(\frac{\gamma^2-1}{\gamma^2}\right)\left(1+\frac{2\lambda q}{D^2}\right)}\right]=\frac{\gamma v_0}{\gamma+1}\left[1-\frac{1}{\gamma}\sqrt{1-\lambda}\right], \quad (7.11)$$

where this (smaller) specific volume root of Eq. 7.8 corresponds to having shocked to the spike point of Fig. 7.5. The larger specific volume root can be interpreted as a deflagration process (see Ref. 18). The pressure through the reaction zone as a function of λ is obtained by use of Eq. 7.11 in the Rayleigh line equation, Eq. 7.4.

We still do not know the reaction-zone structure in space or time [e.g., $P=P(x)$]. To proceed further, one needs to know a chemical-heat-release rate relation

$$\frac{d\lambda}{dt}=r(P,v,\lambda). \quad (7.12)$$

Using what has been found above, Eq. 7.12 can be written as $d\lambda=r(P(\lambda),v(\lambda),\lambda)dt$; integrating gives $\lambda=\lambda(t)$. Given $\lambda=\lambda(t)$, $P=\hat{P}(t)$ and $v=\hat{v}(t)$ are known, since $P(\lambda)$ and $v(\lambda)$ are known. The dependence of P and v on the spatial coordinate (x) is determined as follows. The spatial position of a fluid particle relative to the detonation shock is given by

$$\frac{dx}{dt}=D-u_{\mathrm{p}}. \quad (7.13)$$

We know from Sec. 7.3.4 that $u_{\mathrm{p}}=\sqrt{[P(v_0-v)]}$ and we know $\hat{P}(t)$ and $\hat{v}(t)$ from the work immediately above. Therefore, Eq. 7.13 can be integrated to give a function $t=t(x)$, which, in turn, can be used in $\hat{P}(t)$ and $\hat{v}(t)$ to yield the functions $P(x)$ and $\overline{v}(x)$, i.e., the spatial dependence of P and v through the ZND reaction zone. A typical particle velocity snapshot of the ZND reaction-zone profile is shown in Fig. 7.6 (see also Fig. 7.7).

The most striking physical result shown in Fig. 7.6 is that the particle velocity (and pressure) *falls* through the ZND zone. This result is also implicit in Fig. 7.5; i.e., as one moves across the partial reaction Hugoniots from $\lambda=0$ to 1, the pressure falls.

Consider the ZND picture for a Rayleigh line with $D>D_{\mathrm{CJ}}$. The solution with $P>P_{\mathrm{CJ}}$ is again rejected because the flow at the point of complete reaction is subsonic relative to the following flow. The solution with $P<P_{\mathrm{CJ}}$ cannot be reached from the spike point by moving down the Rayleigh line because of the absence of partial reaction Hugoniots at some points on the path. More general types of the heat-release function can make the point with $P<P_{\mathrm{CJ}}$ accessible (see Sec. 7.4.4.3).

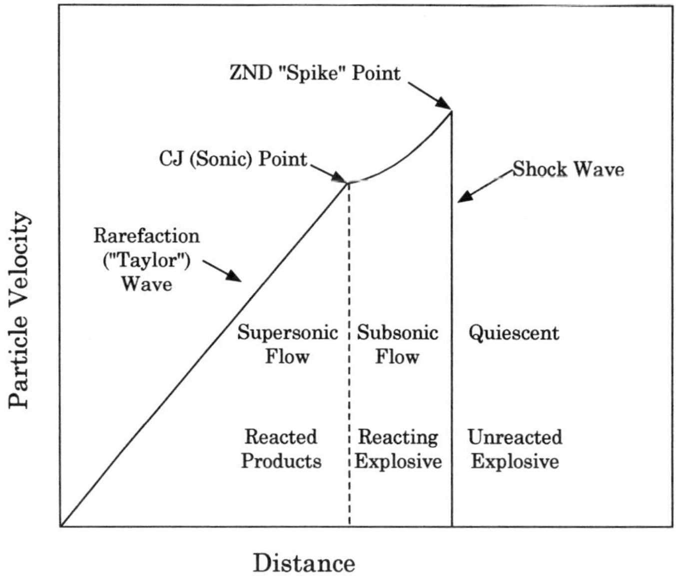

Figure 7.6. Snapshot of the flow in a ZND detonation in the particle-velocity vs. distance plane. The Taylor wave (see Sec. 7.4.4.4) is also shown.

The main features of the ZND model are borne out by experiment; e.g., direct measurement of the particle velocity vs. time in detonating explosives shows a particle-velocity discontinuity (shock front) followed by a zone of decreasing particle velocity (see Fig. 7.7). X-ray radiography of the explosive products has been used to show that the flow behind the shock does become sonic [22]. This feature was examined by tracking strong rarefaction waves introduced into the reaction products. Various other detonation properties can also be measured, e.g., spike pressure, detonation velocity, etc. Methods of making such measurements are discussed in Sec. 7.8.

7.4. Equations Governing Reactive Flow

Above, a description was given of theories of 1D steady planar detonation when the spatial distribution of the chemical-heat addition is either unresolved or is especially simple in form; i.e., one forward exothermic rate process. The essentially algebraic form of the CJ and ZND theories must be generalized when more complex detonation phenomena are considered. Here, complexity can mean, for example, multidimensionality in the flow (e.g., steady two-dimensional detonation) or flow in which the shock wave moves unsteadily (e.g., the shock initiation of detonation).

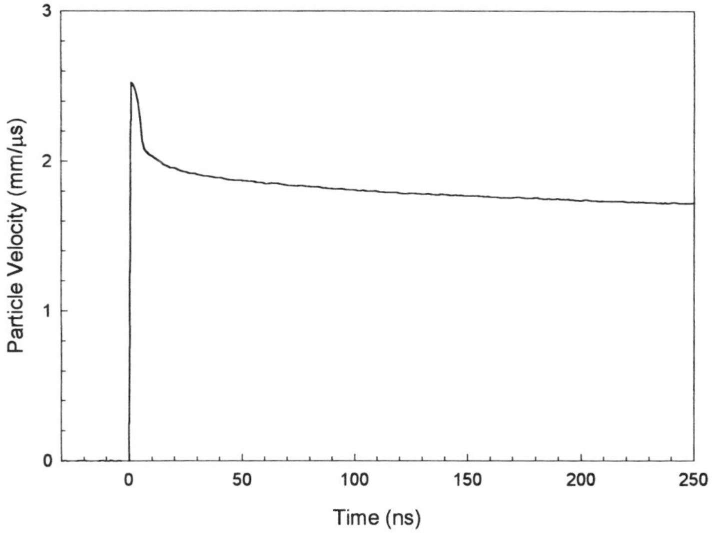

Figure 7.7. Experimentally measured interface particle velocity history showing the ZND reaction-zone structure in detonating nitromethane. The CJ point is at ≈15 ns. See Sec. 7.8 for details of the experiment.

The more general situation is governed by the applicable partial differential equations of fluid mechanics (with initial and boundary conditions) and material constitutive relations. The fluid-mechanical equations governing condensed-phase detonation are usually taken to be the Euler equations of invicid compressible flow—augmented by suitably generalized shock jump conditions if the flow is multidimensional. The Euler equations govern compressible fluid flow in which transport processes are neglected. The equations are statements of the conservation of mass and energy and a dynamical (momentum) equation that governs the particle velocity vector field, $\mathbf{u}$.

As a preliminary to consideration of the Euler equations, it is important to note that in a compressible flow, the time derivative of a flow field variable (say F) on a fluid particle is given by the total time derivative of that quantity evaluated on the particle's path, i.e.,

$$\frac{dF}{dt} = \left[\frac{\partial}{\partial t} + (\mathbf{u}\cdot\nabla)\right]F\ . \tag{7.14}$$

Equation 7.14 is derived by taking the total derivative of F (via the chain rule) and then evaluating the result on a particle path. This result will be useful below where we are interested in defining the behavior of any particular fluid particle.

7.4.1. Fluid-Mechanical Equations

When the notation of Eq. 7.14 is used, the three applicable fluid-mechanical conditions, written in the laboratory reference frame, are as follows. The mass conservation condition is

$$\frac{d\rho}{dt} + \rho \nabla \cdot \mathbf{u} = 0. \tag{7.15}$$

In Eq. 7.15, the second term results from the mass density being dependent on the fluid particle's volume; the $\rho \nabla \cdot \mathbf{u}$ term accounts for dilation of the fluid particle as the flow evolves. The momentum and energy equations, given next, are dependent on the fluid particle mass and, consequently, do not contain dilation terms. The next equation is the expression of Newton's second law for a continuous invicid compressible fluid; i.e.,

$$\rho \frac{d\mathbf{u}}{dt} = -\nabla P, \tag{7.16}$$

where P is pressure at the particle's location. When $\nabla P = \mathbf{0}$, Eq. 7.16 expresses the conservation of linear momentum. Note that Eq. 7.16 is a nonlinear partial differential equation, since the operator d/dt depends on $\mathbf{u}$ (see Eq. 7.14).

The energy conservation condition is

$$\frac{dE}{dt} = -P \frac{dv}{dt}, \tag{7.17}$$

where E and v are the internal energy per unit mass and the volume per unit mass, respectively. Equation 7.17 mirrors the adiabatic nature of the flows being treated. It shows that energy transport occurs only via mechanical work done by or on a particle.

The chemical-energy release in a fluid particle is specified in E via dependence on chemical reaction-progress variables.

7.4.2. Chemical Heat-Release Relations

The evolution of the reaction-progress variable(s), λ, is specified by *rate laws* that govern their time dependence on the state variables of the system. Such a form is

$$\frac{d\lambda}{dt} = r(P, v, \lambda), \tag{7.18}$$

where, in general, λ is an n-dimensional vector of such relationships (see Ref. 23). Note that any particular progress variable (i.e., any component λ_i of the vector λ) does not necessarily correspond to *any* particular microscopic chemical process.

The energy function that drives the fluid mechanics is the *global* heat released (Q) into the flow. Thus, when a number of chemical processes are occurring simultaneously, the pertinent mechanical energy is the sum of the individual energy releases (q_i), i.e.,

$$Q = \sum \lambda_i q_i . \tag{7.19}$$

The difficulty in attempting to probe microscopic energetic events in a detonation via mechanical measurements is inherent in the ambiguities associated with Eq. 7.19. To probe the microscopic processes that produce the global heat-release function Q, one needs to examine events at the microscopic level—e.g., with spectroscopic techniques. This type of investigation is a current area of research in detonation physics and chemistry (see Refs. 24–27).

7.4.3. The Material Equation of State

Equations 7.15–7.17 and 7.18 are $5+n$ differential equations in $6+n$ dependent variables. This system of equations is underdetermined. One further constraint involving the $6+n$ dependent variables is required.

The constraint that closes the system is the equation of state (EOS) of the reactive fluid, e.g., a relationship of the form

$$E = E(P, v, \lambda) \tag{7.20}$$

(see, e.g., Eq. 7.6).

Because one is dealing with mechanical equations, it is not required that the temperature field in the system be known. However, if the chemical-heat-release relations of Section 7.3.2 depend on temperature, this variable must also be known in order to fully define the reactive flow problem. Defining the temperature field in detonations is a current area of research (see Ref. 28).

It is implicit in the above discussion that there is local thermodynamic equilibrium in the system except for the chemical progress variables and the rate laws (Eq. 7.18) governing the kinetics which define their time evolution.

Shock jump conditions and Eqs. 7.15–7.20 (with boundary and initial conditions) define the reactive flow problems found in detonating condensed systems, within the stated approximations. These governing equations are a hyperbolic system; hence, a salient feature of their solutions is transport of energy by wave motion (see Ref. 23).

7.4.4. Qualitative Features of Flow with Reaction

Some insight into the effect of chemical-energy addition on a compressible hydrodynamic flow can be obtained by eliminating dE/dt between the energy conservation condition and the EOS relation. From the EOS relation, we know that

$$\dot{E} \equiv \frac{dE}{dt} = \left(\frac{\partial E}{\partial P}\right)_{\upsilon,\lambda} \dot{P} + \left(\frac{\partial E}{\partial \upsilon}\right)_{P,\lambda} \dot{\upsilon} + \left(\frac{\partial E}{\partial \lambda}\right)_{\upsilon,P} \dot{\lambda}. \tag{7.21}$$

Eliminating $\dot{E}$ from this expression via $\dot{E} = -P\dot{\upsilon}$ gives

$$\dot{P} = \left[\rho^{-2}\left(\frac{\partial E}{\partial P}\right)_{\upsilon,\lambda}^{-1}\left\{\left(\frac{\partial E}{\partial \upsilon}\right)_{P,\lambda} + P\right\}\right]\dot{\rho} + \left[-\left(\frac{\partial E}{\partial P}\right)_{\upsilon,\lambda}^{-1}\left(\frac{\partial E}{\partial \lambda}\right)_{\upsilon,P}\right]\dot{\lambda}. \tag{7.22}$$

By writing the 1D Euler equations in characteristic form, Wood and Kirkwood [23] showed that the coefficient of $\dot{\rho}$ in Eq. 7.22 is the local sound speed ($\equiv c$) squared—at fixed chemical composition. This speed is called the *frozen* (fixed-composition) *sound speed.* The second term in Eq. 7.22 depends on the chemical-heat-release rate $\dot{\lambda}$ ($\equiv r$). Use of partial derivative relations for the differentiation of implicit functions allows the second term in square brackets in Eq. 7.22 to be written as

$$\left(\frac{\partial P}{\partial \lambda}\right)_{E,\upsilon} \dot{\lambda} = P_{\lambda}\dot{\lambda}.$$

If there is more than one chemical reaction, the expression for P_{λ} generalizes to

$$P_{\lambda} = \left(\frac{\partial E}{\partial P}\right)_{\upsilon,\lambda}^{-1} \sum_{i} \left(\frac{\partial E}{\partial \lambda_i}\right)_{P,\upsilon}.$$

With these definitions, Eq. 7.22 can be written as

$$\dot{P} = c^2\dot{\rho} + P_{\lambda}\dot{\lambda}. \tag{7.23}$$

The quantity $P_{\lambda}\dot{\lambda}$ is a measure of the local rate of chemical heat addition. For example, for the energy relationship defined by Eq. 7.6, $P_{\lambda}\dot{\lambda} = [q(\gamma - 1)\rho]\dot{\lambda}$. When $P_{\lambda}\dot{\lambda}$ is positive, Eq. 7.23 shows that, at fixed $c^2\dot{\rho}$, $\dot{P}$ is larger than it would be without chemical energy addition.

7.4.4.1. $\dot{P}$ and Sonic Points in 1D Steady Flow

Equation 7.23 can be used to show an important relationship between the chemical-energy addition rate and the sonic character of chemically driven 1D *steady* planar flow. In this case, Eq. 7.23 can be written as

$$uP_x - c^2 u\rho_x = P_\lambda \dot{\lambda} \tag{7.24}$$

and, for steady 1D flow, Eqs. 7.15 and 7.16 simplify to

$$u\rho_x = -\rho u_x \tag{7.25}$$

$$\rho u u_x = -P_x. \tag{7.26}$$

Rewriting the $u\rho_x$ term in Eq. 7.24, by use of Eqs. 7.25 and 7.26, allows that equation to be written as

$$\dot{P} = \frac{P_\lambda \dot{\lambda}}{1-(c/u)^2}, \tag{7.27}$$

where the sign of the denominator of Eq. 7.27 determines the sonic character of the flow. For subsonic regions (e.g., in the ZND reaction zone), $c^2 > u^2$, and for $P_\lambda \dot{\lambda} > 0$, $\dot{P}$ is negative; that is, the pressure falls through the ZND reaction zone. An important feature of Eq. 7.27 is that at points where the flow is sonic ($u = c$), either P_λ or $\dot{\lambda}$ must be zero—otherwise $\dot{P}$ is infinite. For the simple EOS defined by Eq. 7.6, P_λ cannot be zero and so the only possible sonic point corresponds to complete reaction (i.e., $\dot{\lambda} = 0$, $\lambda = 1$). This is the CJ point defined earlier.

Note that the coincidence of a zero in the heat-release rate and a sonic point in the flow is also true for higher-dimensional flows.

7.4.4.2. Shock-Change Equation

Another valuable application of Eq. 7.23 and the compressible flow equations is to define the evolution of a shock wave in a reactive fluid, given the flow conditions immediately behind the shock wave. For example, we might like to know the rate of change of shock pressure at the shock as the shock wave progresses (see Ref. 9).

The relation to be derived is for a planar shock in steady or unsteady motion. To obtain the desired relationship, we use Eq. 7.23,

$$P_t + uP_x + c^2 \rho u_x = P_\lambda \dot{\lambda}, \tag{7.28}$$

and Eq. 7.16,

$$u_t + uu_x + \rho^{-1} P_x = 0. \tag{7.29}$$

Note the $\dot{\rho}$ term has been reexpressed in Eq. 7.28 by use of Eq. 7.15. Two further equations can be obtained for conditions at the shock by using Eq. 7.14, but evaluating the pressure and particle velocity derivatives along the shock path rather than a particle path. These equations are

$$\left(\frac{dP}{dt}\right)_D = P_t + DP_x \tag{7.30}$$

and

$$\left(\frac{du}{dt}\right)_D = u_t + Du_x, \tag{7.31}$$

where the notation $(d(\cdot)/dt)_D$ has been introduced to denote the total time derivative of a quantity along the shock path.

Equations 7.28–7.31 can be viewed as four equations in the four unknowns P_t, P_x, u_t, and u_x—provided all the other quantities are viewed as parameters. The solution of these equations for u_x is

$$u_x = \frac{1}{\rho\left[c^2 - (D-u)^2\right]}\left[P_\lambda \dot{\lambda} - \left(\frac{dP}{dt}\right)_D + \rho(u-D)\left(\frac{du}{dt}\right)_D\right]. \tag{7.32}$$

The quantities involving the $(d(\cdot)/dt)_D$ terms specify changes along the shock path; it is these quantities that we wish to obtain. Since two such quantities appear in Eq. 7.32, we need to eliminate one of them in order to obtain the other. This is done as follows. The states allowed immediately behind a shock are those defined by the Hugoniot curve. On this curve differential changes in particle velocity and pressure are related by $du = (du/dP)_\mathrm{H}\, dP$. Therefore,

$$\left(\frac{du}{dt}\right)_D = \left(\frac{du}{dP}\right)_\mathrm{H}\left(\frac{dP}{dt}\right)_D, \tag{7.33}$$

where $(du/dP)_\mathrm{H}$ indicates the derivative is evaluated on the appropriate Hugoniot curve of the material. The thermodynamic states of the evolving shock must lie on this curve; note that $(du/dP)_\mathrm{H}$ is a material property and does not depend on the shock evolution process. Using Eqs. 7.1 (mass conservation) and 7.33 in Eq. 7.32 and solution for $(dP/dt)_D$ gives

$$\left(\frac{dP}{dt}\right)_D = \left[P_\lambda \dot{\lambda} - \rho\left[c^2 - (D-u)^2\right]u_x\right]\left[1 + \rho_0 D\left(\frac{du}{dP}\right)_\mathrm{H}\right]^{-1}. \tag{7.34}$$

The denominator of Eq. 7.34 is positive definite for most EOSs; e.g., for a γ-law EOS with $\gamma = 3$, it has the (constant) value of 3/2. Also note that when the quantity $c^2 - (D-u)^2$ in the numerator of Eq. 7.34 is positive (negative), the flow is subsonic (supersonic). We have seen previously that $P_\lambda \dot{\lambda}$ is a measure of the rate of chemical-energy addition to the flow (see Sec. 7.4.4).

A number of interesting results are apparent from Eq. 7.34. For example, it shows that $(dP/dt)_D = 0$ (i.e., steady flow) occurs when

$$P_\lambda \dot{\lambda} - \rho\left[c^2 - (D-u)^2\right]u_x = 0 \tag{7.35}$$

In subsonic flow with exothermic chemistry occurring, $P_\lambda \dot{\lambda}$ and $c^2 - (D-u)^2$ are > 0. In this situation, Eq. 7.35 implies that $u_x > 0$ for the shock to be steady. Thus, the particle velocity must decrease as one moves away from the shock front (see Fig. 7.6).

Another interesting situation is where $(dP/dt)_D > 0$, i.e.,

$$P_\lambda \dot{\lambda} - \rho\left[c^2 - (D-u)^2\right]u_x > 0; \tag{7.36}$$

that is, the shock pressure is increasing with time, as occurs in the initiation of explosives. Suppose that $P_\lambda \dot{\lambda}$ is small near the shock front, so the flow is subsonic and the mechanical term controls the sign of Eq. 7.36—then u_x must < 0 for $(dP/dt)_D$ to be > 0; that is, a compression wave is encroaching on the shock from the rear. This occurs in the initiation of heterogeneous explosives when exothermic chemical reaction in regions nonadjacent to the shock cause pressure increases in the flow.

7.4.4.3. Weak Detonation

In the ZND theory, a single forward rate process was assumed (i.e., chemical-energy addition proceeds monotonically via $Q = \lambda q$, $0 \le \lambda \le 1$, and $q > 0$). In Sec. 7.4.4.1 we showed that, in this case, the only possible location of a sonic point in the flow is at the point of complete reaction.

Here, we consider a more general form of the chemical-energy release function that gives a weak detonation as a solution. We utilize an example due to Fickett and Davis [9], but analyze the problem somewhat differently from them. The heat release is produced by two irreversible chemical reactions—one exothermic and one endothermic; there is no mole change due to the reaction. A γ-law EOS is assumed, so

$$E(P, v, \lambda_1, \lambda_2) = \frac{Pv}{\gamma - 1} - Q(\lambda_1, \lambda_2), \tag{7.37}$$

where $Q(\lambda_1, \lambda_2) = q_1\lambda_1 - q_2\lambda_2$ and $q_1 > q_2 > 0$; i.e., complete reaction is exothermic. We have also assumed that the EOS parameter γ is identical for reactant and products. It is important to note that $\dot{Q}(\lambda_1, \lambda_2) = q_1\dot{\lambda}_1 - q_2\dot{\lambda}_2 = 0$ when $q_1\dot{\lambda}_1 = q_2\dot{\lambda}_2$; that is, there is a point of zero heat addition rate in the flow *before* complete reaction; this point must be a sonic point if $\dot{P}$ is to remain finite (see Sec. 7.4.4 and 7.4.4.1). It is this feature of the heat-release function, Q, that introduces the new physics.

To fully define the problem, we must postulate rate laws for λ_1 and λ_2. We assume that

$$\dot{\lambda}_1 = 1 - \lambda_1 \tag{7.38a}$$

and

$$\dot{\lambda}_2 = \lambda_1 - \lambda_2. \tag{7.38b}$$

Equations 7.38a and 7.38b can be combined to yield

$$\frac{d\lambda_2}{d\lambda_1} = \frac{\lambda_1 - \lambda_2}{1 - \lambda_1}. \tag{7.39}$$

With the initial condition $\lambda_2 = 0$ when $\lambda_1 = 0$, the solution of Eq. 7.39 is

$$\lambda_2 = \lambda_1 + (1 - \lambda_1)\ln(1 - \lambda_1). \tag{7.40}$$

This λ_1–λ_2 relationship defines the *reaction path*; all acceptable (λ_1, λ_2) pairs must lie on this locus. The sonic point condition $q_1(\lambda_1)^{\cdot} = q_2(\lambda_2)^{\cdot}$ can be written as

$$\lambda_2 = \lambda_1 - \frac{q_1}{q_2}(1 - \lambda_1). \tag{7.41}$$

Simultaneous solution of Eqs. 7.40 and 7.41 gives the values of λ_1 and λ_2 at the sonic point/zero-heat-release rate point,

$$\tilde{\lambda}_1 = 1 - \exp\left[-\frac{q_1}{q_2}\right] \tag{7.42a}$$

and

$$\tilde{\lambda}_2 = \tilde{\lambda}_1 - \frac{q_1}{q_2}\exp\left[-\frac{q_1}{q_2}\right], \tag{7.42b}$$

where tildes over variables represent values at the sonic point. The reaction rate values at the sonic point are $(\tilde{\lambda}_1)^{\cdot} = \exp(-q_1/q_2)$ and $(\tilde{\lambda}_2)^{\cdot} = (q_1/q_2)(\tilde{\lambda}_1)^{\cdot}$; i.e., they are *non-zero*. The heat release corresponding to the $(\tilde{\lambda}_1, \tilde{\lambda}_2)$ point is

$$\tilde{Q}(\tilde{\lambda}_1, \tilde{\lambda}_2) = (q_1 - q_2) + q_2\exp\left[-\frac{q_1}{q_2}\right] > Q(1,1) = q_1 - q_2; \tag{7.43}$$

i.e., the $\tilde{Q}$ partial reaction Hugoniot corresponds to a larger exothermicity than the complete reaction Hugoniot.

The condition for the detonation velocity is now obtained from Eq. 7.8, but with $C = (\gamma - 1)(1 + 2\tilde{Q}/D^2)v_0^2/(\gamma + 1)$. A unique solution requires $B^2 - 4AC = 0$, which corresponds to $D^2 = 2(\gamma^2 - 1)\tilde{Q}$. The specific volume through the reaction zone as a function of reaction is

$$v(\lambda_1, \lambda_2) = \frac{\gamma v_0}{\gamma + 1}\left[1 \pm \frac{1}{\gamma}\sqrt{1 - \frac{Q(\lambda_1, \lambda_2)}{\tilde{Q}}}\right]. \tag{7.44}$$

The requirement that we remain on the reaction path (Eq. 7.40) can be used to reduce Eq. 7.44 to a function of one independent variable [e.g., $\bar{v}(\lambda_1)$]. The pressure as a function of λ_1 can be obtained by the use of Eq. 7.44 in the Rayleigh line equation. The dependence of P and v on t and x can be found via the processes as outlined in Eqs. 7.12 and 7.13 of Sec. 7.3.5.

Thus, the weak detonation process summarized on Fig. 7.8 consists of i) a shock from the initial point to the spike point, ii) followed by movement down the Rayleigh line traversing the partial reaction Hugoniots to a sonic point at which the total heat-release rate is zero, but the individual reaction rates are *nonzero*, and iii) then a continued decent down the Rayleigh line traversing the partial reaction Hugoniots to the *weak* point on the fully reacted Hugoniot. The flow in the region after the sonic point is supersonic and, there-

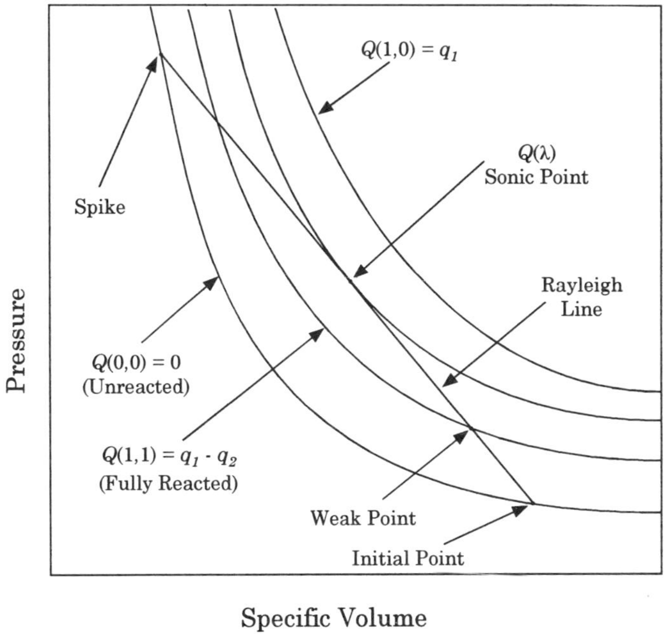

Figure 7.8. Pressure vs. specific-volume plane quantities useful in the discussion of weak detonation.

fore, a growing constant state in x, t space is formed which connects the sonic point to the following flow.

This example shows theoretically how the weak point can be reached. White [21] has observed weak detonations experimentally in gases.

7.4.4.4. Taylor Wave

In Secs. 7.3.4 and 7.3.5, we derived the fluid mechanical state at the point of complete chemical reaction in the CJ and ZND models—i.e., the CJ point values of the flow field. A natural question to ask is: What fluid flow connects the CJ point flow field to the final low-density and pressure state of the expanded reacted gases? The compressible flow equations of Sec. 7.4 can be used to address this question.

First note that in a fluid system describable by the Euler equations (and in which there are no chemical reactions occurring or shock waves present—e.g., the fluid flow following the sonic plane in a CJ or ZND detonation), the thermodynamic states experienced by the initially isentropic fluid must lie on an isentrope of the fluid material. Here, we assume that any fluid particle that has traversed the CJ point can be so described. Further, we assume that the reacted fluid is describable by a γ-law EOS—see Eq. 7.6. The isentropes for this EOS are given by

$$P = C\rho^{\gamma}, \quad C = \text{constant}. \tag{7.45}$$

For such a material in 1D planar time-dependent flow, the mechanical equations governing the density (Eq. 7.15) and momentum (Eq. 7.16), respectively, reduce to

$$\frac{\partial \rho}{\partial t} + u\frac{\partial \rho}{\partial x} + \rho\frac{\partial u}{\partial x} = 0 \tag{7.46}$$

and

$$\frac{\partial u}{\partial t} + u\frac{\partial u}{\partial x} + \gamma C \rho^{(\gamma-2)}\frac{\partial \rho}{\partial x} = 0\ . \tag{7.47}$$

Since there are no pertinent space or time scales in the flow following the detonation reaction zone, Taylor [29] assumed that this part of the flow can be described via the self-similarity variable $z \equiv x/t$. When the partial differential equations, Eqs. 7.46 and 7.47, are expressed in terms of z, they become the ordinary differential equations

$$(u - z)\rho' + \rho u' = 0 \tag{7.48}$$

and

$$(u - z)u' + \gamma C \rho^{(\gamma-2)} \rho' = 0, \tag{7.49}$$

where the prime denotes the ordinary derivative with respect to z. Division of Eq. 7.49 by Eq. 7.48 results in the differential relation

$$u' = \sqrt{\gamma C \rho^{(\gamma-3)}}\, \rho'.$$

This can be integrated to yield

$$u = \frac{D}{\gamma - 1}\left[\frac{2\gamma}{\gamma + 1}\left(\frac{\rho}{\rho_{CJ}}\right)^{(\gamma-1)/2} - 1\right], \tag{7.50}$$

where the known values at the CJ point of

$$u_{CJ} = \frac{D}{\gamma + 1}, \quad c_{CJ} = \sqrt{\gamma P_{CJ} v_{CJ}} = \frac{\gamma D}{\gamma + 1}, \quad \text{and} \quad \rho_{CJ} = \frac{\gamma + 1}{\gamma}\rho_0$$

have been used to evaluate the integration constant and to simplify the expression. Equation 7.50 can be used in Eq. 7.49 to find the dependence of ρ on z. One finds two cases: a solution $\rho' = 0$ and the Taylor-wave solution

$$\rho(x,t) = \rho_{CJ}\left[\frac{(\gamma - 1)x/t + D}{\gamma D}\right]^{2/(\gamma-1)}. \tag{7.51}$$

The $\rho' = 0$ solution [i.e., $\rho(x,t)$ = constant] corresponds to a flow supported at the rear by a piston moving at the CJ particle velocity.

Use of Eq. 7.51 in Eq. 7.50 gives the particle velocity flow field

$$u(x,t) = \frac{1}{\gamma + 1}\left(\frac{2x}{t} - D\right). \tag{7.52}$$

Figure 7.6 shows a schema of the $u(x,t)$ Taylor wave solution. Note that if the back boundary is free, Eq. 7.51 shows that the density and pressure are zero along the line $x/t = -D/(\gamma - 1)$. Thus, for a $\gamma = 3$ material, the speed of free expansion of the reacted gas is one-half the detonation speed.

7.4.4.5. Solutions of the Reactive-Flow Equations for Complex Systems

There are a number of difficulties associated with obtaining solutions of Eqs. 7.15–7.20 for real physical systems. A mathematical obstacle is that the flow equations are nonlinear (in general, partial) differential equations and so the solution techniques available for linear systems are not available. This difficulty is usually avoided by solving the set of equations numerically (see Ref. 30).

A more fundamental physical problem involves Eqs. 7.18 and 7.20. These equations require specification of constitutive properties of dense systems (usually containing many molecular species) at extreme conditions of pressure, density, and temperature. The EOS problem is difficult and is a topic of current statistical mechanical and molecular dynamical research (see, e.g., Ref. 31). Generally, a global rate is used to calculate the state which lies between the unreacted Hugoniot and fully reacted product Hugoniot. An empirical form called the JWL EOS [32] is often used to describe the reaction products and is calibrated by fitting to metal-pushing experiments (see Sec. 7.8). Recently, overdriven detonation experiments have been performed by Fritz et al. [33] on PBX 9501 in which the sound speed at the supported detonation condition was measured. These sound speeds do not agree with those calculated using the JWL product EOS.

Specification of the chemical-heat-release function(s) is in an even more primitive state. Usually, a mathematically simple rate form containing adjustable parameters is *postulated* and its parameters calibrated to the results of experiments on real systems. This calibrated rate can then be used with the other governing equations to make predictions of behavior of systems similar to those used for the calibration. Models of various complexity have been developed by a number of workers, including, e.g., Arrhenius rates, Forest Fire by Mader and Forest [34], DAGMAR by Wackerle et al. [35], Ignition and Growth by Lee and Tarver [36], and JTF by Johnson, Tang, and Forest [37]. These usually depend on the initial shock parameters, such as pressure, and have several adjustable parameters that are calibrated to experimental data for each explosive. Upgrading these models is a current area of research.

It should be remembered that the rate forms being considered above are global rate relations which track the reaction from the unreacted explosive to the fully reacted products. In reality, each explosive has a reaction mechanism that involves a number of interrelated chemical reactions, each with a particular rate function. Because of the difficulty of probing the detonation regime spectroscopically, almost nothing is known about the actual reaction processes that occur. Chemistry of the reaction products is often addressed by minimizing the Gibbs free energy of the assumed product materials (see the discussion in Ref. 16) to get a composition and state. This method has been refined by Hobbs and Baer [38]. It is likely that studying the chemistry of initiation and detonation will become a larger research area in the future as experimental and calculational techniques continue to be improved.

7.5. Initiation of Detonation

Initiation and detonation of explosives depend on the chemical and physical nature of the materials (e.g., molecular structure, density, charge geometry, confining materials, and various other variables). In terms of their physical nature, explosives are generally grouped into homogeneous and heterogeneous materials.

Homogeneous materials are typically liquids or single crystals in which there are a minimal number of physical imperfections (e.g., bubbles or voids) that can cause perturbations in the input shock and the flow behind it. Homogeneous materials viewed with the macroscopic probes characteristic of detonation physics experiments appear uniform.

Heterogeneous explosives are generally all other types; these are usually pressed, cast, machined, or extruded into the shapes or parts desired. Here, heterogeneous means a material that contains any kind of imperfection that can cause fluid-mechanical irregularities (called *hot spots*) when a shock or detonation wave passes over them. Such hot spots cause associated space/time fluctuations in the thermodynamic fields (e.g., the pressure or temperature fields) in the fluid. These thermodynamic variations affect the local chemical-heat-release rate. When averaged over a sufficiently large space scale, such variations convoluted with the underlying chemical rate(s) produce an average heat-release rate that is a combination of chemistry and mechanics. Examples of conditions that cause hot spots are i) void collapse, ii) shock wave propagation through irregular particles that cause complex shock interactions, iii) shock wave interactions between particles and voids that cause jetting, iv) plastic flow involving crystal breakage and shearing, and v) shock impedance mismatch between components of the explosive that cause shock reflections and interactions. Some of these have been modeled using wave propagation computer codes (e.g., see Refs. 30 and 39). Because of these hot spots, initiation and detonation in heterogeneous materials are qualitatively different from that which occurs in homogeneous materials. When explosives are porous (i.e., the density is considerably below the theoretical maximum density [TMD]) or when large amounts of heterogeneities are present, dramatic sensitivity changes can occur (see Sec. 7.5.3).

Since detonation is a self-sustaining, shock-generated chemical reaction, at some point the initiation process must lead to the formation of a shock. The most direct initiation method is to mechanically introduce a shock, e.g., with a thrown plate. If the resultant pressure input is sufficiently large and sustained, the shock will grow to detonation. This process has been labeled a shock-to-detonation transition (SDT).

Shock inputs obviously have a particular duration, depending on how they are generated, and the duration may or may not be relevant to the initiation process. If the shock duration is long compared to the initiation time, it has no relevance. Durations of the same order as the initiation time or shorter can become important because pressure relief from the rear may quench the initiation reactions. Gittings [40] showed that pressure pulse duration is an important parameter (when short pressure pulses are used) and that the threshold for initiation at a given pressure could change as the pressure duration changed. Walker and Wasley [41] reanalyzed her data and found a roughly $P^2\tau$ dependence of the initiation/no initiation threshold; i.e., there is a critical $P^2\tau$ value needed for initiation. (In this expression, τ is a measure of the pressure pulse duration.) This has been studied by a number of workers who have concluded that the $P^2\tau$ relationship should be a $P^n\tau = K$ (constant) relationship where n and K can vary, depending on the explosive.

When the pressure input is less than that required for direct SDT, two other notable processes can take place: i) a burning or deflagration-to-detonation transition (DDT) or ii) what has been called a low-velocity detonation (LVD). These will be discussed in Secs. 7.5.3 and 7.5.4, respectively.

7.5.1. 1D Homogeneous Initiation (SDT)

Homogeneous SDT was studied in detail by Campbell et al. [42] and Chaiken [43]; these studies led to the classical homogeneous initiation model shown in the time–distance diagram of Fig. 7.9. The explosive is shocked and, after an induction period τ_{ind} (that depends on the temperature generated by the initial shock), a thermal explosion occurs at the explosive–driver interface. This explosion produces a superdetonation (detonation in the explosive precompressed by the initial shock) which runs forward and eventually overtakes the initial shock. After the superdetonation overtakes the input shock, an overdriven condition results. After some time, the overdriven detonation wave decays to a steady detonation.

The simplest theory, which assumes Arrhenius kinetics, predicts the induction time (i.e., the time to thermal explosion) to be

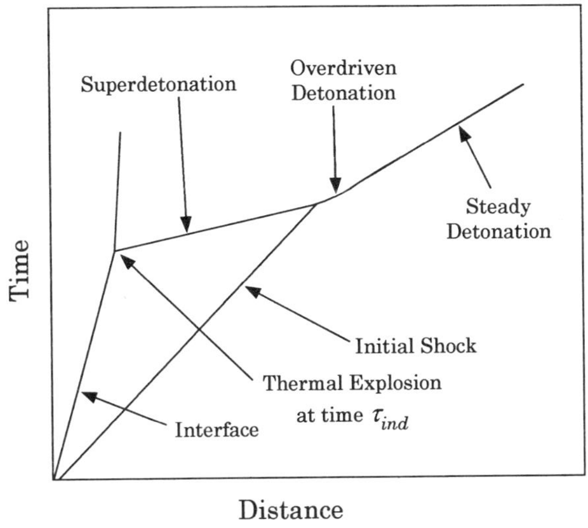

Figure 7.9. Time–distance diagram for a homogeneous explosive initiation process according to Campbell et al. [42] in which a thermal explosion occurs at the driver–explosive interface, where the explosive has been shocked the longest, and a superdetonation (running into precompressed material) immediately develops. The superdetonation velocity is higher than the normal detonation velocity and overtakes the original shock, producing an overdriven detonation in the undisturbed material. Eventually, the overdriven detonation decays to a steady detonation.

$$\tau_{\text{ind}} = \frac{C_v R T^2}{AQE} \exp\left[\frac{E}{RT}\right]. \tag{7.53}$$

Here, C_v is the constant-volume specific heat, R is the gas constant, A is a collision factor, Q is the total chemical heat release, T is the temperature produced by the shock, and E is an Arrhenius activation energy. A number of assumptions, including no heat transfer, lead to this simplified form; a more general treatment has been developed by Frank-Kamenetskii [44]. Because of the exponential dependence of the Arrhenius rate on E, this relation indicates the induction time is a sensitive function of the shock temperature (provided $E \gg RT$); i.e., a small increase in temperature yields a large increase in reactivity and a consequent large decrease in the induction time.

Figure 7.9 indicates that a superdetonation forms immediately after the thermal explosion. This picture was developed without the aid of in situ measurements. In situ magnetic particle-velocity gauge measurements of this growth process in chemically sensitized nitromethane have been made by Sheffield et al. [45] and show the reactive wave develops over relatively long time and distance scales (see Fig. 7.10). When these measurements are transformed into the time–distance plane, the trajectories of the various waves become those shown in Fig. 7.11. This modified homogeneous initiation model consists of the following processes: i) the explosive is initially shocked, raising the pressure and temperature to the point that a chemical reaction starts, ii) this reaction produces waves that coalesce away from the driver–explosive interface and strengthen, iii) a reactive compression wave is formed within the material, iv) this reactive wave builds in strength and steepens into a shock, and v) this reactive shock (which possibly can become a *steady* superdetonation) overtakes the original shock wave, producing an overdriven condition that eventually settles down to a steady detonation [45].

Although this process has not been measured in explosives other than sensitized nitromethane, it is thought to apply to homogeneous liquids and single-crystal solid explosives in general. The space and time scales of the buildup process will differ for different explosives.

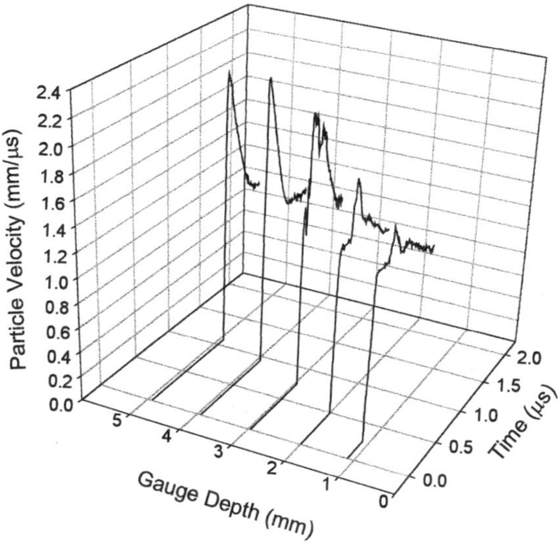

Figure 7.10. In situ particle velocity measurements made in chemically sensitized (sensitized with 5 wt % of the organic base diethylenetriamine) homogeneous nitromethane using magnetic gauges. These waveforms (and those from other similar nitromethane experiments) led to an improved understanding of the reaction process which takes place during the SDT initiation process in homogeneous materials. The reaction produces waves that coalesce into a reactive shock that may grow into a superdetonation before it overtakes the initial wave [45].

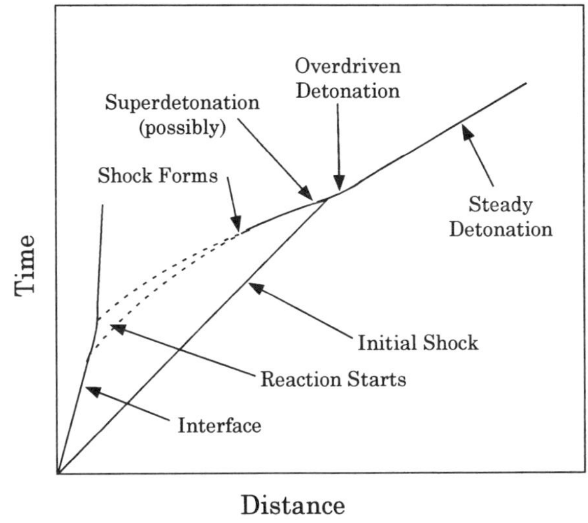

Figure 7.11. Modified time–distance diagram for a homogeneous explosive SDT initiation process. It accounts for the processes observed in in situ particle velocity gauging experiments on sensitized nitromethane [45].

7.5.2. 1D Heterogeneous Initiation (SDT)

Heterogeneous SDT was also studied in detail by Campbell et al. [46], leading to much of the basic understanding that exists today. They showed that wave growth occurs at the front as well as behind the front. This leads to a relatively smooth growth of the initiating shock to a detonation, in contrast to the abrupt changes which occur in the homogeneous case. In this case, no overshoot in the detonation velocity is observed, indicating that the transition to detonation occurs at or near the shock front.

Experiments on heterogenized nitromethane in which in situ magnetic particle-velocity gauge histories were measured [45] are shown in Fig. 7.12. These profiles agree with the heterogeneous initiation model of Campbell et al. [46]. They show a reacting wave growing more at the front than behind the front. The reactive wave reaches a detonation condition between the third and fourth gauges at the shock front.

Other heterogeneous materials (typically plastic-bonded explosives with a few percent voids) have behavior in which the wave grows at the front but more slowly than what is shown in Fig. 7.12. The reactive wave grows behind the front, speeds up as it grows, and finally overtakes the front. After overtake, a steady detonation is rapidly established. The reactive wave does not form a shock until it overtakes the front. An example of this behavior is shown in Fig. 7.13.

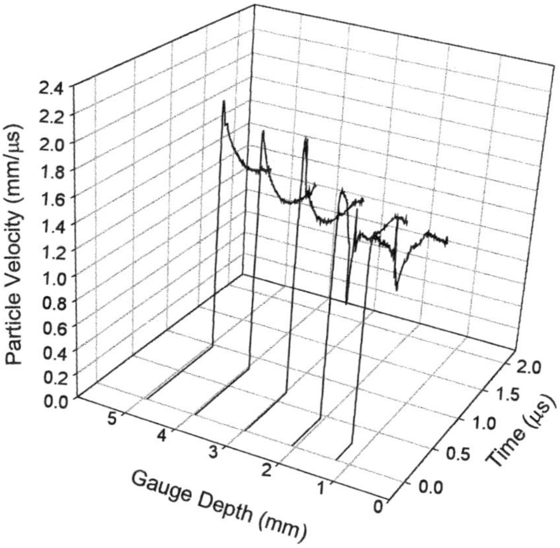

Figure 7.12. In situ particle velocity measurements of a heterogeneous SDT initiation process in heterogenized nitromethane. The heterogenities were silica particles suspended in gelled nitromethane [47,48]; the composition was 92.75/6.0/1.25 wt% nitromethane/silica/guar gum. Input was 7 GPa (see Ref. 45).

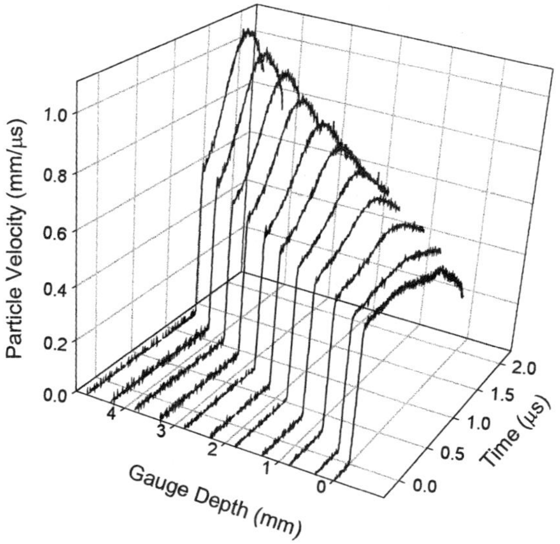

Figure 7.13. In situ particle velocity measurements of a plastic-bonded HMX-based explosive, PBX 9501. The first gauge was at the input interface and the other 10 gauges were from 0.5 to 5 mm into the material [49].

This is an SDT process in an HMX-based plastic-bonded explosive (PBX 9501) [49] which had a magnetic gauge at the input interface and 10 in situ particle velocity gauges. Although the wave never reaches a detonation during the measured buildup process, the growth both at and behind the front can be clearly seen. This PBX 9501 sample had a density of 1.82 g/cm^3, about 98 % TMD (2 % porosity). Because the porosity was so low, it exhibits initiation behavior that falls between the classical homogeneous and heterogeneous SDT processes.

The difference in behavior from that observed in homogeneous explosives is attributed to shock-induced hot spots forming at or near the shock front due to physical imperfections in the explosive and their consequent effect on the chemical-energy release into the flow (see Sec. 7.5). These hot spots can result from any phenomenon that disrupts the smooth nature of the flow. Based on this, one would expect initiation behavior differences when the particle size of the material used to fabricate the plastic-bonded explosive is changed. This is, in fact, observed [50].

Little is known about the detailed mechanisms that produce the shock wave growth other than that the hot spots develop at the inhomogeneities, stimulating chemical-energy release locally and making the explosive initiate at much lower pressure inputs than would otherwise be the case. For example, pure liquid nitromethane can be made to initiate, with a sustained shock input of about 8 to 9 GPa, in approximately 1 μs, whereas the same material with a large number of carborundum heterogeneities will initiate in the same time interval with 2 to 3 GPa inputs (see Ref. 46). An even more striking comparison is that of single-crystal HMX (which requires over 30 GPa to initiate on the microsecond time scale [51]) and porous HMX with a density of 65 % TMD (which initiates with inputs below 1 GPa on similar time scales [52]).

Many experimental studies have looked at hot-spot phenomena. However, the size and nature of the inhomogeneities involved have not been sufficiently experimentally controlled to allow detailed understanding of the underlying phenomena. This is a current area of research.

Because the hot spots produce reaction locally, heterogeneous explosives have much less-sensitive state-dependent chemical-heat-release rates than homogeneous materials. The dependence on the bulk material pressure and temperature produced by the initiating shock (which is all important in the homogeneous case) is secondary. A comparison of the state sensitivity of the heat-release rates for homogeneous and heterogeneous nitromethane is made in Sec. 7.6.

7.5.3. Deflagration-to-Detonation Transition (DDT)

Transition to detonation can take place from much smaller stimuli than those required in a SDT process. When the material and confinement conditions are appropriate, relatively low thermal or pressure inputs start the explosive burning and this can eventually lead to detonation. This process is called deflagration-to-detonation transition (DDT). Whereas SDT takes place on a time scale of a few microseconds, DDT takes place on a time scale 10 to 1000 times slower. DDT has been observed in both cast and pressed solid explosives. Heavy confinement is usually required for DDT. Experimental and theoretical research relating to DDT phenomena is active because of the safety concerns associated with small stimuli leading to accidental detonations. DDT was discussed extensively by a number of researchers in a recent detonation symposium [53].

DDT in low-density pressed HMX has been experimentally studied and the data interpreted to include the following phenomena [54]. After the explosive starts to burn, the heavy confinement causes the pressure to increase with a consequent increase in the burning rate. The hot product gases move out into the pores of the unburned explosive and convective burning develops. This burning drives a compaction wave into the material ahead of the burning; this makes it more difficult for the gases to penetrate the porous bed of material. This compaction wave compresses the material to ~90% of TMD. As pressure builds up, stress waves are formed and coalesce, compacting the explosive to nearly 100% of TMD; i.e., a plug is formed in the material. As the combustion process gets near the plug, pressures increase and a strong shock forms at the front of the plug. When the shock amplitude increases sufficiently, a SDT process takes place in the 90% TMD material ahead of the plug. DDT in cast materials is thought to involve similar phenomena with the addition that the burning and compaction processes cause crushing of the material with resultant enhanced convective burning. From a theoretical standpoint, it has proven difficult to adequately simulate the plug formation process. It has not yet been determined whether this plug formation process is material specific; i.e., it only occurs in HMX.

Baer and Nunziato [55] have had success in modeling the DDT process by using a two-phase flow model. Although many of the phenomena they calculate are similar to the experimental observations, they do not calculate a plug formation. Because of the many different processes involved in DDT, more experiments are needed to accurately calibrate the parameters of their model. Baer has recently used this model [56] to calculate new experiments on porous HMX

[52] and other materials. Some of the compaction and reaction parameters have been recalibrated with these data. There is still much discussion about the exact nature of the transition process; it is possible that the exact process is material-specific rather than general. Bdzil et al. have also been addressing the DDT modeling process in detail [57,58].

In addition to heavy confinement, many other explosive properties are important in the DDT process (e.g., density, porosity, compressibility and material strength, grain size, shock sensitivity, reaction rates, heats of combustion and detonation). Essentially no work relating to the detailed chemistry involved in the DDT process has been attempted.

7.5.4. Low-Velocity Detonation (LVD)

In some cases, condensed-phase explosives undergo *low-velocity detonation* (LVD), a process completely different from that of steady 1D detonation. In fact, to call LVD a detonation is a misnomer, but since this is the way it is referred to in the literature, we follow the same pattern. There are two kinds of LVD that are known to occur: i) a non-1D process in which the waves in the confinement prepare the material for a low-level reaction [59] and ii) a low-velocity wave resulting from the reaction of one of the most sensitive components in a mixture (such as nitroglycerin in dynamite) while the other energetic materials do not react [60]. These waves are, in a sense, constant speed and require rather stringent conditions (in most cases) for their production and propagation. They depend on i) the properties and preparation of the explosive, ii) the confinement diameter and wall thickness, and iii) the input shock conditions. The LVD waves die out or transition to a true detonation when sufficiently perturbed. However, constant-velocity LVD waves have been observed in charges as long as several tens of centimeters. We discuss the two types of LVD briefly below.

The first type of LVD is usually associated with liquid explosives. Initiation inputs are in the few tenths of a GPa regime. LVD propagation velocities are at or a little above sonic velocity, i.e., about 2 km/s, compared to high-order detonation velocities which are approximately three to four times higher. Reaction in an LVD wave releases only a small portion of the available chemical energy. The pressures in such waves are approximately 1.5 GPa or lower. Nevertheless, LVD can lead to considerable energy release and to fatal accidents. Many accidents involving nitroglycerin have been attributed to LVD.

In the case of liquids, confining materials with a sound speed higher than that of the liquid are required in order that hydrodynamic waves in the confinement wall travel faster than waves in the liquid explosive. This leads to 2D wave interactions in the liquid that can cause bubble formation (cavitation) ahead of the main reactive wave; the wave in the confinement can lead the reactive wave by as much as tens of microseconds. The cavitation bubbles are compressed by the shock in the liquid and form hot spots when the pressure wave in the liquid explosive passes; this leads to enhanced chemical-energy release which supports the further propagation of the waves. This process has been described in detail by Watson, et al. [59].

The second type of LVD have been observed in dynamites without confinement at diameters just below failure diameter with low pressure input shocks. Voids or imperfections (in one case, glass beads) produce the hot spots and enough subsequent chemical reaction to sustain the process. Reaction is thought to occur in the most sensitive ingredients without the other energetic materials reacting.

Other solid explosives such as PETN, HMX, Tetryl, and TNT (pressed and cast) have been observed to undergo LVD when the charge diameters, particle size, density, confinement, and pressure input are just right (see Ref. 12). The elastic-plastic nature of the initial compressive wave (0.5 to 1.5 GPa in stress) is thought to have some bearing upon whether LVD occurs. There is still much to learn about the nature of LVD and, because it is often hard to make repeatable measurements, understanding comes slowly.

7.6. 2D Steady Detonation in Homogeneous and Heterogeneous Materials

An experimentally realizable system with some conceptual simplicity is *steady* two-dimensional detonation. Such detonations are observed in explosive charges that have an axis of symmetry. The most common experimental configuration is a right circular cylinder of explosive detonated in air (see Fig. 7.14; this experiment will be discussed in more detail in Sec. 7.8). A detonation wave propagating in such a charge reaches a steady state after an initiation transient. In high-density explosive charges, experiment shows that a steady two-dimensional shock followed by a steady chemical reaction zone is closely approached after the detonation wave has propagated approximately four charge diameters. Thereafter, the steady 2D flow in the chemical reaction zone is determined by the interaction of the chemical-energy-release rate and the fluid mechanics of the divergent flow (see Fig. 7.15). At the steady state, the mode of initiation has been forgotten.

The hydrodynamic equations governing steady axisymmetric (2D) detonation in the steady reaction zone are functions of two space variables and independent of time (in the Galilean frame traveling with the detonation shock).

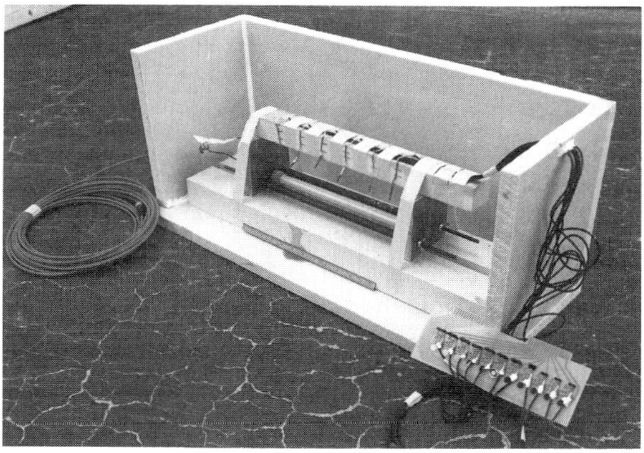

(a)

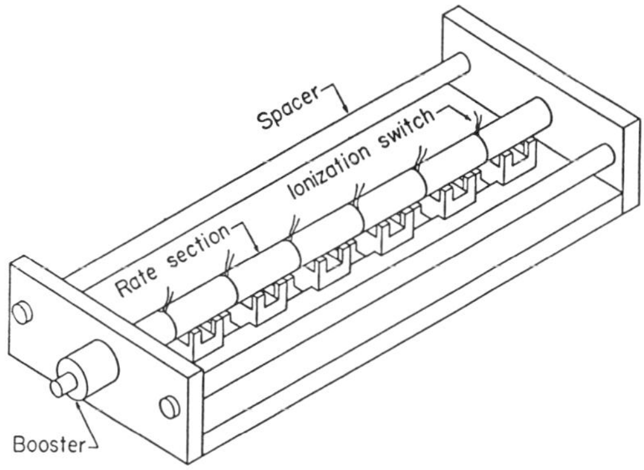

(b)

Figure 7.14. Photograph of a rate-stick assembly with which accurate measurements of steady 2D detonation velocity are made is shown in (a). The important features of the assembly are shown in schema (b). Fine copper wires (2 mils in diameter) are shorted as the detonation wave moves across them. Signals produced by electrical circuitry actuated by the shorting give the time history of the detonation wave motion. Measurements made in this way yield wave speeds accurate to a few m/s out of 8000 m/s.

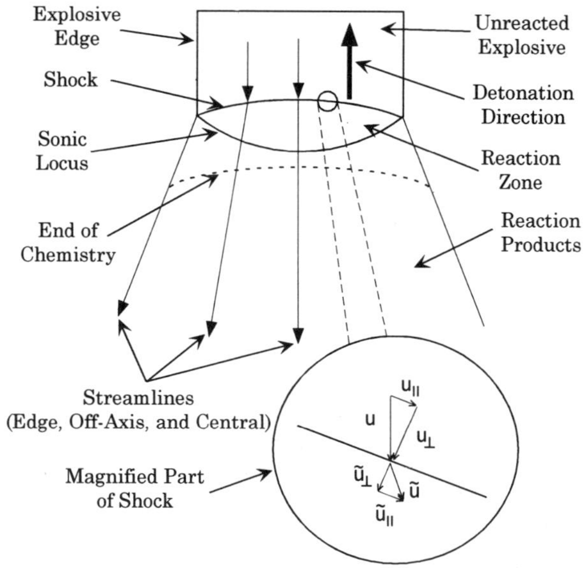

Figure 7.15. Quantities of interest in steady 2D detonation. The enlarged section of the shock (inset) shows how shock curvature deflects the streamlines in the flow.

It is experimentally known that a steadily propagating 2D detonation front, as discussed above, is curved. The shock on the symmetry axis leads the remainder of the shock wave; the qualitative form of such a detonation is shown on Fig. 7.15. A number of features of steady 2D detonation can be obtained from the fact that the detonation shock is curved.

To deduce these features, we need the shock jump conditions for fluid motion that is oblique (i.e., nonperpendicular) to the shock. Suppose that one rides in the shock reference frame and that the fluid velocity observed from this frame is **u**. As shown on the inset of Fig. 7.15, the velocity vector **u** can be resolved into a component, $u_{\perp}$, that is perpendicular and another component, $u_{\parallel}$, that is parallel to the local shock surface. The conditions that relate the values of the components of **u** in front of and in back of the shock wave are obtained as follows. The component $u_{\parallel}$ is continuous across the shock; $u_{\perp}$ is governed by Eqs. 7.1–7.3 with U_s and $U_s - u_p$ replaced by the values of $u_{\perp}$ in front of and behind the shock wave.

One sees from these considerations that, other things being equal, shock obliquity results in lower mass density, pressure, and internal energy jumps across the shock wave than occurs in normal incidence. Further, as can be seen from the inset to Fig. 7.15, the curvature of the steady 2D detonation shock causes divergence of the flow streamlines away from the central streamline as material crosses the shock. This divergence has a number of effects. First, it introduces mechanically based terms into the 2D Euler equations which have the appearance of an endothermic chemical reaction (see Ref. 9). This mechanical endothermicity is due to some diversion of chemical heat to producing lateral fluid motion. Because of this, the detonation speed is reduced from that observed in steady 1D detonation. A second, more subtle, effect is that this mechanical endothermicity produces a *zero* in the total heat-release rate before the full chemical heat release is completed. As indicated in Sec 7.4.4.1, a zero in the heat-release rate must be accompanied by a sonic point in the flow. The chemical heat released behind this sonic point cannot affect the shock and so it is lost in this sense. This effect also contributes to the reduction in the speed of the steady 2D detonation shock wave relative to the planar case.

A question of interest is: How does the steady 2D detonation speed depend on the lateral charge size? This is answered experimentally by measuring the detonation speed as a function of the lateral charge dimension. Such measurements produce an explosive's diameter–effect curve as shown on Fig. 7.16. Diameter–effect curves are usually plotted in the detonation velocity vs. 1/(lateral dimension) plane; this choice allows one to estimate the steady detonation speed by an extrapolation to zero. For cylindrical charges, the pertinent lateral dimension is the charge radius, R.

A striking feature of diameter–effect curves (see Fig. 7.16) is that they *end* at a *finite* lateral size. In cylindrical charges, this lateral dimension is called the *failure diameter* and denoted by D_f. It is an experimental fact that it is *impossible* to propagate a steady 2D detonation in a cylindrical charge whose lateral size is less than D_f. The failure diameter D_f is dependent on the molecular structure of the explosive, its density, the material surrounding the explosive (confinement), the temperature, and other variables. Cylindrical charges fired with lateral dimension smaller than D_f result in shock waves that attenuate rapidly and completely. It is thought the failure phenomenon is a result of detonation waves in charges near failure being catastrophically hydrodynamically unstable to local perturbations.

The 2D steady detonation of homogeneous and heterogeneous materials exhibit strong differences from each other. This parallels the behavior discussed in Secs. 7.5.1 and 7.5.2 relating to the initiation of such materials, and the difference in behavior can be traced to the same origin; that is, the absence or presence of hot-spot phenomena.

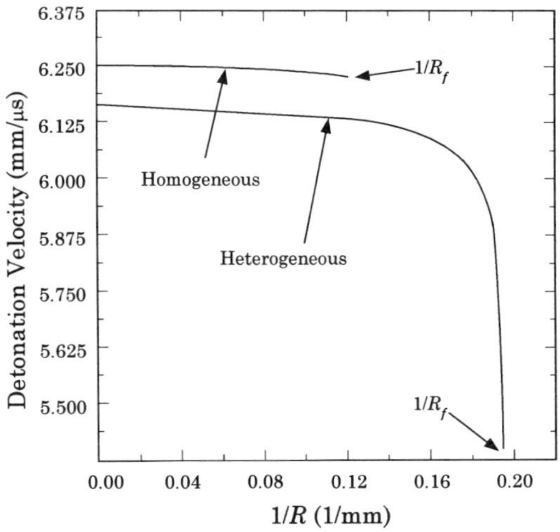

Figure 7.16. Typical diameter–effect curves for homogeneous and heterogeneous explosives. Steady 2D detonation velocity is plotted vs. the reciprocal of charge radius (R). These curves apply to homogeneous and heterogeneous nitromethane [48].

We discuss the homogeneous case (see Sec. 7.5.1) first, because it is simpler. The diameter–effect curves of homogeneous explosives are well represented by a line in the D vs. $1/R$ plane (see Fig. 7.16). A second characteristic is that the detonation velocity decrease from the steady 1D value (D_∞) to the speed at failure [$D(D_f)$] is small; a value of $[D(D_\infty) - D(D_f)] / D(D_\infty) = +0.01$ is typical.

Heterogeneous explosives (see Sec. 7.5) have diameter–effect curves that are qualitatively different from those of homogeneous materials; they are strongly downward concave and have a value of approximately $D(D_\infty) - D(D_f)] / D(D_\infty) = +0.1$ (see Fig. 7.16). Note that for heterogeneous explosives with more than a few percent voids, even larger velocity deficits have been measured [60].

It should be pointed out that the slope of the diameter–effect curve, at large charge diameters, is often said to be related to the reaction-zone thickness. However, in general, one knows the reaction-zone length involves interweaving of chemical kinetics and the EOS of the reacting mixture. Thus, it is unlikely that simply comparing the diameter–effect slopes for chemically very different explosives can be used to obtain relative reaction-zone lengths.

Nonideal heterogeneous explosives (e.g., those containing large amounts of aluminum, ammonium nitrate, or ammonium perchlorate) have diameter–effect curves that have larger velocity deficits [61,62] than those discussed above for heterogeneous materials. For these materials, the extrapolated infinite–diameter detonation velocity is well below what is calculated using the equilibrium thermodynamic codes [63,64]. Because of the interest in the relationship between the diameter–effect curves and the reaction zone, this is a current area of research.

It has been experimentally shown in a particular system that the downward concave portion of a diameter–effect curve is caused by flow field fluctuations (hot spots). Experimental curves for the explosive nitromethane exist for the pure liquid and for the liquid heterogenized by the addition of chemically inert silica particles (see Fig. 7.16 and Refs. 47 and 48). The addition of heterogeneities produces the expected downward concave region and the much enlarged value of the velocity deficit at failure of detonation relative to that of the homogeneous material. An analysis of these nitromethane data has produced global heat-release rates for the liquid and heterogenized materials as shown in Fig. 7.17. The most important feature of Fig. 7.17 is that the heat-release rate for the homogeneous material is much more sensitive to the thermodynamic state than is the rate for the heterogenized material (see Refs. 48 and 65). In this figure, $\dot{Q}_1$ is the heat-release rate function for homogeneous nitromethane and $\dot{Q}_2$ is the same function for heterogenized nitromethane.

These results offer a more complete view of the heat-release rates needed to produce the qualitatively different detonation phenomena seen in homogeneous and heterogeneous materials than is allowed by the hot-spot picture alone. A more encompassing view is that homogeneous (heterogeneous) behavior is observed in explosives that have strong (weak) dependence of the heat-release rate on the thermodynamic state. Within this picture, e.g., a physically homogeneous material could exhibit heterogeneous detonation behavior provided the thermodynamic state dependence of the underlying microscopic chemical rates is weak enough. Such behavior is rare enough that the partitioning of materials as homogeneous and heterogeneous as discussed above is a useful viewpoint.

7.7. Properties of High Explosives

Explosives are generally classified in accordance with their sensitivity as judged from various experimental tests. Primary explosives are the most sensitive, followed by secondary, and, finally, insensitive explosives. The chemical properties of many explosives have been measured; however, only a few materials have been studied to the extent that a complete set of initiation and detonation information is available. In addition, the data reported are often unreliable or de-

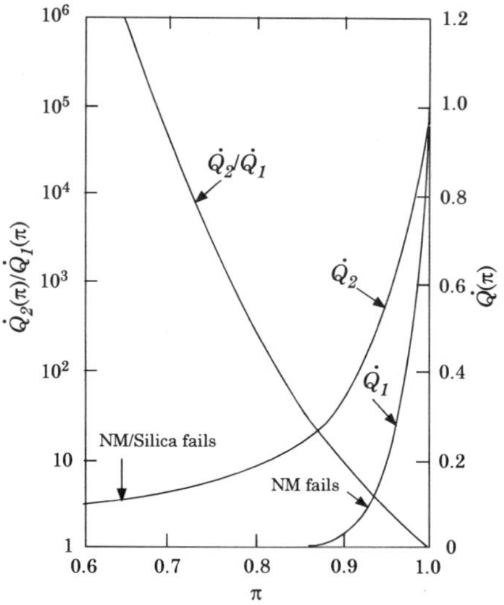

Figure 7.17. Global heat-release rates for a homogeneous material (liquid nitromethane) and a heterogeneous material (nitromethane with silica particles added). The right ordinate applies to the individual heat-release rates and the left one to the ratio of the two rates; π is the pressure normalized to the CJ pressure. The heat-release rate for the homogeneous material is much more sensitive to changes in the pressure than is the heterogeneous rate.

pendent on exactly how the test was done. Reaction-zone structure, von Neumann spike, and CJ state properties are particularly difficult to measure.

Based on the ZND detonation theory (Sec. 7.3.5), if the unreacted explosive and the reaction-product Hugoniots are well known, the von Neumann spike and CJ point can be determined. In practice, the unreacted Hugoniot is not well known at higher pressures because accurate measurements are not possible; high-pressure input shocks cause prompt reaction. This makes an extrapolation from the lower-pressure data necessary to obtain the higher-pressure portions of the unreacted Hugoniot. Similar measurement problems are encountered with determining the reaction-product Hugoniot. Estimates are usually obtained via a 2D reactive computer code which is used to do parameter variation studies, thereby calibrating a particular EOS form to experimental data. Because of this and difficulties in making direct measurements, reaction-zone parameters (e.g., spike and CJ pressures and reaction-zone length) are not sharply defined for any explosive. In contrast, detonation velocity and shock sensitivity are relatively easily measured.

Nearly all explosive properties depend on an explosive's mass density, so it must be accurately known to properly specify an experiment. In addition, some properties, such as shock sensitivity, are dependent on the explosive's particle-size distribution; this distribution may be altered during processing, e.g., by pressing.

Blasting agents, which are composite mixtures of ammonium nitrate, fuel, gelling agents, metals, etc., are difficult to define in a general sense since they are usually formulated for a particular situation and application. Because of this, we have chosen to discuss only molecular explosives.

7.7.1. Chemical Explosive Classifications

Explosives are usually classified according to three broad sensitivity classifications: primary, secondary, and insensitive. Primary explosives can be initiated with rather small inputs of mechanical shock, heat, flame, or spark and are used in initiators to start an explosive train. Examples of primary explosives are lead azide, lead styphnate, mercury fulminate, and tetrazene. Small amounts of these materials are located in initiators; electrical or mechanical input will reliably initiate them. They, in turn, reliably initiate the output charges, e.g., secondary explosive booster pellets, gun powder, etc. Extreme care is required when handling primary explosives to prevent accidental initiation.

Secondary explosives are considerably less sensitive to the same inputs and can be handled with less concern about accidental initiation. These materials are reliably initiated by moderate- to high-pressure shock inputs. They are used in high input initiating devices (exploding bridgewire or slapper detonators), output pellets and boosters, and main charges in explosive assemblies. Examples are TNT, HMX, pentaerythritol tetranitrate (PETN), cyclo-1,3,5-trimethylene-2,4,6-trinitramine (RDX), and trinitro-2,4,6-phenylmethylnitramine (Tetryl). Hundreds of explosives fit in this category.

Insensitive explosives are difficult to initiate and shock inputs for reliable initiation must be quite large; i.e., careful planning is usually required to eliminate failures, particularly when there can be environment changes, e.g., low ambient temperature. Examples of materials in this category are 1,3,5-triamino-2,4,6-trinitrobenzene (TATB), nitroguanidine (NQ), ammonium nitrate (AN), ANFO, and many AN-based slurry and emulsion blasting agents.

Some explosives are borderline and could be placed in either of two categories, depending on the particular sensitivity property being used as the criterion. For instance, nitroglycerin might be put into the secondary explosive category because it requires a rather high-level shock input to initiate it in the absence of heterogeneities. However, when these are present (which is almost always the case, because of NG's relatively high viscosity), it becomes very shock sensitive, so it is generally classified as a primary explosive.

7.7.2. Chemical and Detonation Properties for Selected Explosives

Chemical and physical properties of a selected group of molecular explosives (some from each sensitivity class) are presented in Table 7.1. Most of these are organic molecules containing nitro groups (NO_2). All of the materials produce gases upon reaction; i.e., the product Hugoniot lies above the unreacted Hugoniot (see Fig. 7.5). (Other organic materials react under the action of shock waves, e.g., carbon disulfide and acrylonitrile, but since the reaction products are more dense than the reactants; they would not be expected to detonate [66].) Product gases are typically N_2, CO, CO_2, and H_2O; very little is known about the reaction mechanisms by which the explosive is transformed into reaction products. Also, very little is known about how the chemical reactions (and gas production) feed energy to the detonation front to sustain the leading shock wave.

In Table 7.1, the heat of formation is the difference in enthalpy between the molecule and its elements (at their reference state), and the heat of detonation is the enthalpy difference between the explosive molecule and the product molecules that result from the reaction. Comparison of the heat of detonation gives a relative measure of the expected output; higher heats of detonation correspond to more energetic explosives, as is the case for HMX, RDX, and PETN. (Data for Table 7.1 were taken from Refs. 67 and 68.)

Research on the synthesis of new explosive materials continues. A current emphasis of this work is in developing explosive molecules that contain more nitrogen atoms. A pure nitrogen-containing molecule, N_8 cubane, has been theoretically studied by Engelke and Stine [69] and found to have potential properties that would make it a very important explosive (estimated detonation velocity of ≈14 km/s and CJ pressure of ≈130 GPa). However, this molecule has not been synthesized and questions exist about its kinetic stability. Recent theoretical and chemical synthesis work is aimed at understanding this and other high-nitrogen compounds.

Table 7.2 is a tabulation of some of the detonation properties for the explosives listed in Table 7.1. (Data were taken from Refs. 67 and 70). Densities are given for each entry because the values are density

Table 7.1. Chemical properties of selected molecular explosives

Common name	Chemical name[a]	Chemical formula	Molecular weight	Theoretical density (g/cm^3)	Melting point (°C)	Heat of formation, ΔH_f (kcal/mol)	Heat of detonation[b] ΔH_{det} (kcal/g)
Lead azide	Lead azide	$Pb(N_3)_2$	291.3	4.8	dec.[c]	+112	0.37
NG	Nitroglycerin or glycerol trinitrate	$C_3H_5N_3O_9$	227.1	1.60	13.2	−88.6	1.48
PETN	Pentaerythritol tetranitrate	$C_5H_8N_4O_{12}$	316.2	1.78	140	−128.7	1.51
Tetryl	Trinitro-2,4,6-phenyl-methyl-nitramine	$C_7H_5N_5O_8$	287.0	1.73	130	+4.67	1.45
RDX	Cyclo-1,3,5-trimethylene-2,4,6-trinitramine	$C_3H_6N_6O_6$	222.1	1.81	205[d]	+14.7	1.48
HMX	Cyclotetramethylene-tetranitramine	$C_4H_8N_8O_8$	296.2	1.90	285	+17.9	1.48
HNS	Hexanitrostilbene	$C_{14}H_6N_6O_{12}$	450.3	1.74	316[d]	+18.7	1.36
TNT	2,4,6-trinitrotoluene	$C_7H_5N_3O_6$	227.1	1.65	80.9	−16.0	1.29
NM	Nitromethane	CH_3NO_2	61.0	1.13	−29	−27	1.36
AN	Ammonium nitrate	NH_4NO_3	80.7	1.73	169	−87.3	0.38
TATB	1,3,5-triamino-2,4,6-trinitro-benzene	$C_6H_6N_6O_6$	258.2	1.94	dec.[c]	−36.9	1.08
NQ	Nitroguanidine	$CH_4N_4O_2$	104.1	1.78	257	−22.1	0.88

Note: Selected data from Refs. 67 and 68.

[a] Common chemical names are used rather than the names used in the *Chemical Abstract Index Guide* (see Ref. 67).
[b] Heats of detonation are calculated (rather than experimental) and assume H_2O is a gas.
[c] Decomposes before melting.
[d] Melts with some decomposition.

Table 7.2. Detonation properties of selected molecular explosives (densities are in parentheses)

Material	Theoretical density (g/cm^3)	Detonation velocity (mm/μs)	Estimated CJ pressure (GPa)	Estimated spike pressure (GPa)	Estimated reaction-zone length (mm)	Failure radius (mm)	Plate-dent depth[a] (mm)
Lead azide	4.80	5.5 (3.80)	23 (4.0)	—	—	—	—
NG	1.60[b]	7.7	25	—	0.2	1.1[c]	—
PETN	1.78	8.26 (1.76)	30 (1.67)	—	—	—	9.8 (1.66)
Tetryl	1.73	7.85 (1.71)	26 (1.71)	—	—	—	8.1 (1.68)
RDX	1.81	8.7 (1.77)	34 (1.77)	—	—	—	10.4 (1.75)
HMX	1.90	9.11 (1.89)	39 (1.89)	—	—	—	10.1 (1.73)
HNS	1.74	7.00 (1.70)	23 (1.74)	—	—	—	—
TNT	1.65	6.93 (1.64)	21 (1.63)	28 (1.64)	0.6 (1.64)	7.5 (1.62)[d]	7.0 (1.63)
NM	1.13[b]	6.35	13	20	≈0.1[e]	7.6[c]	4.1
AN	1.73	5.27 (1.30)	22 (1.73)	—	—	> 50 (1.4)[c]	—
TATB	1.94	7.76 (1.88)	29 (1.88)[f]	38 (1.88)[f]	2.1 (1.88)[f]	4.5 (1.88)[d,f]	8.3 (1.87)
NQ	1.78	7.93 (1.62)	25 (1.63)	—	—	—	—

Note: Data from Refs. 67 and 70.
[a] Charge diameter = 41.3 mm; cold rolled steel dent plates.
[b] Liquid.
[c] Confined in glass.
[d] Unconfined or paper tube confinement.
[e] Experiments are underway to define this more accurately (see Fig. 7.7).
[f] PBX 9502 which is 95/5 wt% TATB/Kel-F.

dependent. Detonation velocity is the best measured detonation parameter so there are considerable data available, including initial density and temperature dependence. High detonation velocities are associated with high outputs (higher CJ state properties) which is apparent in the table. The CJ pressure is more difficult to measure and, hence, these values are less reliable. The von Neumann spike and reaction-zone length data are only estimated for a few materials because of the measurement difficulties. Reaction-zone estimates were made by the authors, based on the literature. Improved measurements are now being made by laser velocity interferometry and in situ gauge techniques (see Sec. 7.8.8), so better estimates are forthcoming. TATB has one of the longer reaction zones at about 2 mm and NM has one of the shorter ones at about 0.1 mm. It is interesting to note that reaction-zone estimates for nitromethane can be found in the literature that range from approximately 0.3 to 600 μm.

Explosives are often mixed with polymeric materials (to make them easier to process—cast, press, mold, machine, etc.) and then are called plastic-bonded explosives (PBX). Examples of this are i) PBX 9404, which has a composition of 94/3/3 wt% of HMX, nitrocellulose, and chloroethylphosphate and ii) PBX 9502, which has a composition of 95/5 wt% TATB and Kel-F 800. Many PBX materials are quite strong, easy to machine, and, therefore, easy to build into experimental assemblies to make precise measurements. For this and other reasons, a relatively large amount of data are available on some of these materials.

Many detonation properties depend on initial temperature; e.g., liquid TNT has a failure diameter of 63 mm at an initial temperature of 80°C and 4 mm at 240°C (see Ref. 71). When relying on published data, it is always important to know the initial temperature at which the experiments were done.

7.7.3. Sensitivity Properties

Explosive materials vary widely in their sensitivity to different inputs. Because experience has shown it necessary to understand explosive hazards, a rather large number of sensitivity tests have been developed (some are peculiar to a particular operation or hazard). Several test methods are listed in Table 7.3, categorized by input types, e.g., shock, friction and crushing, heat, and spark or static electricity.

Gap tests, of which there are several types, are widely used as a measure of shock sensitivity. In this test, a standard explosive driver system (donor) produces a shock that is attenuated to different levels

Table 7.3. Explosive sensitivity tests

Type of experiment	Sensitivity property	Measurement method[a]	Sensitivity indicator	Less sensitive explosive indicated by[b]
			Shock input tests	
Gap test	Shock	G,P	Attenuator thickness	Small thickness
Minimum priming charge	Shock	G,P	Primer charge mass	Larger primer charge
Plate impact (critical $P^n\tau$ value)	Input shock parameter, such as $P^n\tau$	G,P	Value of shock parameter ($P^n\tau$); depends on flyer material, velocity, thickness	Larger input values ($P^n\tau$); higher-impedance materials, thicker plates, higher velocity
			Shock, friction, and crushing input tests	
Drop weight (various anvils and surfaces used)	Impact (crushing)/friction	G,A,P	Drop height for given weight to reaction	Greater height for a given weight
Skid test (angles and height varied)	Impact/friction	G,A,P	Sample height & surface angle	Greater height
			Heating tests	
Differential thermal analysis	Thermal input	C	Endotherm, exotherm temperature	Higher exotherm temperature
Time-to-explosion testing	Thermal input	T	Time to explosion in hot metal bath or hot metal enclosure	Longer times to reaction
			Static electricity input tests	
Spark test	Spark	G,A,P	Spark energy to reaction	Higher spark energy

Note: Tests representative of those done to characterize explosives; see Refs. 67 and 70.

[a] Methods to measure reaction: A : active measurement of noise, etc., indicating reaction. G : go–no go test.
P : postmortem of debris from test. C : differential temperature monitoring of endotherms and exotherms for constant heating rate.
T : time to explosion measured after sample is put in a hot metal environment.

[b] Less sensitive means the test parameter condition where an explosive is relatively safer.

by varying the thickness of an inert material between the donor and the sample explosive. The thickness is varied until the attenuator thickness at which the explosive sample is just initiated is determined. This is the gap thickness reported. A large gap indicates a more sensitive material, since, with a larger amount of the inert material present, the input pressure to the explosive sample is attenuated to a lower value.

The minimum priming charge test is similar, except the explosive-driver mass is varied and there is no attenuator. In flyer-plate impact experiments, the flyer material and its thickness and velocity can be varied (changing the input pressure pulse) to determine a critical initiation input parameter, such as $P^n\tau = K$. The quantity K has been shown to be relatively constant over a range of input pressures and pulse widths for a particular explosive (see Sec. 7.8.5).

Impact sensitivity is important because impacts are the type of input that occur in many accident situations. Drop weight setups (of which there are several with different drop heights, weights, anvils, and anvil surfaces) are used to determine impact sensitivity. Samples are usually in the form of powders or pellets. The samples are placed on an anvil and then the drop weight is allowed to fall onto the sample. The minimum height of free fall that causes a reaction is the value reported. Evidence of reaction is determined by both active measurement and postmortem examination of the debris from the test.

In the skid test, a piece of explosive is subjected to a particular drop environment in order to simulate dropping a large charge. The angle of the surface (as well as the character of the surface itself) on which the explosive piece is dropped can be modified to vary the surface interaction. Evidence of reaction is determined by both active measurement and postmortem observations.

Heating tests indicate the sensitivity to thermal input. A DTA test involves a continuous comparison of an explosive sample's temperature with that of an inert standard when both are heated at a constant rate. Using this method, energy excursions can be determined, i.e., the temperatures at which there are endotherms and exotherms. In the time-to-explosion test, an explosive sample is confined (so product gases cannot escape) and subjected to a constant temperature metal bath or metal enclosure with the time to explosion determined. Using the thermal explosion analysis developed by Frank-Kamenetskii [44], Arrhenius rate parameters can be determined.

Sensitivity to spark input (such as might be produced by release of a static charge on the human body) is of interest because of the possibility of electrical initiation of an explosive material during handling. Critical-spark-energy values from this test are test specific; i.e., there are several different test setups.

Because threshold reaction conditions are being measured, probabilistic data come from many of the above tests. Therefore, it is necessary to run several tests at each input condition to establish statistically the lowest input reaction condition.

Sensitivity information from some of the above tests is given in Table 7.4 for selected molecular explosives. Information about the particular test setup that was used to obtain the data is given in the table footnotes. A detailed discussion of these tests and references to more information relating to each test are given in Refs. 67, 70, and 72. Large differences in sensitivity to the various inputs are apparent between lead azide (a primary explosive) and the secondary materials (even PETN—which is considered one of the more sensitive secondaries). Differences between the secondary and insensitive explosives are less apparent. The order of the materials in the table (from top to bottom) is from more to less sensitive, but there are order reversals for some tests. This indicates that, depending on material properties, different explosives can be more or less sensitive to certain inputs. TATB is quite obviously a very insensitive explosive.

A notable exception to the ordering is nitromethane, which is a liquid. It initiates homogeneously to shock inputs (see Sec. 7.5.1) while all the other materials are heterogeneous and initiate heterogeneously (see Sec. 7.5.2); this applies to both the critical energy and gap test data. In addition, the drop test is much more difficult to do with liquids because the normal crushing and friction which occurs with solid explosive particles is not present. Data on nitromethane have been included to demonstrate these differences.

7.8. Initiation and Detonation Measurement Techniques

Because of the harsh environment produced by the detonation and initiation processes, measurements of certain phenomena are extremely difficult, e.g., measurement of the detailed chemistry occurring in an initiation or detonation. However, considerable progress has been made in measuring the mechanical aspects of initiation and detonation phenomena. A wide variety of experimental techniques is employed to make such measurements, including various gauging techniques, laser velocity interferometry, streak cameras, framing cameras, flash x-ray, ionization, and piezoelectric pins that measure

Table 7.4. Sensitivity of selected high explosives (densities are in parentheses when appropriate)

Material	Theoretical density (g/cm³)	Gap test[a] (mm)	Gap test[b] (mm)	Drop weight[c] (m)	Spark input[d] (J)	DTA exotherm (°C)	Critical energy[e] (cal/cm²)
			Primary Explosive				
Lead azide	4.80	27.2 (3.66)	—	—	0.007	—	0.03 (1.26)
			Secondary explosives				
PETN	1.78	14.4 (1.58)	5.21 (1.76)	0.11	0.19	190	4. (1.6)
RDX	1.81	7.9 (1.72)	5.18 (1.74)	0.28	0.21	200	~16(1.55)
HMX	1.90	8.7 (1.81)	4.27 (1.79)	0.33	0.23	260	—
TNT	1.65	6.25 (1.56)	0.33 (1.63)	0.80	0.46	250	34 (1.64)
NM	1.13	—	0.18–0.43 (1.13)	>3.2[f]	—	—	405 (1.13)
			Insensitive explosives				
TATB	1.94	1.12 (1.88)	0.13 (1.87)	>1.77	4.25	340	72–88 (1.76)
NQ	1.78	2.72 (1.27)	no-go (1.58)	>1.77	—	240	—

Note: Data from Refs. 67, 70, and 72.

[a] NSWC (Naval Surface Warfare Center) Gap Test.

[b] Los Alamos (LANL) small-scale gap test.

[c] Performed using a 5-kg weight and Type 12 tooling (sandpaper surfaces).

[d] Highest electrostatic-discharge energy at 5000 V without ignition; sample confined.

[e] Critical energy is defined as $P^2 \tau / \rho_0 U_s$ in Ref. 67.

[f] Performed using a 2.5-kg weight and Type 12 tooling (sandpaper surfaces), modified tooling for liquid.

time-of-arrival of a shock or material surface, photodiode light measurement, various spectroscopic techniques, and integrated output tests such as plate-dent testing. Table 7.5 lists a number of experimental methods used to measure various attributes involved in initiating and detonating explosives. Although this listing is not exhaustive, it indicates the wide variety of techniques being used. Most of the tests provide time-resolved data relating directly to the physics of the processes occurring. However, a few are integrated tests used mostly for comparison purposes.

Streak camera wedge and interface measurements, framing camera visualizations, ionization pins, and plate-dent output tests are used extensively to provide the major part of the detonation physics and explosive output properties data available. In situ gauging experiments provide a great deal of information per experiment, but are complicated to perform. Velocity interferometry has become more important because of the high time resolution with which interface particle-velocity histories are measured. Photodiode and spectroscopy measurements are becoming more important because of the unique information they can provide, e.g., chemical species data and temperature.

The measurement methods of Table 7.5 have been incorporated into Table 7.6 to indicate the methods used to obtain particular initiation/detonation information. Articles in the various Detonation Symposia [73] contain more detailed discussions of these methods.

Wave profiles can be measured by in situ particle velocity or pressure gauges; however, one should realize that gauge presence can be perturbing, particularly on the time and space scales present in the reaction. Interface velocimetry measurements of the particle velocity at explosive–window interfaces provide the highest-time-resolution reaction-zone data. Initiation and detonation shock-shape experiments, coupled with computer modeling, also provide estimates of global reaction-rate parameters. These methods are routinely used in energetic material research as discussed below.

In situ gauging (i.e., low-perturbation gauges located in the flow field) is becoming important. These gauges give direct information on wave profiles in the flow and contain implicit information on how the chemical heat release is occurring. The particle-velocity waveforms shown in Figs. 7.10 and 7.12 are examples of in situ gauging measurements in which the initiation process has been measured in nitromethane-based materials. These measurements provided the insight necessary to modify the long-held homogeneous-initiation model as discussed in Sec. 7.5.1. In situ manganin-gauge pressure measurements are also providing important insights into these processes.

An example of in situ gauging is a magnetic particle velocity gauging experiment in which the particle velocity of an initiating explosive is measured. These experiments are becoming more important to provide the data used in developing and calibrating global reaction-rate models. Figure 7.18 shows how such an experiment (done with a gas gun) using a solid explosive sample is put together. A gauge membrane 60 μm thick is carefully epoxied between two precisely machined pieces of explosive. It is important to make sure the gauge ends are parallel with the top of the sample. The gauge is usually installed at an angle of 30 degrees. A cut is then taken off the top of the glued assembly to ensure a smooth flat surface at the projectile–target impact face. The target is carefully placed in the magnetic field so that the gauge ends are perpendicular to the field lines. After impact, the shock moves past each gauge and a voltage signal proportional to the gauge length, magnetic field strength, and gauge speed vs. time is produced; this voltage is recorded by a digitizer. Figure 7.13 shows data recently obtained in an experiment on PBX 9501. The large amount of information contained in a single experiment is obvious.

Laser velocity–interferometry measurements also provide wave-profile information as an initiation or detonation wave accelerates an explosive–window material interface from which a laser beam is reflected. An important advantage (e.g., over in situ gauging) is that extremely high-time-resolution measurements are possible even at very high pressure. A disadvantage is that the wave profiles are perturbed by the mechanical impedance mismatch at the window through which the observations are made. This is complicated by the fact that the impedance of the unreacted explosive is different from that of the reaction products so it is impossible to have a window match the impedance of the reacting explosive through the entire reaction process. There are several types of laser interferometry used to make measurements in both inert and energetic materials. These are VISAR [75], Fabry–Perot [76], and ORVIS [77]; each of them have undergone improvements. A notable improvement in VISAR due to Hemsing was the push–pull VISAR system [78]. Time resolution of streak VISAR [79] and ORVIS can be as good as 200–300 ps. VISAR, with photomultiplier tubes and digitizers for recording the fringes, has a time resolution of 1–2 ns. Goosman et al. [80] have recently described a system involving 10 Fabry–Perot interferometers that can be used to probe a surface at 10 different places during a single experiment.

All these interferometers (VISAR, ORVIS, and Fabry–Perot) operate on the same principle; laser light (usually 514.5 nm wavelength) is Doppler shifted by a moving surface. This Doppler-shifted

Table 7.5. Experimental methods used to study initiation and detonation [a]

Type of experiment (designation[b])	Property measured	Location of measurement	Comments
	Gauging techniques		
Manganin gauge (G-M)	pressure	in situ	thick gauge package
Electromagnetic gauge (G-E)	particle velocity	in situ	nonconductors required
Quartz gauge (G-Q)	stress	interface	4 GPa limit
PVDF gauge (G-PV)	stress rate	in situ	10 GPa limit
	Velocity interferometry		
VISAR/digitizers (I-VD)	particle velocity	interface	1–3 ns resolution
VISAR/streak camera (I-VS)	particle velocity	interface	0.2 ns resolution
Fabry-Perot (I-FP)	particle velocity	interface	8–10 ns resolution
ORVIS (I-O)	particle velocity	interface	0.3 ns resolution
	Streak camera record (rotating mirror or electronic camera)		
Wedge test (SC-W)	1st wave arrival	interface	requires light source
Cylinder test (SC-C)	wall velocity vs. time	wall position	requires back light
Wave/interface by light generation (flashers) (SC-F)	1st wave arrival	interface	requires flashers or mirror destruction
	Framing camera records		
Rotating mirror camera (F-RMC)	position/light	2D area	10s of pictures, 200 ns resolution
Electronic camera (F-E)	position/light	2D area	up to 16 pictures, 50 ns time resolution, poor spatial resolution
Image-converter camera (F-IC)	position/light	2D area	one picture, 10–20 ns resolution, good spatial resolution, (multiple exposure)
	Flash x-ray records		
Single head (X-1)	density variation	thru expt.	one picture, poor spatial resolution
Multiple head (X-M)	density variation	thru expt.	multiple exposure, measure motion

Table 7.5. (cont.) Experimental methods used to study initiation and detonation[a]

Type of experiment (designation[b])	Property measured	Location of measurement	Comments
	Emitted light history		
Photodiodes unfiltered (P-UF)	light output	line of sight/ interface	few ns resolution
Several photodiodes filtered (P-F)	light output at filtered freq.	line of sight/ interface	few ns resolution
	Spectroscopy techniques		
Absorption (S-A)	intensity vs. wavelength	line of sight	absorbed spectrum, 10–20 ns resolution
Emission (S-E)	intensity vs. wavelength	line of sight	emitted spectrum, 10–20 ns resolution
Spontaneous Raman (S-R)	intensity vs. Raman shift	line of sight	weak signal; 20 ns resolution
Spontaneous Raman–Stokes/anti-Stokes (S-S)	intensity vs. Raman shift	line of sight	weak signal; 20 ns resolution
Coherent anti-Stokes Raman (S-CARS)	intensity vs. Raman shift	in situ, point	analysis is difficult; few ns resolution
Time resolved mass spectra (S-M)	intensity vs. mass	samples a small area	samples chemical species vs. space and time; ~10 ns resolution
	Time of wave arrival techniques		
Ionization (shorting) pins and switches (A-I)	1st wave arrival	interface, point	few ns resolution.
Piezoelectric pins (A-PZ)	1st wave arrival	interface, point	rough pressure measurement
Continuous resistance wire (A-CW)	1st wave motion	interface, line	low resolution
	Integrated explosive output tests		
Plate dent (I-PD)	deformation	interface	plate material dependent
Lead block (I-LB)	volume change	inside block	standardized test
Ballistic pendulum (I-BP)	arm swing	—	standardized test
Upsetting test (I-U)	deformation	—	standardized test

[a] Information relating to most of these tests can be found in Refs. 10, 12, 15, 16, 67, 68, 70, and 73,

[b] These designations will be used in Table 7.6.

Table 7.6. Initiation and detonation measurement experiments

Attribute measured	Method	Specific information obtained	Comments
Detonation profile (1-D) Particle velocity Pressure	 G-E, I-VD,-VS,-FP,-O G-M	Estimate of von Neumann spike, reaction time and/or wave shape.	Change in slope at CJ point not apparent (Fig. 7.7).
2D detonation shock shape (from arrival time)	SC-F,X-1	2D wave shape, reaction information.	Requires modeling to obtain global reaction data.
1D initiation/growth profile In situ gauging Interface gauging Emitted light	 G-M,-E I-VD,-VS,-FP P-UF	Unreacted Hugoniot data, type of growth,where growth occurs, global reaction information.	Requires modeling to obtain global reaction rate (Figs. 7.10, 7.12, 7.13).
Run distance- or time-to-detonation	SC-W	Run distance and time to detonation vs. input pressure.	See data plotted in Fig. 7.20.
Shock/detonation velocity	SC-W, X-M, A-I,-PZ,-CW	Unreacted Hugoniot data, detonation velocity, diameter-effect curves.	Velocity often accurate to a few m/s (Figs. 7.14, 7.16).
Detonation product EOS	SC-C, I-VD,-FP	Cylinder wall expansion history at one or more positions.	Calibrates reaction-product EOS, requires modeling.
Detonation product EOS (overdriven detonation)	P-UF, SC-F	Sound speed at supported pressure.	Sound speed is a differential quantity for product EOS.

Table 7.6. (cont.) Initiation and detonation measurement experiments

Attribute measured	Method	Specific information obtained	Comments
Detonation failure diameter	I-PD (postmortem), A-CW	Failure diameter for steady detonation in a specific confinement.	Critical diameter related to reaction parameters.
Detonation corner turning	SC-F, F-IC, X-1, X-M	Detonation wave front spreading ability.	Related to reaction parameters.
Detonation/initiation overall event visualization	F-RMC, -E, -IC	Several pictures of the detonation wave interacting with it confines.	Helpful in understanding entire event (Figs. 7.1 and 7.2).
Detonation output	I-PD, -LB, -BP, -U	Relative measure of output.	Related to CJ properties.
Initiation/detonation chemistry	S-A, -E, -R, -S, -CARS, -M	Time-resolved chemical spectra indicating chemistry occurring.	Difficult measurements, optical opacity a problem.
Initiation/detonation temperature	P-F, S-S	Intensity of certain lines; Stokes and anti-Stokes Raman shift data.	P-F requires emissivity assumptions.

Note: Experimental methods are defined in Table 7.5.

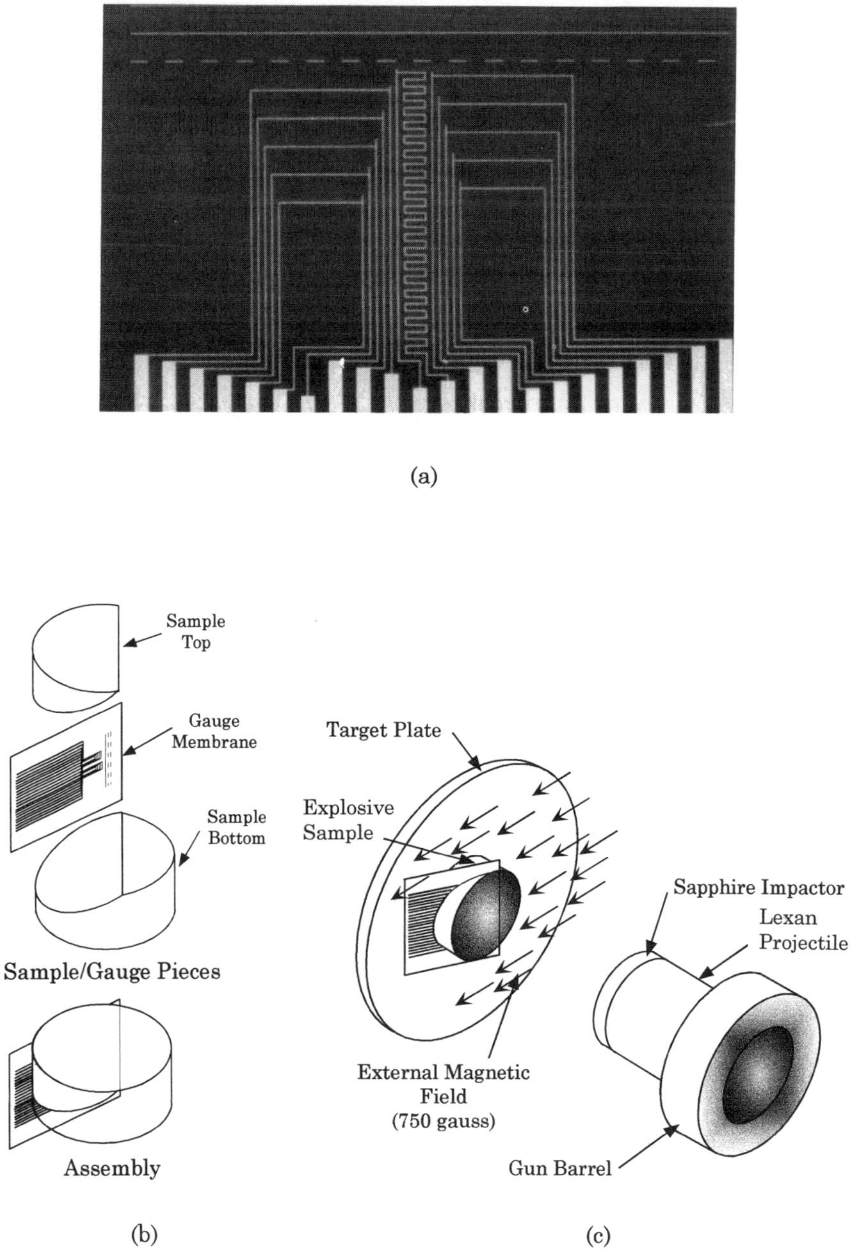

Figure 7.18. Various aspects of a magnetic particle velocity gauge experiment in an explosive sample. (a) shows the gauge pattern—10 particle velocity gauges and a shock tracker in the center. (b) shows how the gauge is installed in the sample. (c) shows how the target is situated in the gun. The technique was developed by Vorthman and Wackerle [74].

light is routed through an optics system that beats Doppler-shifted light at one time with that Doppler shifted at a later time. This produces fringes, the number of which is directly proportional to the velocity change of the surface. The time delay between the two Doppler-shifted beams can be controlled and this, in turn, controls the velocity per fringe of the instrument (fringe constants are typically between 0.1 and 1.8 km/s/fringe). The fringe constant is usually set to record an expected surface velocity as a several-fringe change. Because shock jumps are so fast that fringes cannot be tracked by the instrument (one or more fringes can be lost), it is often advantageous to use two instruments with different fringe constants to aid in data reduction.

Figure 7.19 depicts a modern VISAR system. Data shown in Fig. 7.7 were obtained using a similar VISAR system. The data in this figure show the measured detonation front in nitromethane [81]. This experiment consisted of a glass tube 24.5-mm inside diameter by 127-mm long filled with nitromethane. It was initiated by detonating PBX 9404 pellets which, in turn, were initiated by projectile impact. The detonation wave in the nitromethane interacted with a PMMA window having a very thin (submicron thick) layer of aluminum between the nitromethane and PMMA. PMMA is a good mechanical impedance match with nitromethane, so perturbations of the detonation wave profile were small. These data provided the reaction-zone-length estimate for nitromethane in Table 7.2. Similar measurements using an ORVIS system [82] provided the reaction-zone estimates for TATB and TNT also included in Table 7.2.

Detonation velocity is precisely measured in rate-stick experiments as shown in Fig. 7.14. Ionization switches (in some cases consisting of 2-mil-diameter copper wires) are installed between precisely measured pieces of the explosive and provide arrival time information from which the detonation velocity is obtained.

Diameter-effect curves (see Sec. 7.6 and Fig. 7.16) are generated when the charge diameter is varied in a set of rate-stick experiments. These measurements become more difficult when the initial temperature is changed from ambient because of the different thermal expansion (or contraction) of the various parts. Nevertheless, accurate velocity measurements are made under these conditions.

Wedge experiments yield run distance- or time-to-detonation data from sustained shock input experiments. A precisely machined explosive wedge is mounted on a plane wave explosive driver–attenuator system as shown in Fig. 7.20a. The wedge is illuminated by a light source (usually explosively-compressed argon) and a streak camera is used to monitor thelight changes as a function of time along

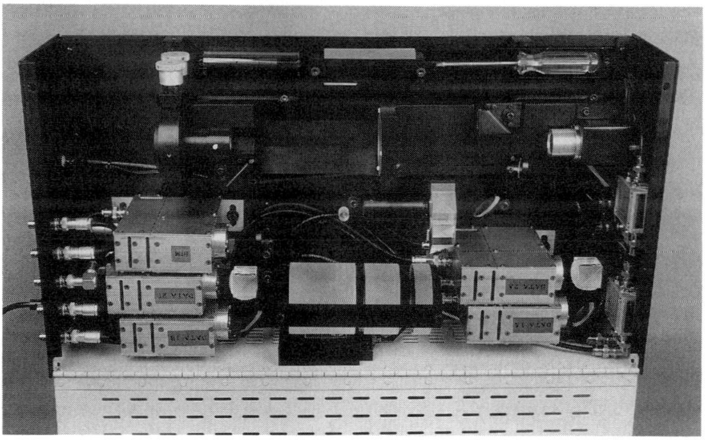

(a)

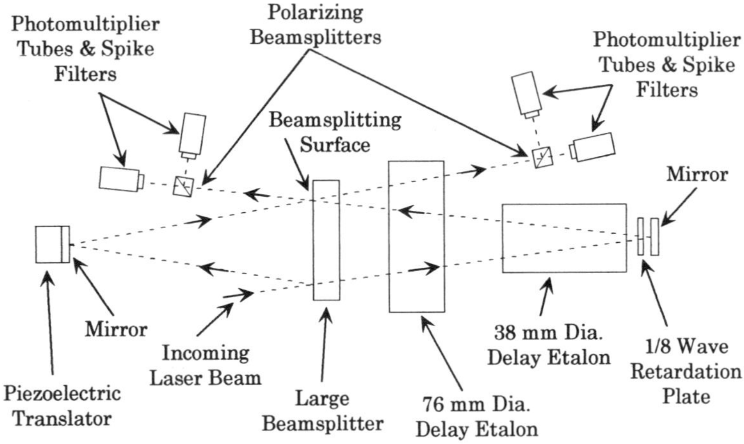

(b)

Figure 7.19. Modern VISAR system setup. (a) is a photograph of a Valyn VISAR system; (b) is a schema of the optics involved in a Hemsing-type push-pull VISAR system. (Photograph and schema courtesy of Lynn Barker, Valyn International.)

the wedge surface as the shock interacts with the surface; such a trace is shown in Fig. 7.20b. The point at which the distance–time streak record has a large change in slope is the distance- or time-to-detonation for that particular pressure input. This is clearly shown in Fig. 7.20c in which the slope (shock velocity) is plotted versus distance of shock travel into the wedge. By doing several experiments, in which the input shock pressure is varied, a set of pressure versus distance- or time-to-detonation data is compiled. A plot of these data in the log(pressure) vs. log(run distance) plane is typically a straight line [83], as shown in Fig. 7.21. Note that for the explosives on this plot, PETN is the most sensitive and TATB and NQ are the least sensitive materials.

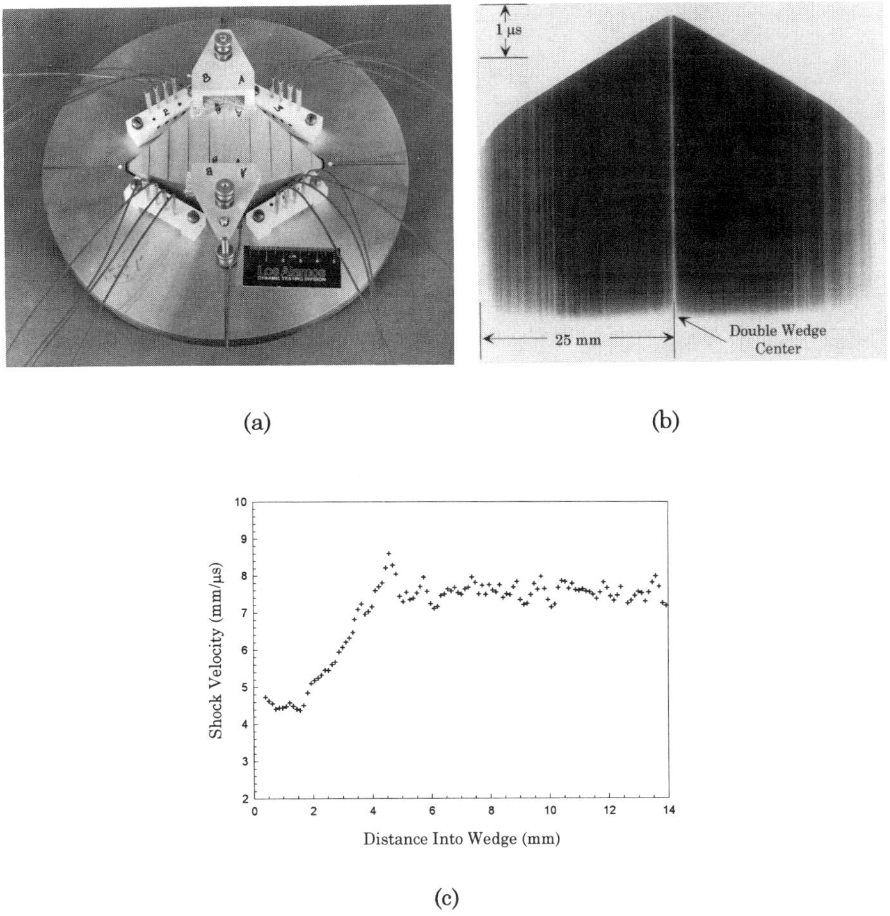

(a) (b) (c)

Figure 7.20. (a) The configuration for an explosively driven wedge experiment. The 16 wires are for electrical pins to measure the input to the explosive wedge. (b) A streak-camera record showing shock contact with the wedge surface. Shock contact reduces the reflectivity and therefore the light intensity. The dark area in the streak record photograph is the region of high reflectivity. (c) The shock velocity plotted versus distance into the wedge obtained from the space-time trajectory of the shock contact with the wedge surface. The distance-to-detonation point is about 4 mm.

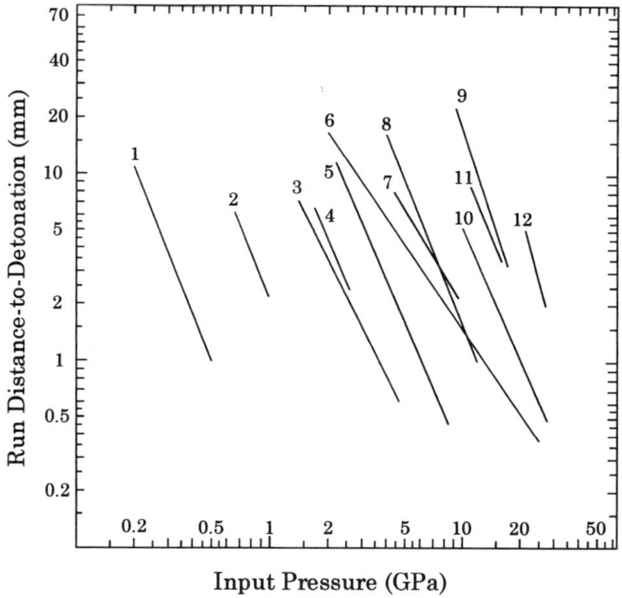

Figure 7.21. Plot of sustained input shock pressure vs. run distance-to-detonation (in the log-log plane) as observed from wedge tests, for a number of explosives. Explosives (with densities in g/cm³) are (1) PETN ($\rho_0 = 1.0$); (2) PETN ($\rho_0 = 1.4$); (3) PBX 9407 (94/6 wt % RDX/Exon, $\rho_0 = 1.6$); (4) PETN ($\rho_0 = 1.75$); (5) Tetryl ($\rho_0 = 1.70$); (6) PBX 9404 (94/3/3 wt% HMX/nitrocellulose/chloroethylphosphate $\rho_0 = 1.84$); (7) HMX ($\rho_0 = 1.891$); (8) pressed TNT ($\rho_0 = 1.63$); (9) cast TNT ($\rho_0 = 1.635$); (10) superfine particle TATB ($\rho_0 = 1.806$); (11) TATB ($\rho_0 = 1.876$); (12) NQ ($\rho_0 = 1.688$). Data were taken from Ref. 70.

Rotating-mirror framing cameras and electronic image-converter cameras are used to make stop-action pictures of detonation events. Exposure times range from microseconds to nanoseconds, depending on the camera used. For example, Fig. 7.2 was taken using a rotating-mirror framing camera with the frames 5 μs apart (the two frames shown are 10 μs apart). This figure shows the progress of the detonation wave along a charge contained in copper and expanding in air. Estimates of the reaction-product Hugoniot of the explosive can be obtained from this type of data by calibrating computer models to the experiment. A related test involves the measurement of the wall expansion of a precisely machined and explosively loaded copper cylinder when the explosive is detonated. This experiment is conducted with a streak camera and is called a *cylinder test*. Information obtained from this test is also used for calibrating reaction-product Hugoniots (see Sec. 7.4.4.5).

Failure radii for propagation of steady detonation waves are determined by making a series of shots at progressively smaller charge radius until detonation failure occurs. Failure is determined by the condition of dent blocks or postshot examination of the debris. The failure radius is reduced by heavier confinement; e.g., NM has a failure diameter of $\simeq$ 16 mm in Pyrex glass and $\simeq$ 2.5 mm in brass.

Integrated output from a detonating charge is useful in understanding the ability of an explosive to do work. Higher detonation velocities and pressures are not always as important as the gas volume produced in determining output. This is particularly true of blasting explosives.

Other measurement methods listed in Tables 7.5 and 7.6 are discussed in articles in the various Detonation Symposia [73].

7.9. Summary

Developments in condensed-phase explosive technology in the last 150 years have led to substantial increases in understanding of the initiation and detonation process. A planar steady detonation is a chemical-reaction-supported shock. The shock takes the unreacted material to a high pressure (the von Neumann spike); this is followed by decreasing pressure in the chemical-reaction zone in which the explosive is transformed into gaseous reaction products. In the simplest case, the chemical-reaction zone terminates at a sonic point (the CJ state). In situ and interface velocity measurements of the reaction zone support this picture of 1D steady planar detonation theory. Spectroscopic measurements that attempt to interrogate the reaction-zone chemistry are a current area of research.

Explosives are classified according to sensitivity as primary (the most sensitive), secondary (which have intermediate sensitivity), and insensitive explosives (the least sensitive). Sensitivity is measured by a number of well-defined test methods in which sensitivity to shock, impact, heat, and electrical inputs are measured.

The initiation and 2D steady detonation processes depend qualitatively on whether the explosive is homogeneous or heterogeneous. Homogeneous explosives are typically liquids or single crystals; their initiation and detonation behavior depend on the bulk temperature and pressure generated by an input shock. Heterogeneous explosives have voids, grain boundaries, etc., which perturb the flow so that spatially localized hot spots are produced; these hot spots are energy concentrations where the chemical-reaction rate is greatly increased. These regions control the initiation and detonation behavior of heterogeneous explosives.

Important areas of explosive research have been neglected in our exposition. Examples of this are i) the study of the hydrodynamic stability of detonations to perturbations, ii) studies concerning numerical solution of the partial differential equations governing multidimensional time-dependent initiation and detonation, and iii) detailed discussion of the theoretical developments in DDT as it relates to two-phase flow. Discussion of these and other important topics can be found in the more detailed discussions of detonation phenomena in Refs. 9–16 and in the Detonation Symposia [73].

We have shown that, in contrast to the intuitive feeling that explosives are disorderly, irreproducible, and chaotic processes, they can be quite orderly and are governed by the conservation equations of compressible fluid mechanics along with material constitutive relations.

Evidence has been discussed showing that there are explosives with widely varying characteristics. These include, e.g., explosives with different sensitivities to accidental insult, levels of power and pressure generation during detonation, and cost. Because of this variation, one can choose an explosive appropriate to the application. Further work is being done to increase this variety of choice. An important example of this is the study of insensitive explosives for use in nuclear weapons, the goal being to produce a situation where it is virtually impossible to produce detonation of the explosive, except by design.

7.10. Glossary

CJ (Chapman-Jouguet) Model: The simplest model of planar steady detonation.

DDT (Deflagration-to-Detonation Transition): The mechanical compaction and reaction buildup processes an explosive goes through to make the transition from deflagration to detonation.

Diameter-Effect Curve: The relation between steady 2D detonation velocity and the lateral charge size.

EOS (Equation of State): The functional relationship between internal energy, pressure, molecular composition, and volume for a material in thermodynamic equilibrium.

Failure Diameter: The diameter of a cylindrical explosive charge below which one cannot propagate a steady detonation.

Heterogeneous Explosive: A material in which the release of chemical heat is controlled by local high-temperature and high-pressure regions near flow irregularities ("hot spots").

Homogeneous Explosive: A material in which the release of chemical heat is controlled by the (uniform) bulk temperature and pressure.

Hot Spots: Localized regions in a shocked explosive where energy is concentrated by fluid flow in the vicinity of physical imperfections.

Ideal Explosives: Molecular or mixture explosives in which essentially all the heat of explosion is released before the sonic surface is traversed.

LVD (Low-Velocity Detonation): A low-level reactive wave in an explosive; LVD is not a true detonation.

Molecular Explosive: An explosive material in which both the fuel and oxidizer are contained in a single molecule but are isolated by chemical bonds.

Nonideal Explosive: Mixture explosives (usually containing aluminum or other metals) in which a considerable amount of the energy release (and chemical reaction) takes place after the sonic plane.

Rankine–Hugoniot Conditions: The three mathematical conditions that guarantee conservation of mass, momentum, and energy across flow discontinuities (e.g., shocks); they also apply between any two locations in 1D steady planar flow.

SDT (Shock-to-Detonation Transition): The reaction buildup process an explosive goes through to make the transition from shock to detonation.

Superdetonation: A detonation which occurs in an explosive precompressed by an initial shock.

Taylor Wave: The pressure relief wave (rarefaction) following an unsupported planar steady detonation.

TMD (Theoretical Maximum Density): The maximum density a material would have if all voids were removed from it.

Thermal Explosion: A rapid chemical reaction that occurs when the temperature in an explosive gets above a threshold value.

von Neumann Spike: The leading (high pressure) point in the ZND model of planar detonations (sometimes referred to as the chemical peak).

ZND (Zeldovich, von Neumann, Doering) Model: The simplest model of planar steady detonation with a resolved chemical reaction zone.

Acknowledgment

Preparation of this manuscript was supported by the United States Department of Energy.

References

[1] M. Berthelot and P. Vielle, *C. R. Hebd. Sceances Acad. Sci.* **93**, p. 18 (1881).

[2] M. Berthelot and P. Vielle, *C. R. Hebd. Sceances Acad. Sci.* **94**, p. 149 and **94**, p. 882 (1882).

[3] E. Mallard and H. Le Chatelier, *C. R. Hebd. Sceances Acad. Sci.* **93**, p. 145 (1881).

[4] D.L. Chapman, *Philos. Mag.* **47**, p. 90 (1899).

[5] E. Jouguet, *J. Math. Pure Appliq.* **1**, p. 347 (1905).

[6] Ya.B. Zeldovich, *Sh. Eksp. Teor. Fiz.* **10**, p. 542 (1940). (English Translation: NACA TM 1261, 1960.)

[7] J. von Neumann, *Progress Report on the Theory of Detonation Waves,* OSRD Report No. 549 (1942); in *John von Neumann Collected Works, Vol. 6,* Pergamon Press, New York, pp. 203 (1963).

[8] W. Doering, *Ann. Phys.* **43**, p. 421 (1943).

[9] W. Fickett and W.C. Davis, *Detonation,* University of California Press, Berkeley (1979).

[10] A.N. Dremin, S.D. Savrov, V.S. Trofimov, and K.K. Shvedov, *Detonation Waves in Condensed Media* (1970). Translation from Russian, Nat. Tech. Info. Service AD-751417, Springfield, VA.

[11] R.A. Strehlow, *Combustion Fundamentals,* McGraw-Hill Book Co., New York (1984).

[12] C.H. Johansson and P.A. Persson, *Detonics of High Explosives.* Academic Press, New York (1970).

[13] Picatinny Arsenal/Large Caliber Weapons Systems Lab., *Encyclopedia of Explosives and Related Items* (ed. B.T. Fedoroff et al.), in 10 volumes, Dover, NJ (1960 to 1983).

[14] Ya.B. Zeldovich and Yu.P. Raizer, *Physics of Shock Waves and High-Temperature Hydrodynamic Phenomena,* Academic Press, New York (1966–67); Ya.B. Zeldovich and A.S. Kompaneets, *Teoriya Detonatsii (Theory of Detonations),* Gostekhizdat, Moscow (1955). (English Translation: Academic Press, New York, 1960.)

[15] P.W. Cooper, *Explosives Engineering,* VCH, New York (1996).

[16] R. Chéret, *Detonation of Condensed Explosives,* Springer-Verlag, New York (1993).

[17] M.B. Boslough and J.R. Asay, in *High-Pressure Shock Compression of Solids* (ed. J.R. Asay and M. Shahinpoor), Springer-Verlag, New York, p. 7 (1993).

[18] P.A. Thompson, *Compressible-Fluid Dynamics,* McGraw-Hill, New York, pp. 306–311 (1972).

[19] J. Rayleigh, *Proc. Roy. Soc. London* **84**, p. 247 (1910).

[20] P.H. Hugoniot, *J. École Polytech.* (1887–1889), 57th and 58th cahiers.

[21] D.R. White, *Phys. Fluids* **4**, p. 465 (1961).

[22] R. Engelke and W.C. Davis, Los Alamos National Laboratory, unpublished data (1972).

[23] W.W. Wood and J.G. Kirkwood, *J. Appl. Phys.* **25**, p. 395 (1957).

[24] N.R. Greiner and N. Blais, in *Proceedings of the Ninth Symposium (Intl.) on Detonation,* Office of Naval Research OCNR 113291-7, pp. 953–961 (1989).

[25] N.R. Greiner, H.A. Fry, N.C. Blais, and R. Engelke, in *Proceedings of the Tenth International Detonation Symposium,* Office of Naval Research ONR 33395-12, pp. 563–569 (1993).

[26] R. Engelke and N.C. Blais, *J. Chem. Phys.* **101**, p. 10961 (1994).

[27] R. Engelke, D.R. Pettit, and S.A. Sheffield, *J. Phys. Chem. A* **101**, p. 1696 (1997).

[28] C.S. Yoo and N.C. Holmes, in *High Pressure Science and Technology—1983* (ed. S.C. Schmidt, J.W. Shaner, G.A. Samara, and M. Ross), American Institute of Physics, p. 1567 (1993).

[29] G.I. Taylor, *Proc. Roy. Soc. A* CC, pp. 235–247 (1950).

[30] C.L. Mader, *Numerical Modeling of Detonations*, University of California Press, Berkeley (1979).

[31] F.H. Ree and M. Van Thiel, in *Proceedings of the Eight Symposium (Intl.) on Detonation*, NSWC MP 86-194 U.S. GPO, Washington, DC, pp. 501–512 (1985).

[32] E.L. Lee, H.C. Hornig, and J.W. Kury, *Adiabatic Expansion of High Explosive Detonation Products*, technical report UCRL-50422, Lawrence Livermore National Laboratory (May 1968).

[33] J.N. Fritz, R.S. Hixson, M.S. Shaw, C.E. Morris, and R.G. McQueen, *J. Appl. Phys.* **80**, p. 6129 (1996).

[34] C.L. Mader and C.A. Forest, *Two-Dimensional Homogeneous and Heterogeneous Detonation Wave Propagation*, Los Alamos Scientific Laboratory Report LA-6259, p. 60 (App. C) (June 1976). See also C.A. Forest, *Burning and Detonation*, Los Alamos Scientific Laboratory Report LA-7245, p. 22 (App. B) (July 1978).

[35] J. Wackerle, R.L. Rabie, M.J. Ginsberg, and A.B. Anderson, in *Proceedings of the Symposium on High Dynamic Pressures*, Paris, France, p. 127 (1978).

[36] E.L. Lee and C.M. Tarver, *Phys. Fluids* **23**, p. 2362 (1980).

[37] J.N. Johnson, P.K. Tang, and C.A. Forest, *J. Appl. Phys.* **57**, p. 4323 (1985).

[38] M.L. Hobbs and M.R. Baer, in *Proceedings of the Tenth International Detonation Symposium*, Office of Naval Research ONR 33395-12, pp. 409-418 (1993); M.L. Hobbs and M.R. Baer, *Shock Waves* **2**, p. 177 (1992).

[39] J.W. Nunziato, M.E. Kipp, R.E. Setchell, and E.K. Walsh, *Shock Initiation in Heterogeneous Explosives*, technical report SAND81-2173, Sandia Natl. Labs., Albuquerque, NM (1982).

[40] E.F. Gittings, in *Proceedings of the Fourth Symposium (Intl.) on Detonation*, Office of Naval Research Department of the Navy Report ACR-126, Washington D.C., pp. 373–380 (1965).

[41] F.E. Walker and R.J. Wasley, *Explosivstoffe* **1**, p. 9 (1969).

[42] A.W. Campbell, W.C. Davis, and J.R. Travis, *Phys. Fluids* **4**, p. 498, (1961).

[43] R.F. Chaiken, *The Kinetic Theory of Detonation of High Explosives*, M.S. Thesis, Polytechnic Institute of Brooklyn (1958).

[44] D.A. Frank-Kamenetskii, *Diffusion and Heat Exchange in Chemical Kinetics,* Princeton University Press, Princeton, NJ (1955).

[45] S.A Sheffield, R. Engelke, and R.R. Alcon, in *Proceedings of the Ninth Symposium (Intl.) on Detonation,* Office of Naval Research OCNR 113291-7, pp. 39–49 (1989).

[46] A.W. Campbell, W.C. Davis, J.B. Ramsay, and J.R. Travis, *Phys. Fluids* **4**, p. 511 (1961).

[47] R. Engelke, *Phys. Fluids* **22**, p. 1623 (1979); and **23**, p. 875 (1980).

[48] R. Engelke and J.B. Bdzil, *Phys. Fluids* **26**, p. 1210 (1983).

[49] R.L. Gustavsen, S.A. Sheffield, and R.R. Alcon, Los Alamos National Laboratory, unpublished data (1996).

[50] H. Moulard, in *Proceedings of the Ninth Symposium (Intl.) on Detonation*, Office of Naval Research OCNR 113291-7, pp. 18–24 (1989).

[51] A.W. Campbell and B.G. Craig, Los Alamos National Laboratory unpublished data.

[52] S.A. Sheffield, R.L. Gustavsen, and M.U. Anderson, in *High-Pressure Shock Compression of Solids IV* (ed. L. Davison, Y. Horie, and M. Shahinpoor), Springer-Verlag, New York (1997).

[53] Section on DDT, in *Proceedings of the Ninth Symposium (Intl.) on Detonation*, Office of Naval Research OCNR 113291-7, pp. 259–376 (1989).

[54] J.M. McAfee, B.W. Asay, and A.W. Campbell, in *Proceedings of the Ninth Symposium (Intl.) on Detonation*, Office of Naval Research OCNR 113291-7, pp. 265–279 (1989).

[55] M.R. Baer and J.W. Nunziato, in *Proceedings of the Ninth Symposium (Intl.) on Detonation*, Office of Naval Research OCNR 113291-7, pp. 293–305 (1989).

[56] M.R. Baer, in *High-Pressure Shock Compression of Solids IV* (ed. L. Davison, Y. Horie, and M. Shahinpoor), Springer-Verlag, New York (1997).

[57] A.K. Kapila, S.F. Son, J.B. Bdzil, R. Menikoff, and D.S. Stewart, submitted to *Phys. Fluids*, July 1996.

[58] J.B. Bdzil, S.F. Son, R. Menikoff, A.K. Kapila, and D.S. Stewart, in preparation for submittal to *Phys. Fluids*, 1997.

[59] R.W. Watson, C.R. Summers, F.C. Gibson, and R.W. Van Dolah, in *Proceedings of the Fourth Symposium (Intl.) on Detonation*, Office of Naval Research ACR-126, pp. 117–125 (1965).

[60] David Kennedy, ICI Australia, personal communication, 1995.

[61] G.A. Leiper and J. Cooper, in *Proceedings of the Tenth International Detonation Symposium*, Office of Naval Research ONR 33395-12, pp. 267–275 (1993).

[62] D.L. Kennedy and D.A. Jones, in *Proceedings of the Tenth International Detonation Symposium*, Office of Naval Research ONR 33395-12, pp. 665–674 (1993).

[63] J. Forbes, Lawrence Livermore National Laboratory, personal communication, January 1997.

[64] R. Bernecker, Naval Surface Warfare Center, Indian Head, Maryland, personal communication, January 1997.

[65] J.B. Bdzil, *J. Fluid Mech.* **108**, p. 195 (1981).

[66] S.A. Sheffield, *Bull. Am. Phy. Soc.* **35**, p. 697 (1990); see also S.A. Sheffield, R.L. Gustavsen, and R.R. Alcon, in *Shock Waves in Condensed Matter—1995* (ed. S.C. Schmidt and W.C. Tao), American Institute of Physics, New York, p. 771 (1996).

[67] B.M. Dobratz and P.C. Crawford, *LLNL Explosives Handbook*, UCRL-52997, Lawrence Livermore National Laboratory, Livermore, CA (1985).

[68] R. Meyer, *Explosives*, VCH Publishers, New York (1987).

[69] R. Engelke and J.R. Stine, *J. Phys. Chem.* **94**, p. 5689 (1990).

[70] T.R. Gibbs and A. Popolato, *LASL Explosive Property Data*, University of California Press, Berkeley (1980).

[71] A.F. Belyaev and R.Kh. Kurbangalina, *Russ. J. Phys. Chem.* **34**, p. 285 (1960); see also B.M. Dobratz and P.C. Crawford, *LLNL Explosives Handbook*, UCRL-52997, Lawrence Livermore National Laboratory, Livermore, CA, p. 8–34 (1985).

[72] L.C. Smith, *LANL Explosives Orientation Course: Sensitivity and Sensitivity Tests*, Los Alamos National Laboratory Report LA-11010-MS (1987).

[73] Proceedings of the First through the Tenth (International) Symposia on Detonation, sponsored by the Office of Naval Research, U.S. Government Printing Office, Washington, DC (1951 to 1993).

[74] J.E. Vorthman and J. Wackerle, in *Shock Waves in Condensed Matter—1983* (ed. J.R. Asay, R.A. Graham, and G.K. Straub), Elsevier Science Publishers, Amsterdam, p. 613 (1984).

[75] L.M. Barker and R.E. Hollenbach, *Rev. Sci Instrum.* **36**, p. 1617 (1965).

[76] M. Durand, P. Laharrague, P. Lalle, A. Le Bihan, J. Morvan, and H. Pujols, *Rev. Sci. Instrum.* **48**, pp. 275–278 (1977)

[77] D.D. Bloomquist and S.A. Sheffield, *J. Appl. Phys.* **54**, p. 1717 (1983).

[78] W.F. Hemsing, *Rev. Sci. Instrum.* **50**, p. 73 (1979).

[79] W.F. Hemsing, in *Proceedings of the Eighth Symposium (Intl.) on Detonation*, Naval Surface Weapons Center NSWC MP 86-194, pp. 468–472 (1985).

[80] D. Goosman, G. Avara, L. Steinmetz, C. Lai, and S. Perry, in *Proceedings of the 22-nd International Congress on High Speed Photography and Photonics*, SPIE Vol. 2869, SPIE Press, Bellingham, WA (1997).

[81] S.A. Sheffield, R. Engelke, and R.R. Alcon, Los Alamos National Laboratory, unpublished data.

[82] S.A. Sheffield, D.D. Bloomquist, and C.M. Tarver, *J. Chem. Phys.* **80**, p. 3831 (1984).

[83] J.B. Ramsay and A. Popolato, in *Proceedings of the Fourth Symposium (Intl.) on Detonation*, Office of Naval Research ACR-126, p. 233 (1965).

CHAPTER 8

Analysis of Shock-Induced Damage in Fiber-Reinforced Composites

F.L. Addessio and J.B. Aidun

8.1. Introduction

Under the conditions of high strain rates and large deformations, engineered composites can experience large mean- and shear-stress states. Accurate numerical simulations for the mechanical response of these anisotropic materials under these conditions must include the effects of nonlinear elasticity and inelastic phenomena such as plasticity and damage. Modeling composite materials involves the additional complexity of requiring constitutive models for the interfaces as well as the constituents. Interfacial debonding of the fiber and matrix materials and delamination between the layers of a laminate provide two important failure mechanisms within composite structures. Physically based material models are desirable for simulating the thermomechanical response of composites. Computational simulations are useful for reducing the number of candidate designs for laminated, fiber-reinforced composites and for interpreting the deformation mechanisms observed in both controlled experiments and integrated tests. Furthermore, the availability of good computational models can reduce the costs of validation experiments, which are necessary for the design of engineering structures.

Computational expediency requires material models that are numerically robust and computationally efficient. Because the length scales necessary to model the detailed response of the constituents and the interfaces within a composite are much smaller than the length scales of engineering structures, it is impractical to resolve numerically the composite microstructure. Consequently, it is necessary to model composites as equivalent homogeneous materials.

A homogenization technique for unidirectional fiber-reinforced and layered composites is reviewed in this development. It is shown how the details of the micromechanical response of the constituents and the interfaces of a composite are included in the analysis. However, this approach also offers the advantage of eliminating the need to provide a detailed description of the composite at the level of its microstructure. The homogenization technique has been extended to include phenomena inherent to the high-strain-rate loading conditions of impact events.

Numerical simulations for the dynamic uniaxial-strain loading of fiber-reinforced composites and laminates are presented. The simulations illustrate the deformation mechanisms which have been captured using the micromechanical homogenization approach to modeling these composites.

8.2. Background

There is considerable interest in using engineered composite materials to develop lighter structures that are strong under adverse conditions, including high-temperature environments, thermal and mechanical cycling, vibrations, and chemical corrosion. Mechanical attributes that can be tailored to provide high strength and stiffness, as well as light weight, abrasion resistance, improved damage tolerance, and inexpensive fabrication requirements have established composites as ideal materials for many structural applications. Aerospace applications include aeroframes, fan blades, and blade containment systems. The utilization of composites for automotive applications includes drive trains, brakes, and crash-mitigating structural members. Recently, composites have been considered for the design of munitions, armor, and weapons systems. The conditions of high strain rates and large deformations are encountered in many of these applications.

The ability to design composite structures relies on the availability of constitutive models that provide the thermomechanical response for complex loading paths. Research leading to the application of composite materials can benefit from numerical simulations to assist in the interpretation of experiments used to characterize them, as well as for designing composite structures. The value of having the capability to predict the response of an engineering system to dynamic loads with a high degree of confidence reduces the need for expensive field or laboratory experiments. Furthermore, dependable computational tools can expedite the design of engineering systems. A continuum analysis of the dynamic response of solids to large deformations requires a mathematical description for the material behavior. These material models are provided in terms of the properties for material strength and failure.

A physically based, as opposed to an empirical, material model description is desirable. It is expected that a physically based model will be capable of extrapolation outside of the regime of direct experimental verification. Also, when sufficient data are not available, the parameters required by physically based constitutive models may be estimated more easily than those of empirical models. A thermodynamically consistent continuum formulation is necessary. Viable computational constitutive models also should be compatible with the incremental continuum formulation inherent to large deformation numerical descriptions. Furthermore, developing material descriptions that are numerically robust and computationally efficient is desirable to allow large-scale simulations.

Two approaches have been pursued for modeling composite materials. In the micromechanical approach, material descriptions for the constituents and the interfaces within the composite are provided. The microstructure of the composite is idealized by a periodic representative volume element (RVE). The macromechanical response then is obtained as an appropriate average of the response of the constituents and interfaces over the RVE. Numerical investigations have been conducted using a finite-element description [1] for the RVE. However, a finite-element simulation [1] requires a fine computational mesh to resolve the material microstructure. Consequently, although a finite-element approach is useful for investigating micromechanical phenomena, it is too costly for pursuing the response of composite structures. Instead, simpler homogenization techniques have been invoked to obtain the average composite response. In a macromechanical approach, the composite is modeled as an equivalent homogeneous, anisotropic material. Therefore, details of the microstructure are ignored. Consequently, the macromechanical description is convenient and may be implemented easily into continuum computer methods. Unfortunately, classical principles such as plastic incompressibility, convexity of the yield surface, and normality of the plastic strain rate, which may apply to the constituents, are not generally valid for the composite. Therefore, macromechanical descriptions are dependent on obtaining a large experimental database.

The Method of Cells (MoC) is a homogenization technique, which has demonstrated versatility [2–5]. It is applicable to general loading conditions, compatible with a wide variety of constitutive relations for the constituent materials and interfaces, and easy to implement as the material model in existing continuum analyses. The MoC has been employed to calculate the thermomechanical response of a variety of engineered composites with good success. The method is applicable to any ordered composite that can be idealized with a microstructural RVE which generates the entire composite through periodic repetition. The macroscopic stress state calculated for the RVE is taken to be the stress state at the corresponding location in the equivalent homogeneous continuum. The homogenized stress state is obtained as a solution for the mechanical response of the RVE to an applied macroscopic strain, knowing the response of the constituent materials.

The analysis of engineering structures subjected to loads producing high-strain-rate, large-deformation responses is modeled using numerical continuum-mechanics analyses [6]. For high strain rates, a time-explicit formulation of the governing conservation equations is used. Typically, finite-difference and finite-element formulations are used, although particle methods [7] are becoming more widely available. The solution to the continuum equations provides the constitutive model with strain rates (in the form of velocity gradients) and the internal energy field. The material model uses this information to update the stress field. For large-scale continuum-mechanics simulations, it is impractical to provide a detailed resolution for the microstructure. Furthermore, only the macroscopic response of the composite is required by the continuum analysis. The MoC provides a solution technique which can be conveniently implemented into a continuum analysis. The robustness of this approach is dependent on the constitutive representation used to describe the response of the constituent materials and the interfaces within the composite. Therefore, the computational cost is dependent on the details chosen to be modeled.

A model is presented for the analysis of composite structures. A sample composed of a fiber-reinforced, metal-matrix composite is used for demonstrative purposes. Quantitative predictions of the mechanical response may be obtained only to the extent that sufficient microstructural phenomena have been incorporated into the model. Inelastic deformation of the constituents, including plasticity and damage, is included. The nonlinear elastic deformation of the constituents, which is necessary for large-strain phenomena, also is treated. Failure phenomena are incorporated as dependent on the micromechanical stress state in the material. The failure model provided in this development is phenomenological. Consequently, this model has a limited predictive capability. It relies on a far-field stress state to predict the state of damage. This is consistent with a continuum damage formulation. Lacking data, the present damage model is useful for establishing a framework for representing material failure. It is useful for illustrating the versatility of the MoC homogenization technique. An simple interface model is provided for the fiber–matrix interface to facilitate computational simulations. Finally, an approach for modeling layers of fiber-reinforced laminae is developed. An interface model is provided in the layered problem, allowing investigations to include delamination phenomena.

8.3. Micromechanical Model

A detailed development of the micromechanical model is provided in the Refs. 2–5. In this section, the first-order theory is reviewed for both continuous fiber-reinforced and layered geometries. The development proceeds by i) identifying an idealized representative volume element (RVE), ii) defining strains and material models on the micromechanical level, iii) imposing continuity of displacements and tractions across interfaces in an average sense, and iv) providing suitable averages, which results in the macromechanical response of the RVE. A typical RVE comprises a number of subcells. In general, each subcell is associated with one of the constituents of the composite. Constitutive models are specified for each of the constituent materials. The model provides only a quasi-static response for the representative volume. The solution to the continuum conservation equations provides the transient response of the structure. However, this is sufficient for applications with spatial resolutions much larger than the microstructural dimensions and time resolutions longer than wave transit times through the RVE. Many structural and material responses meet these restrictions.

Two separate coordinate systems are used in the development of the homogenization technique. A local coordinate system (y_i^α) is used to describe the heterogeneous material and the RVE. A second coordinate system (x_i) is used to describe the macroscopically equivalent homogeneous continuum. Microscopic variables, which provide the detailed state of the material within a RVE, are expressed as functions of the local coordinates (y_i^α) and time (t), and are parameterized in the continuum coordinates (x_i). For example, the micromechanical stress field is $\sigma_{ij}^{\alpha\beta}(y_i^\alpha, t; x_i)$. In the development, Latin subscripts denote Cartesian components and Greek indices identify subcells within the RVE. Summation is always implied by repeated Latin indices, but never for Greek indices. The parameterization in x_i results from the relation between the two coordinate systems; that is, an entire RVE maps into a point in the continuum coordinate system. Known macromechanical conditions at a continuum location provide loading conditions for determining the thermomechanical state of the corresponding RVE through the microstructural model. The resulting state of the RVE, when mapped into the continuum, completes the determination of the macromechanical state at a given location.

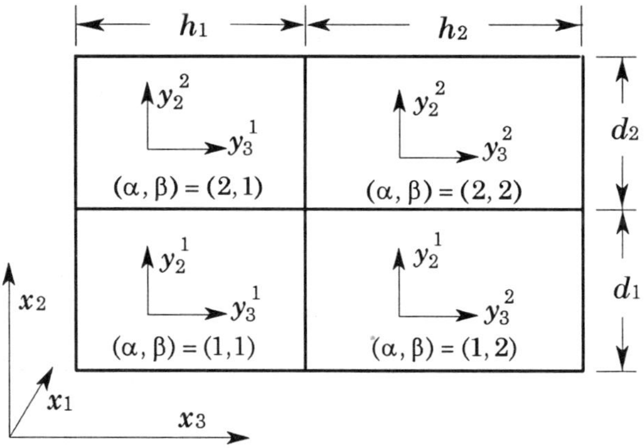

Figure 8.1. Representative volume element for a unidirectional fiber-reinforced composite.

8.3.1. Fiber-Reinforced Model

For a continuous fiber-reinforced composite, an idealized RVE can be identified as a rectangular volume containing four rectangular subcells (Fig. 8.1). One of the subcells ($\alpha = 1,\ \beta = 1$) corresponds to the fiber cross section and the other three to the matrix material. The composite is assumed to be periodic in the x_2 and x_3 directions. The RVE is continuous and infinite along the fiber direction (x_1). Two-dimensional local coordinates (y_2^α, y_3^β) are defined within each subcell. The subcell dimensions are d_α and h_β along the y_2^α and y_3^β directions, respectively. The dimensions of the RVE are $d = d_1 + d_2$ and $h = h_1 + h_2$. For the first-order description, the velocity field of the continuous-fiber model is bilinear within each subcell,

$$\dot{u}_i^{\alpha\beta}(y_i^{\alpha\beta}, t; x_i) = \dot{w}_i^{\alpha\beta}(x_i, t) + \dot{\phi}_i^{\alpha\beta}(x_i, t)\, y_2^\alpha + \dot{\psi}_i^{\alpha\beta}(x_i, t)\, y_3^\beta\ , \tag{8.1}$$

where $\phi_i^{\alpha\beta}$ and $\psi_i^{\alpha\beta}$ are the model parameters and $u_i^{\alpha\beta}$ are the subcell displacements.

For each subcell, the rate-of-deformation tensor may be written

$$\dot{\varepsilon}_{ij}^{\alpha\beta} = \frac{1}{2}\left(\frac{\partial \dot{u}_i^{\alpha\beta}}{\partial y_j} + \frac{\partial \dot{u}_j^{\alpha\beta}}{\partial y_i} \right), \tag{8.2}$$

where the differentiation in the y_2 and y_3 directions are with respect to the local coordinates (y_2^α and y_3^β). The subcell coordinate y_1 is identical with x_1, the nonperiodic coordinate. Subcell rotations are neglected within the development. These approximations are acceptable, provided planes normal to the fiber direction remain planar and the interfaces between the constituents do not deviate substantially from being parallel to the local coordinate axes. From Eqs. 8.1 and 8.2, the rate-of-deformation tensor is constant within each subcell:

$$\begin{aligned} \dot{\varepsilon}_{11}^{\alpha\beta} &= \dot{w}_{1,1} & 2\dot{\varepsilon}_{23}^{\alpha\beta} &= \dot{\phi}_3^{\alpha\beta} + \dot{\psi}_2^{\alpha\beta} \\ \dot{\varepsilon}_{22}^{\alpha\beta} &= \dot{\phi}_2^{\alpha\beta} & 2\dot{\varepsilon}_{13}^{\alpha\beta} &= \dot{\psi}_1^{\alpha\beta} + \dot{w}_{3,1} \\ \dot{\varepsilon}_{33}^{\alpha\beta} &= \dot{\psi}_3^{\alpha\beta} & 2\dot{\varepsilon}_{12}^{\alpha\beta} &= \dot{\phi}_1^{\alpha\beta} + \dot{w}_{2,1} \, , \end{aligned} \tag{8.3}$$

where the assumptions [8]

$$\begin{aligned} \dot{w}_{i,2}^{1\beta} &= \dot{w}_{i,2}^{2\beta} = \dot{w}_{i,2} \\ \dot{w}_{i,3}^{\alpha 1} &= \dot{w}_{i,3}^{\alpha 2} = \dot{w}_{i,3} \end{aligned} \tag{8.4}$$

have been used. The comma subscript denotes differentiation with respect to the x_i component identified by the trailing index. The variables $\dot{w}_{i,j}$ are the macromechanical velocity gradients, which are obtained from solutions to the global conservation equations.

The description of the approximate mechanical response of a RVE requires satisfying the continuity of tractions and displacements across the subcell interfaces. In the context of modeling the constitutive behavior, it only is necessary to consider static equilibrium at a point. The equilibrium conditions are satisfied by the formulation [9]. The model parameters corresponding to this static solution are completely determined from the macroscopic velocity gradients through the imposition of continuity of displacement and tractions in an average sense. The macroscopic quantities, however, are governed by the conservation equations for the homogenized continuum. Consequently, momentum conservation at the subcell level is considered as extraneous. For this reason, the static solution of the first-order MoC relations is appropriate.

The subcell stresses are uniform, to be consistent with the constant subcell strain rate given by Eq. 8.3. Therefore, the rate form of the traction continuity conditions is

$$\begin{aligned} \dot{\sigma}_{2j}^{1\beta} &= \dot{\sigma}_{2j}^{2\beta} \\ \dot{\sigma}_{3j}^{\alpha 1} &= \dot{\sigma}_{3j}^{\alpha 2} \, . \end{aligned} \tag{8.5}$$

Accounting for the macroscopic variations that occur around the continuum location x_i is of fundamental importance in developing the displacement continuity expressions. This is referred to as "the transition to the continuum." It introduces macroscopic strain information into the micromechanics, which ensures that the solution for the mechanical response of the RVE is appropriate to the prevailing conditions at x_i. In particular, the macroscopic displacement gradients provide loading conditions on the approximate solution for the RVE response.

Displacement conditions are derived by considering the local jump in the displacement field across the interfaces

$$\begin{aligned} u_i^{1\beta}(y_2 = d_1/2, y_3, t; x_i) - u_i^{2\beta}(y_2 = -d_2/2, y_3, t; x_i) &= f_i^{1\beta}(\sigma) \\ u_i^{\alpha 1}(y_2, y_3 = h_1/2, t; x_i) - u_i^{\alpha 2}(y_2, y_3 = -h_2/2, t; x_i) &= f_i^{\alpha 1}(\sigma), \end{aligned} \tag{8.6}$$

where $f_i^{\alpha\beta}$, the interfacial constitutive relations, are provided in terms of the tractions on the interface. In the current formulation, the displacement conditions are satisfied on an average basis for each of the interfaces within the RVE. The averaged form for Eq. 8.6 may be written [8]

$$\begin{aligned} d_1 \dot{\phi}_i^{1\beta} + d_2 \dot{\phi}_i^{2\beta} + 2\dot{\Delta}_{2i}^{1\beta} &= d \frac{\partial \dot{w}_i}{\partial x_2} \\ h_1 \dot{\psi}_i^{\alpha 1} + h_2 \dot{\psi}_i^{\alpha 2} + 2\dot{\Delta}_{3i}^{\alpha 1} &= h \frac{\partial \dot{w}_i}{\partial x_3} \end{aligned} \tag{8.7}$$

for $i = 1, 2$, or 3. In Eq. 8.7, the quantities $\Delta_{ij}^{\alpha\beta}$ are the average interfacial jumps in the displacement field between the fiber and matrix constituents. It is assumed that the matrix–matrix interfaces remain perfectly bonded. Consequently, $\Delta_{ij}^{\alpha\beta}$ are nonzero only for $\beta = 1$ and $\alpha = 1$ in the first and second equations, respectively.

Once the constitutive models are defined for the constituents, Eqs. 8.5 and 8.7 provide the relations necessary to determine the 24 microvariables $\phi_i^{\alpha\beta}$ and $\psi_i^{\alpha\beta}$. The macromechanical stress state then may be determined from the definitions of appropriate averaged quantities. The average stresses within the subcells are defined as

$$\bar{\sigma}_{ij}^{\alpha\beta} = \frac{1}{d_\alpha h_\beta} \iint \sigma_{ij}^{\alpha\beta} \, dy_2^\alpha \, dy_3^\beta \,. \tag{8.8}$$

For the first-order theory, the local and average subcell stresses are identical. Finally, the average stresses for the RVE are

$$\sigma_{ij} = \frac{1}{dh} \sum_{\alpha\beta} \bar{\sigma}_{ij}^{\alpha\beta} d_\alpha h_\beta \,. \tag{8.9}$$

8.3.2. Layered Model

The equations necessary to define the micromechanical and macromechanical stress states for homogeneous layers may be developed in a manner similar to the previous section. An ideal RVE for a layered composite is provided in Fig. 8.2. It is assumed that each layer is a fiber-reinforced composite whose constituents are identical. Only the fiber orientation is allowed to vary among the layers. The homogenization technique provides an estimate of the overall stress in a computational element that contains the multiple stacking sequences. In continuum-mechanics simulations of thick structures composed of laminated composite plates, it is impractical to resolve the lamina microstructure or the layers. Furthermore, only the macroscopic response is required. Consequently, the macroscopic stresses in the laminate are appropriate and sufficient for many computational simulations. The laminate is periodic in the x_3 direction. The RVE is continuous and infinite along the x_1 and x_2 directions. A one-dimensional, local coordinate system (y_3^γ) is defined within each layer. The dimension of each layer is l_γ and the dimension of the laminate is $l = l_1 + l_2 + \cdots + l_n$.

Consistent with a first-order development, the velocities within each layer are written

$$\dot{u}_i^\gamma(y_i^\gamma, t; x_i) = \dot{w}_i^\gamma(x_i, t) + \dot{\zeta}_i^\gamma(x_i, t) y_3^\gamma , \tag{8.10}$$

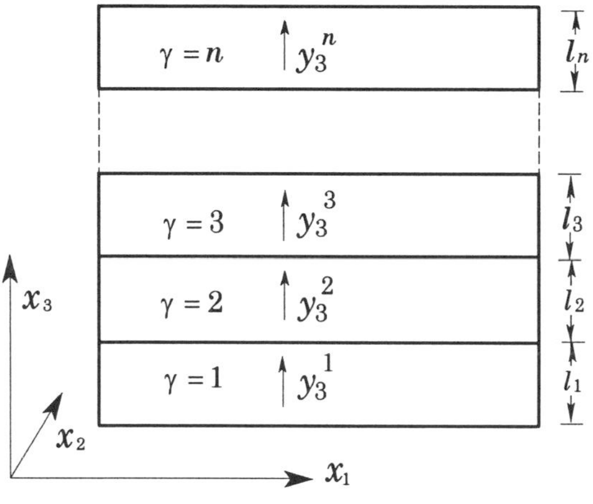

Figure 8.2. Representative volume element for a layered composite.

where ζ_i^γ are the model parameters, and u_i^γ are the subcell displacements. The local rate of deformation again is defined by Eq. 8.2. Consequently, the local rates for each subcell may be written

$$\begin{aligned} \dot{\varepsilon}_{11}^{\gamma} &= \dot{w}_{1,1} & 2\dot{\varepsilon}_{23}^{\gamma} &= \dot{\zeta}_2^{\gamma} + \dot{w}_{3,2} \\ \dot{\varepsilon}_{22}^{\gamma} &= \dot{w}_{2,2} & 2\dot{\varepsilon}_{13}^{\gamma} &= \dot{\zeta}_1^{\gamma} + \dot{w}_{3,1} \\ \dot{\varepsilon}_{33}^{\gamma} &= \dot{\zeta}_3^{\gamma} & 2\dot{\varepsilon}_{12}^{\gamma} &= \dot{w}_{1,2} + \dot{w}_{2,1}\,. \end{aligned} \tag{8.11}$$

The assumption [8]

$$\dot{w}_{i,j}^{\gamma} = \dot{w}_{i,j} \tag{8.12}$$

has been used to simplify Eq. 8.11. The variables $\dot{w}_{i,j}$ are the macromechanical velocity gradients, which are obtained from solutions to the continuum conservation equations.

Because the subcell stresses are uniform for the first-order theory, the rate form of the traction continuity conditions may be written

$$\begin{aligned} \dot{\sigma}_{3i}^{1} &= \dot{\sigma}_{3i}^{2} \\ \dot{\sigma}_{3i}^{2} &= \dot{\sigma}_{3i}^{3} \\ &\vdots \\ \dot{\sigma}_{3i}^{n-1} &= \dot{\sigma}_{3i}^{n}\,. \end{aligned} \tag{8.13}$$

The displacement relations are

$$\sum_{\gamma=1}^{n} [l_\gamma \dot{\zeta}_i^{\gamma} + 2\dot{\Delta}_i^{\gamma}(\sigma)] = \frac{\partial \dot{w}_i}{\partial x_3}\,, \tag{8.14}$$

where the interfacial displacements (Δ_i^γ) are functions of the tractions in the neighboring constituents and are provided by the interfacial constitutive relations.

Once the constitutive relations for each of the layers are defined, Eqs. 8.13 and 8.14 may be solved for the $3n$ micromechanical variables ζ_i^γ. Once a solution for the micromechanical variables has been obtained, the average response of the RVE is determined from the definitions of the average stresses. The average layer stresses are

$$\bar{\sigma}_{ij}^{\gamma} = \frac{1}{l_\gamma} \int \sigma_{ij}^{\gamma}\, dy_3^{\gamma}\,. \tag{8.15}$$

For the first-order theory, the local and average subcell stresses are identical. Finally, the average stresses for the laminate are

$$\sigma_{ij} = \frac{1}{l} \sum_{\gamma=1}^{n} \bar{\sigma}_{ij}^{\gamma} l_{\gamma} . \tag{8.16}$$

8.4. Constitutive Models

Once the constitutive relations for the constituents and the interfaces have been defined, the response of an equivalent homogeneous continuum may be obtained by solving the micromechanical models. As noted in the previous sections, one of the strengths of the micromechanical approach is that the material response is defined for each of the constituents within the composite rather than for the overall composite.

8.4.1. Constituent Material Model

Within this section, a general framework is provided for the constitutive models for each of the constituents of the composite. Many of the materials used for the fiber reinforcement of composites, such as graphite, are anisotropic. Furthermore, for high-strain-rate, shock-driven deformations, nonlinear elastic phenomena must be included. Inelastic effects, including plasticity and damage, also must be considered for problems involving large deformations. Employing strain-rate decomposition, these phenomena may be accounted for by the relation

$$\dot{\sigma}_{ij} = \hat{C}_{ijkl}(\dot{\varepsilon}_{kl} - \dot{\eta}_{kl}) - C_{ijkl}\dot{\xi}_{kl} , \tag{8.17}$$

where C_{ijkl} and $\hat{C}_{ijkl}$ are the elastic stiffness tensor and the degraded stiffness, respectively. Also, $\dot{\varepsilon}_{ij}$ is the rate-of-deformation tensor, $\dot{\eta}_{ij}$ is the plastic strain rate, and ξ_{ij} is a damage strain rate. Ductile failure, which is characterized by a plastic flow surface and porosity, for example, may be included in the plastic strain rate [10]. Brittle failure may be characterized by a crack surface energy (i.e., the stress intensity factors) and a crack size, which may be modeled by a crack strain [11]. All damage mechanisms are included in the damage strain rate in the model provided by Eq. 8.17.

Numerous approaches are available for defining the plastic strain rate ($\dot{\eta}_{ij}$). Classically, it is assumed that the plastic strain rate is incompressible ($\dot{\eta}_{kk} = 0$). Therefore, it only is necessary to model the deviatoric part of the plastic strain ($n_{ij} = \eta_{ij} - \eta_{kk}\,\delta_{ij}$). Many flow surfaces [12–14] are available to define the plastic deformation. Classically, the plastic flow surfaces are written in terms of the second invariant of the deviatoric stress

$$J_2 = \tfrac{1}{2} s_{ij} s_{ij} \tag{8.18}$$

and a flow stress (Y), which is a function of the equivalent plastic strain (e^p)

$$F(s_{ij}, e^p) = \tfrac{3}{2} s_{ij} s_{ij} - Y^2(e^p) = 0. \tag{8.19}$$

The associative flow rule is then applied to obtain the plastic strain rate

$$\dot{\eta}_{ij} = \frac{\partial F}{\partial s_{ij}} \dot{\lambda}, \tag{8.20}$$

where $\dot{\lambda}$ is a parameter which is determined from the condition that the stress state is constrained to remain on the flow surface ($\dot{F} = 0$). This approach for determining the plastic strain rate requires the solution of a system of ordinary differential equations, which can become mathematically stiff. Another approach, which yields an equivalent result while remaining numerically robust, is the normal return method. This technique initially computes an elastic prediction for the stress state. The second invariant of the stress deviator provided by the elastic prediction is then compared to the flow stress. If the second invariant is less than the flow stress, then the deformation is elastic and the solution for the stress state is complete. However, if the second invariant is greater than the flow stress, then the stresses are determined by returning normally to the flow surface, Eq. 8.19. Finally, the plastic strain rates are obtained from the constitutive relation, Eq. 8.17. Unfortunately, the normal return method requires two passes through the micromechanical model for the composite problem: once to obtain an elastic prediction and again following the determination of the plastic strain rates for each subcell. Within this development, the plastic strain rate is obtained from a unified viscoplastic model [15]. For this approach, the plastic strain rate is written

$$\dot{\eta}_{ij} = s_{ij} \dot{\lambda}, \tag{8.21}$$

similar to Eq. 8.20. The parameter $\dot{\lambda}$ is then obtained from the relations

$$\dot{\lambda} = \frac{D_0}{J_2} \exp\left(-\frac{n+1}{n} \frac{Z^2}{3J_2} \right) \tag{8.22}$$

and

$$Z = Z_1 + (Z_0 - Z_1) \exp\left(-\frac{m w^p}{Z_0} \right), \tag{8.23}$$

where D_0, n, m, Z_0, and Z_1 are material parameters and w^p is the plastic work

$$w^p = 2\dot{\lambda} J_2 \,. \tag{8.24}$$

Stress states with a high mean stress often occur under dynamic loading conditions. Mean stresses in the range of tens to hundreds of MPa can produce significant deviations from the linear stress–elastic strain relation. To adequately simulate this aspect of the dynamic response of composites to the high-strain-rate loadings typical of crash safety, armor, and munitions applications, nonlinear elastic effects must be included. Nonlinear elasticity provides a convenient means of accounting for the common increase of the bulk modulus with compression. Consequently, it allows for the possibility of shock waves to develop in the material. For this reason, nonlinear elasticity is an essential aspect of the constitutive response. This extension modifies the calculated response throughout the deformation. Its effects are not limited to the strain interval of purely elastic response because even during inelastic deformations, the stress depends explicitly on the elastic contribution to the total strain. An investigation including the quadratic terms, which represent the leading order of nonlinearity in the stress–elastic strain relation, has demonstrated the importance of this extension for composite materials [8].

An approximate approach for including nonlinear elasticity, which provides sufficient accuracy for pure constituents, is the implementation of an equation of state (EOS). Because shear stress is limited by material strength, the predominant contribution of nonlinear elasticity is through the volumetric strain. In isotropic materials, the volumetric strain affects only the mean stress. Consequently, the mean and deviatoric stresses can be computed independently of one another. Elastic nonlinearity is approximated and accurately accounted for by computing the mean stress from an EOS. In each subcell of the RVE, the mean stress is described by a volume- and energy-dependent EOS. In this development, a polynomial representation is used for the EOS

$$P = a_1 \varepsilon + a_2 \varepsilon^2 + a_3 \varepsilon^3 + a_4 \varepsilon^4 + (b_0 + b_1 \varepsilon) e \,, \tag{8.25}$$

where P, the pressure, is the trace of the stress ($P = -\sigma_{kk}/3$), $\varepsilon = 1 - (\rho/\rho_0)$ is the volumetric strain, and e is the internal energy change from an initial reference state. The material density is ρ. The parameters a_i and b_i are material constants. A polynomial approximation for the equation is chosen to facilitate the implementation of the EOS into the numerical algorithm. The material being considered may satisfy a Mie–Grüneisen EOS; for example,

$$P = H(\varepsilon)\left(1 + \tfrac{1}{2}\Gamma \varepsilon\right) + \rho_0 (1 - \varepsilon) \Gamma e \,, \tag{8.26}$$

where

$$H(\varepsilon) = \frac{\rho_0 c^2 \varepsilon(\varepsilon - 1)}{[1 + \varepsilon(s-1)]^2} \; . \tag{8.27}$$

The quantity Γ is the Grüneisen constant, ρ_0 is a reference density, c is a sound speed, and s is the slope of the linear shock velocity versus particle velocity response. The coefficients of the polynomial approximation are chosen to be consistent with Eqs. 8.26 and 8.27. The resulting values are

$$\begin{aligned}
a_1 &= -\rho_0 c^2 \\
a_2 &= \rho_0 c^2 [2s - 1 + \tfrac{1}{2}\Gamma] \\
a_3 &= -\rho_0 c^2 [3s^2 - 4s + 1 - \tfrac{1}{2}(2s-1)\Gamma] \\
a_4 &= \rho_0 c^2 [4s^3 - 9s^2 - 2s + 1 - \tfrac{1}{2}(3s^2 - 4s + 1)\Gamma] \\
b_0 &= -b_1 = -\rho_0 \Gamma \, .
\end{aligned} \tag{8.28}$$

For large deformations, material damage, including fiber breakage and matrix cracking, should be included in the constitutive response. Damage mechanisms such as shear banding, crack propagation, and the growth of porosity occur on a scale much smaller than the stress and strain fields used to characterize the continuum description of deformations. However, solutions for the growth of cracks [16] and voids [17] have been obtained based on the far-field distributions of the stress state. Consequently, acceptable descriptions of a damage state may be obtained for a continuum description without increasing the resolution of the analysis. The addition of damage to a material model also adds numerical complexities such as localization phenomena and considerations of ill-posedness of the governing equations under the conditions of material softening [18]. In an effort to mitigate these difficulties, a rate-dependent approach to modeling damage is provided [10].

A phenomenological approach has been taken to model damage in the constituents. In this manner, both brittle and ductile damage mechanisms are approximated with the same model. This approach establishes a framework for representing failure and illustrating the versatility of the homogenization technique. Similar to the flow surface used to compute the plastic strain rates, a damage surface is defined as

$$F_d(\sigma^0_{ij}, \alpha_{ij}, Y_d) = (\sigma^0_{ij} - \alpha_{ij})(\sigma^0_{ij} - \alpha_{ij}) - Y_d^2 \, , \tag{8.29}$$

where the α_{ij} are equivalent to a back stress and allow the material to respond differently under tensile and compressive conditions, for example. The back stress and the damage strength are allowed to degrade quadratically as damage progresses

$$\alpha_{ij} = d^2 \alpha_{ij}^0 \tag{8.30}$$

and

$$Y_d = d^2 Y_{d0}, \tag{8.31}$$

where α_{ij}^0 and Y_{d0} are material constants and d is a scalar measure of damage. The damage is taken to be a function of the damage strain

$$d = 1 - \frac{\xi}{\xi_0}, \tag{8.32}$$

where

$$\dot{\xi}_{ij} = \frac{1}{\mu}(\sigma_{ij} - \sigma_{ij}^0) \tag{8.33}$$

and

$$\dot{\xi} = \sqrt{\tfrac{3}{2}\dot{\xi}_{ij}\dot{\xi}_{ij}}\ . \tag{8.34}$$

In Eq. 8.32, ξ_0 is a material parameter and μ is the damage viscosity. As the deformation is allowed to proceed, Eqs. 8.29 through 8.34 allow the damage state to progress from an initially undamaged condition ($d = 1$) to a fully damaged state ($d = 0$). As a result, the stress state is relaxed to zero. For a computational cell, which is loaded in tension to a fully damaged state, the inability to support a stress field if the cell is allowed to reload in compression could result in a collapsed computational element. Consequently, the damage variable (d) is allowed to degrade to a nonzero value under compressive conditions. For path reversals, the damage is allowed to increase (heal) when a computational element is compressed from a damaged state where the damage variable is less than the minimum compressive damage limit. More general multidimensional approaches to path reversals are available [19].

The response of pure aluminum to a uniaxial loading/unloading strain path using the constitutive model is provided in Fig. 8.3 for a strain rate of $10^3\,s^{-1}$. The material variables used in the simulation are provided in Table 8.1, with the exception of the damage viscosity, which was set to the large value of 15 Mbar μs to exhibit better the damage process. In Fig. 8.3a, the solid line is the stress along the direction of the applied strain. The dashed line is the stress transverse to the applied strain. The plastic and damage strain rates are provided by the solid and dashed lines, respectively, in Fig. 8.3b. For the loading path, plastic yielding (point P) is reached at 0.7% strain. The stress continues to increase until the damage surface is reached

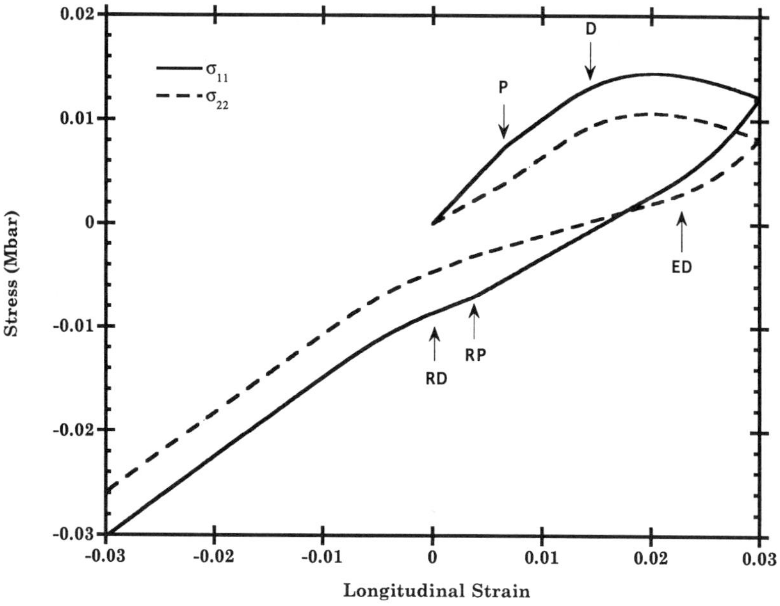

(a)

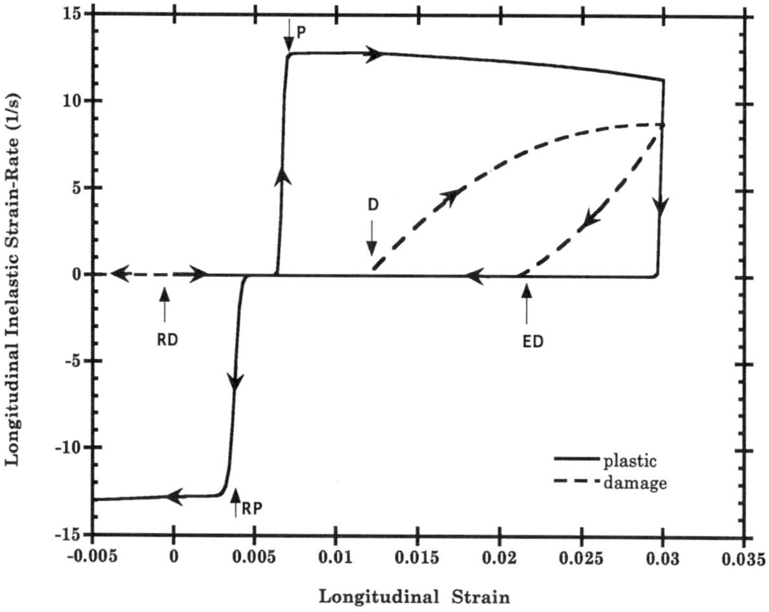

(b)

Figure 8.3. Uniaxial strain cycle (a) stress and (b) inelastic strain.

Table 8.1. Material properties

	Aluminum	Alumina	Quartz[b]	PMMA
ρ (gm/cm^3)	2.699	3.972[a]	2.65	1.184[c]
E (Mbar)	0.721	3.987		
G (Mbar)	0.268[a]	1.613[a]		
ν	0.345[a]	0.236[a]		
Z_0 (Mbar)	0.00445			
Z_1 (Mbar)	0.00550			
n	50.0			
m	5.5			
D_0 (μs^{-1})	0.355			
Γ	1.34[a]	1.5	0.635	1.0
c (cm μs^{-1})	0.535	0.797	0.636	0.276[c]
s	1.34	1.60	1.36	1.43[c]
ξ_0	0.02			
μ (Mbar μs)	0.02			
Y_{d0} (Mbar)	0.06			
α^0_{kk} (Mbar)	– 0.0255			

[a] Ref. 28; [b] Ref. 35; [c] Ref. 36

at 1.2% strain (point D), when damage begins to accumulate and the stress decreases. The loading path is reversed at 3% strain and the plastic strain rate immediately drops to zero. Damage, however, continues to accumulate up to 2.2% strain because the stress state remains above the damage surface. Between 2.2% and 0.5% strain, the aluminum unloads elastically but at a degraded modulus. At 0.5% strain, reverse plastic yielding (point RP) occurs. Upon attaining a compressive strain, the material is allowed to heal (point RD).

8.4.2. Interface Model

Interfaces within composite materials transfer the thermomechanical loads between the fiber and matrix constituents as well as between the layers within the composite structure. Furthermore, damage propagation is influenced by the interfaces. Because interfaces have a strong effect on the response of the composite, they should be included in the composite model. The tractions (T_i) across the interface as a function of the jump in the displacement ($\Delta_i = [\![u_i]\!]$) are provided by the interface constitutive model. In this development, $\Delta_i > 0$ provides an increasing separation distance. In general, the normal displacement (Δ_n) initially may be greater than zero, implying that the constituents bounding the interface are not in contact. The response of both the fiber–matrix interface (debonding) and the interface between laminae (delamination) is provided by the interface model. There is no jump in the displacement field across the interface ($\Delta_i = 0$) for a perfect bond. For an imperfect interface, as a load is applied to the structure, the tractions in the materials adjacent to the interface increase and the interface displacement increases. The jump in the displacement field becomes large ($\Delta_i \to \infty$) as the interface fails. The response of the interface is described by the strength and stiffness in directions normal and tangential to the interface. Phenomena including friction, chemical bonds, and surface imperfections may be included in the model.

Interface constitutive relations have been provided in terms of an interface potential ϕ [20–25]:

$$T_i = -\frac{\partial \phi}{\partial \Delta_i} . \tag{8.35}$$

A simple linear elastic [20,22] model

$$\phi(\Delta_n, \Delta_t) = -\frac{1}{2}\left(\frac{\Delta_n^2}{R_n} + \frac{\Delta_t^2}{R_t} \right) \tag{8.36}$$

has been used to investigate the influence of an imperfect interface response for composite structures. In Eq. 8.36, Δ_n and Δ_t are the normal and tangential (shear) components of the displacement jump, respectively. The interface parameters R_i are related to the interfacial stiffness.

For a more realistic interface model (Fig. 8.4), the tractions across the interface initially increase with the surface displacement. The interface stiffness is defined as the slope of the traction versus displacement response at $\Delta_i = 0$. In the absence of friction, as the displacement field is increased, the tractions in the adjacent materials reach a maximum value, which is defined as the strength of the interface. A further increase in the displacements results in a decrease in the load-carrying capacity of the interface, leading ultimately to failure ($T_i \to 0$). Extensions of Eq. 8.36, using polynomial interface potentials [23], have been used to model this more realistic interface response.

Atomistic studies suggest that an exponential expression for the normal response of the interface may provide a more physically based model. Consequently, exponential potential formulations also have been used to model the interfaces between the constituent materials in composite structures [24–26]. Consider, for example [25],

$$\phi(\Delta_n,\Delta_t)=\frac{9}{16}\sigma_0\delta\left\{1-\left[1+z\frac{\Delta_n}{\delta}-\frac{1}{2}\alpha z^2\left(\frac{\Delta_t}{\delta}\right)^2\right]\exp\left[-z\frac{\Delta_n}{\delta}\right]\right\}, \quad (8.37)$$

where z is a interface constant, δ is a characteristic length, and α specifies the ratio of shear to normal interface stiffness. Interfacial imperfections are included [25] by providing a relation for the interfacial strength (σ_0). Additional modifications to include Coulombic friction and to allow perfect bonding under the conditions of interfacial compression also have been included [27]. When friction is included, the shear response unloads to a frictional resistance level when the shear strength is greater than the frictional resistance. A "perfectly plastic" response may result when the frictional resistance is greater than the chemical bond strength [27].

The interfacial models provided by Eqs. 8.35 through 8.37 apply locally at every point located along the interface. The jump in the displacements contained in the displacement continuity conditions for the unidirectional composite (Eq. 8.7), however, are averages of the displacement taken over the interfaces between the constituent materials. Consequently, an averaged interface model, which provides the functional relation between the average surface tractions and the average displacements must be provided. Finite-element sim-

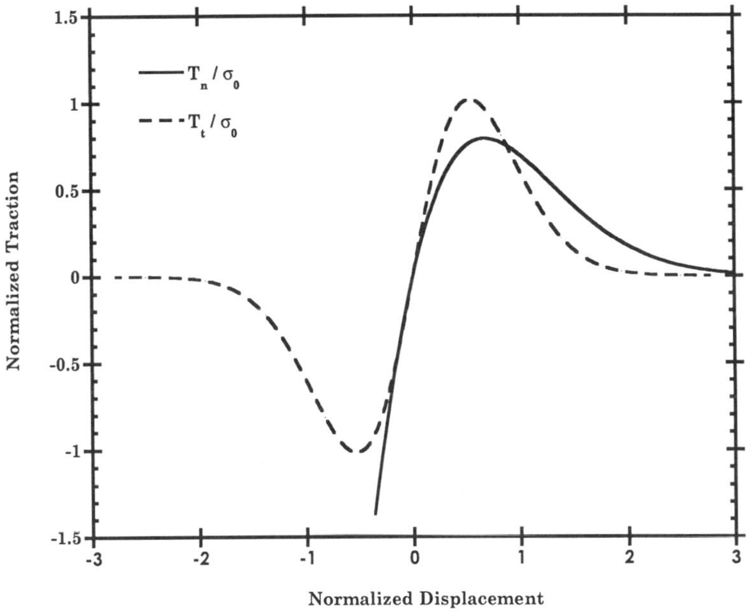

Figure 8.4. Interfacial constitutive model.

ulations [27], using local interfacial constitutive models, have been pursued in an effort to investigate average or global interfacial models, which may be appropriate for use in the average displacement conditions. It was concluded that the global models exhibited functional relations similar to the local models with modified interfacial parameters.

It may be observed from Eqs. 8.35 through 8.37 that the normal and shear responses of the interface are coupled. Consequently, the use of more general interface constitutive models would require a coupled solution for the normal and shear responses of the composite. In order to mitigate this complication, a simpler model has been chosen to provide the normal and shear responses for the interfaces in this investigation:

$$T_n = \frac{2}{\pi}\sigma_0 \sec^{-1}\left(\frac{\Delta_n}{\delta_n} + 1\right) \tag{8.38}$$

and

$$T_t = \frac{2}{\pi}\tau_0 \tan^{-1}\left(\frac{\Delta_t}{\delta_t}\right) . \tag{8.39}$$

This simplified model allows an investigation, including interface phenomena associated with debonding and delamination, without the additional complexities demanded by the more general interface models.

8.5. Numerical Implementation

The Method of Cells, as presented in the previous sections, is suitable for implementation in a continuum analysis. For solutions to large-deformation, high-strain-rate problems, time-explicit, finite-difference, or finite-element numerical techniques typically are employed. For each time cycle of the problem, the continuum state of each computational element is obtained from solutions to the equations of conservation of mass, momentum, and energy as well as a constitutive model for the stress-strain response of the material in each element. Solutions to the macromechanical conservation equations provide the material constitutive model with the macromechanical velocity gradients or rates of deformation and the internal energy. Both the micromechanical and macromechanical stress states, which are consistent with the macromechanical continuum state, are obtained from the solution to the homogenization technique. The macromechanical stress state is then returned to the continuum analysis and the solution for another time cycle is pursued. Details of the solution to the equations governing the response of the composite material are provided in this section. Materials composed of a unidirectional fiber-reinforced composite as well as symmetric lay-ups of fiber-reinforced layers whose orientations vary among the layers are considered. Each of the layers in this investigation is composed of a 6061-T6 aluminum (Al) matrix and alumina (Al_2O_3) fibers. Both materials are assumed to be isotropic and the alumina is assumed to respond elastically. Plasticity and damage are included in the response of the aluminum. The properties used for these materials are provided in Table 8.1. The nonlinear elastic properties are available in the literature [28].

8.5.1. Unidirectional Composites

The rate-of-deformation tensor, which is obtained from solutions to the global conservation equations, is first rotated from the global reference frame of the computational element into the orientation of the unidirectional composite. The rotation tensor, which is updated for each element at each time step [29,30], is used for this purpose. For the analyses in this investigation, the rotation tensor is approximated with the spin tensor

$$r_{ij} \approx \omega_{ij} = \frac{1}{2}\left(\frac{\partial \dot{u}_i}{\partial x_j} - \frac{\partial \dot{u}_j}{\partial x_i}\right). \tag{8.40}$$

A solution for the micromechanical variables ($\dot{\phi}_{ij}^{\alpha\beta}, \dot{\psi}_{ij}^{\alpha\beta}$), which is consistent with the macromechanical strain field, is obtained as a solution to Eqs. 8.5 and 8.7 with the constitutive model for each of the constituents provided by Eq. 8.17. Equations 8.38 and 8.39 are used for the average jump in the displacement ($\Delta_{ij}^{\alpha\beta}$) across the fiber–matrix interface in Eq. 8.7. The solution for the normal and shear stresses, as provided by the Method of Cells, is uncoupled. Once the microvariables have been obtained, the subcell stress state ($\sigma_{ij}^{\alpha\beta}$) can be calculated from the constitutive model for each subcell. The macromechanical stress state (σ_{ij}) is then determined as a volume average over the subcells (Eq. 8.9). Finally, the macromechanical stresses are rotated back into the reference frame of the computational element.

The uniaxial tensile response for a single computational element of a unidirectional composite composed of an aluminum matrix and an alumina fiber is provided in Fig. 8.5 for a strain rate of 10^3 s^{-1}. The applied strain is along the fiber direction. It may be seen that the micromechanical model provides useful information concerning the response of the composite. In Fig. 8.5a, the stress in the direction

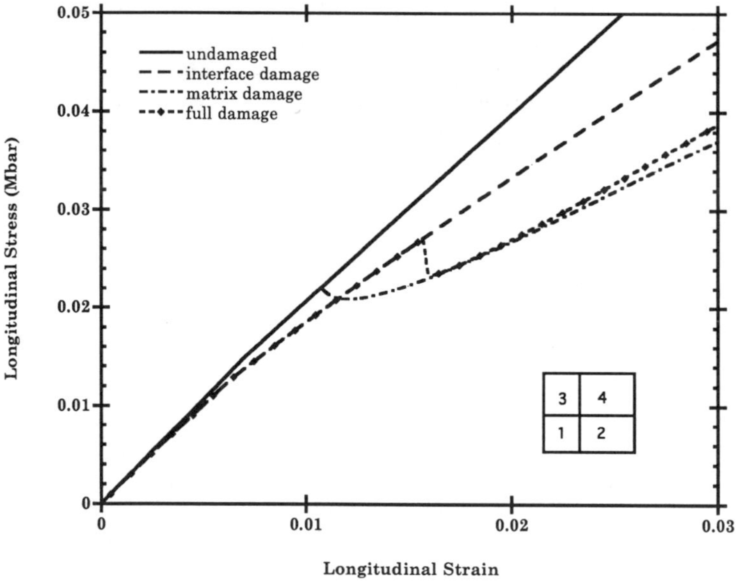

(a)

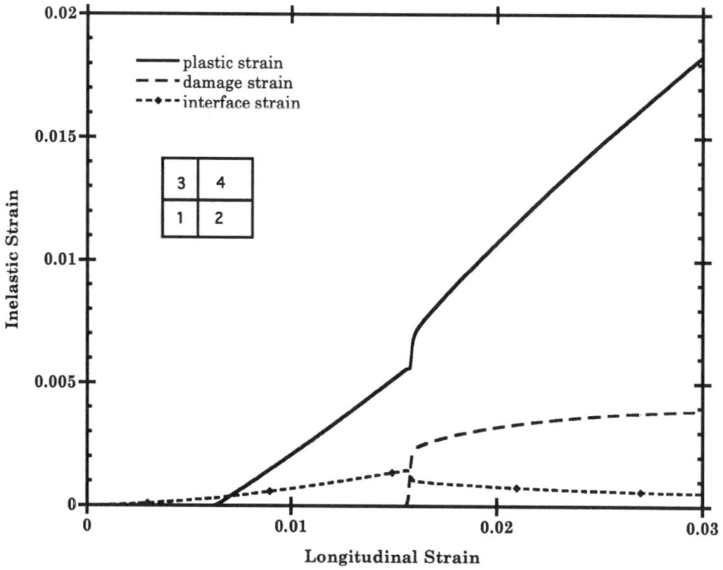

(b)

Figure 8.5. Unidirectional fiber-reinforced composite loaded in uniaxial strain parallel to the fiber direction. (a) longitudinal stress and (b) inelastic strain.

of the applied strain is provided for four cases: i) no damage, ii) matrix damage, iii) interface debonding, and iv) both matrix damage and debonding. It may be seen that, because a threshold stress state is necessary before constituent damage is initiated, matrix damage is delayed in the deformation process. Matrix damage proceeds quickly after initiation because of the small value for the damage viscosity, which results from comparison of the damage model with spall data [28]. Interface damage, however, is initiated immediately with the load and progresses gradually during the deformation. Because there is no fiber damage, the longitudinal stress relaxes to a nonzero value when the composite is strained along the fiber direction. However, the stress would approach zero if the composite were strained transverse to the fiber direction. When only matrix damage is modeled, the subcells adjoining the fiber (subcell 1) initially fail because of the greater stiffness of the fiber. For a strain along the fiber direction, matrix subcells 2 and 3 (Fig. 8.5) initially fail.

The accumulation of the inelastic strains with combined matrix and interface damage is provided in Fig. 8.5b for longitudinal loading. An effective inelastic strain is used as the measure for the plastic and damage strains. The second invariant of a strain measure based on the interfacial displacements is used for the interfacial strain. The gradual accumulation of interfacial damage from the initiation of loading is evident. This is accompanied by a gradual increase in the plastic flow of the matrix once matrix yielding occurs. After matrix damage is initiated, there is a rapid decrease in the interface displacements and a rapid increase in the plastic flow of the undamaged matrix subcells. Matrix damage occurs in the subcell that does not share a side with the fiber (subcell 4) when both damage mechanisms are present.

8.5.2. Multiple Layers

A laminate is constructed of layers of a unidirectional fiber-reinforced composite. Typically, these layers are fabricated in a symmetric stacking sequence. Each of the layers or laminae is assumed to contain the same matrix and fiber materials. However, the orientation of each of the lamina varies among the layers. A simple RVE for the laminate is provided in Fig. 8.6. To facilitate the numerical solution to the RVE, the micromechanical analysis is divided into two coupled solutions: one for the laminate and one for each of the unidirectional layers. The macromechanical response is obtained by first determining a solution for the laminate which is consistent with the continuum macromechanical strain state. This solution results in the strain states for each of the lamina and is obtained from the solution

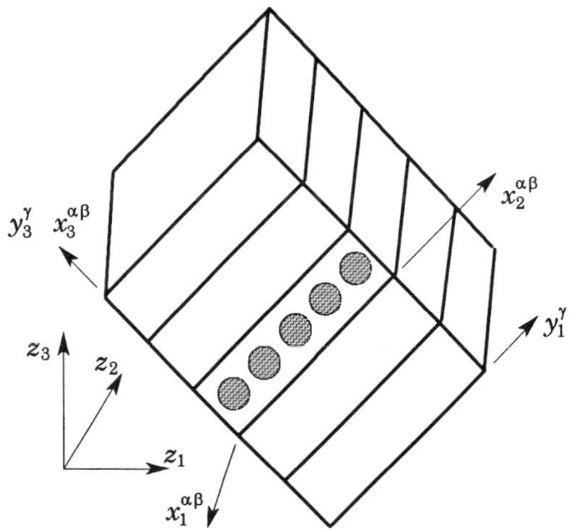

Figure 8.6. Laminate geometry.

to the one-dimensional layered problem provided in Sec. 8.3.2. The solution to the laminate problem requires the average properties for each layer. The response of each layer is obtained as a solution to the unidirectional, fiber-reinforced RVE, which was provided in Sec. 8.3.1. The stress state for each layer is determined to be consistent with the strain state obtained from the solution to the laminate problem. Results for the layered solution provide the average properties of each layer. This approach of decoupling the laminate and layer solutions results in a convenient computational algorithm for a numerical continuum analysis.

Again, the macromechanical rate of deformation and energy of each computational element are available from solutions to the global conservation equations. These gradients are rotated from the reference frame of the element (z_i) into the reference frame of the laminate (y_i^γ), as shown in Fig. 8.6. The constitutive response for each of the fiber-reinforced layers of the laminate is expressed as

$$\dot{\sigma}_{ij}^{\gamma} = C_{ijkl}^{\gamma} \dot{\varepsilon}_{kl}^{\gamma} - \dot{\Gamma}_{ij}^{\gamma}, \tag{8.41}$$

where Γ_{ij}^{γ} provides the inelastic response of each layer (γ). Currently, constituent and interface damage are not included in the solution for each layer. Therefore, the stiffness (C_{ijkl}^{γ}) tensor is constant. Knowing the rate of deformation in the laminate coordinate system (y_i^γ), Eq. 8.41 coupled with Eqs. 8.13 and 8.14, and using the layer strain rates as defined by Eq. 8.11, provides a system of equations for the layer microvariables ζ_1^γ, once the elastic stiffness (C_{ijkl}^{γ})

and inelastic contributions (Γ_{ij}^{γ}) are known. Delamination between the layers is modeled. Equations 8.38 and 8.39 are used for the jump in displacements between each layer. The solution to the laminate problem provides the strain rates ($\dot{\varepsilon}_{ij}^{\gamma}$) for each of the fiber-reinforced layers. A solution for the response of each layer next is pursued by rotating the strain rates ($\dot{\varepsilon}_{ij}^{\gamma}$) into the reference frame of each layer ($x_i^{\alpha\beta}$) and solving for the layer microvariables ($\dot{\phi}_{ij}^{\alpha\beta}$, $\dot{\psi}_{ij}^{\alpha\beta}$) as discussed in Sec. 8.5.1. The solution for each layer results in the layer stress rates ($\dot{\sigma}_{ij}^{\alpha\beta}$). The average layer stress rates then may be determined using Eq. 8.9. The inelastic strain rates for each layer ($\dot{\eta}_{ij}^{\alpha\beta}$ and $\dot{\xi}_{ij}^{\alpha\beta}$) are updated as part of the layer solution. The layer stress rates are then rotated back into the reference frame of the laminate (y_i^{γ}) and the inelastic rates ($\dot{\Gamma}_{ij}^{\gamma}$) for each layer are updated using Eq. 8.41. The average laminate stresses are obtained using Eq. 8.16. Finally, the average stress state is rotated back into the reference frame of the computational element.

The uniaxial tensile response for a single element of a quasi-isotropic laminate is provided in Fig. 8.7. A four-layer laminate is modeled. The relative orientation of the fibers in successive layers is 0°/45°/90°/–45°. The layering direction is taken to be along x_3 and the applied strain is at 45° to the x_1 and x_3 directions. Only damage due to delamination is modeled. The macromechanical stresses normal to the laminate are provided in Fig. 8.7a. Normal stresses for both the intact and damaged laminate are shown. For the intact laminate, σ_{11} is large because the fibers are partially aligned to the x_1 direction. The σ_{22} stress is smallest because the x_2 direction is transverse to the applied load. All three normal stresses are degraded as delamina-ion progresses. The stress along the layering direction (σ_{33}) is preferentially reduced. The laminate shear stress (σ_{13}) is shown in Fig. 8.7b for four cases: the intact laminate, shear interface damage, normal interface damage, and for both shear and normal interface damage. It may be seen that when both modes of damage are included, shear delamination dominates for small values of strain. As the strain increases, the combined response is dominated by normal interface damage. Shear delamination has only a minor effect on the response of the laminate because the shear stresses are limited by the yield strength of the matrix.

8.6. Computational Simulations

The composite model, provided in the preceding sections, has been implemented into an explicit, finite-difference, one-dimensional continuum analysis [31]. The analysis includes all three displacement components. Numerical simulations of plate-impact experiments are

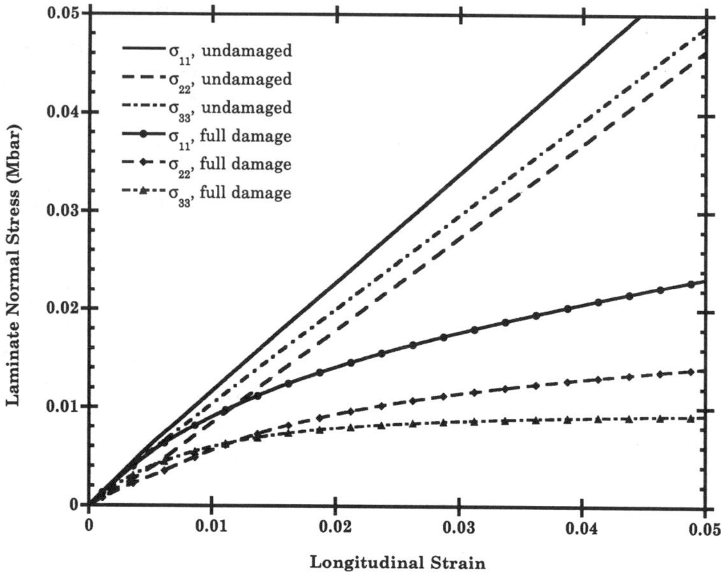

(a)

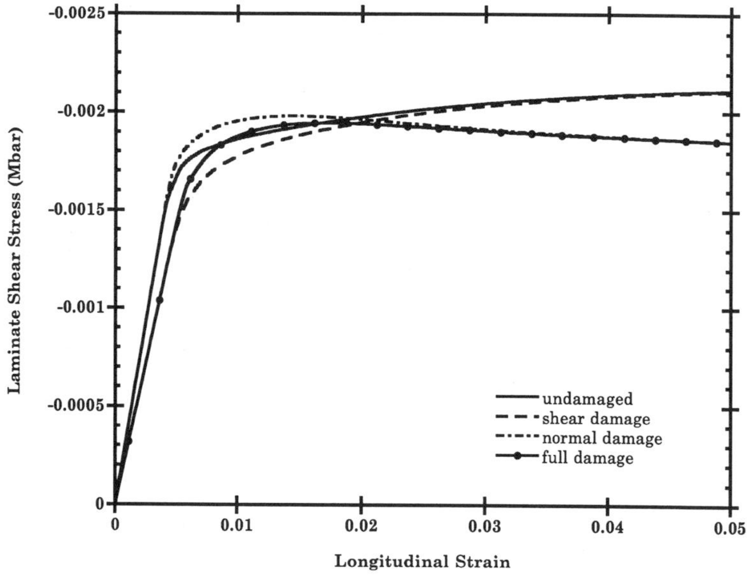

(b)

Figure 8.7. Laminated composite loaded in uniaxial strain at 45° to the layered direction. (a) laminate normal stresses and (b) laminate shear stress.

presented to demonstrate the versatility of the micromechanical approach for modeling high-strain-rate problems. A schematic of the plate-impact problem is provided in Fig. 8.8. This geometry has been used to pursue Hugoniot, dispersion, and spall information for composite materials [28]. A composite which has a matrix material of 6061-T6 aluminum and an alumina fiber is used as the target. In the experiments, a VISAR is used to measure the particle velocity on the back of the target material. Spallation measurements are conducted with a stress-free back surface. Wave profiles are measured at the target–window interface when a window is attached to the back of the target material. Similar to experimental investigations, a Z-cut quartz impactor, which is backed with PMMA, is used as the projectile in the simulations. A fiber volume fraction of 30% is assumed. The elastic properties for the constituent materials are provided in Table 8.1. The initial conditions of the simulations are shown in Table 8.2.

Data for plate-impact experiments are available [28] for pure 6061-T6 aluminum. Data for the aluminum target backed with a LiF window and with a free surface are provided in Fig. 8.9. The wave-profile data are used to determine the parameters for the plastic response of the aluminum. A comparison of the data with a calculation

Table 8.2. Plate-impact simulations

	Thickness, mm				
	PMMA	Quartz	Sample	Window	Impact velocity, m/s
Aluminum wave profile [28]	1.521	1.521	3.920	3.92	505
Aluminum spall [28]	2.700	2.700	3.866	0.00	550
Composite spall	2.700	2.700	3.866	0.00	550

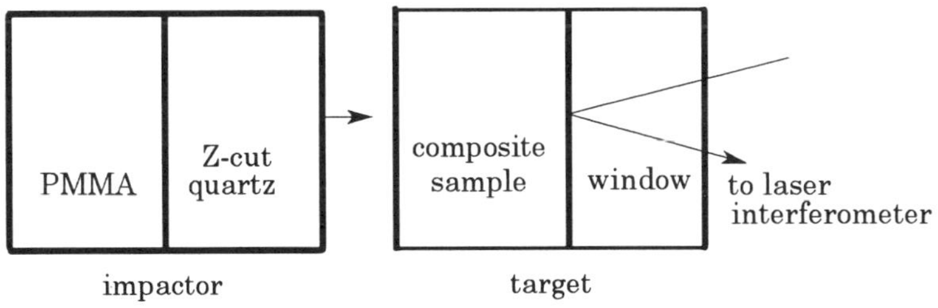

Figure 8.8. Plate-impact geometry.

using the Bodner–Partom viscoplastic model is provided in Fig 8.9a. Also provided are comparisons using a nonlinear (Eq. 8.25) and linear EOS (i.e., $a_i = 0$ for $i > 1$ in Eq. 8.25). The spallation measurement was used to obtain the parameters used in the damage model, which is discussed in Sec. 8.4.1. A comparison of the calculation and the data for the spallation experiment is provided in Fig. 8.9b. The model parameters for the inelastic response of the aluminum also are provided in Table 8.1. Currently, no experimental data are available for the fiber-reinforced aluminum/alumina composite material.

8.6.1. Unidirectional Composite Simulations

Computational simulations of spall experiments for a unidirectional fiber-reinforced composite are shown in Fig. 8.10. In Fig. 8.10a, surface velocity histories are provided for four cases: no damage, matrix damage, debonding, and both matrix and interface damage, when the velocity of impact is 90° to the fiber direction. At 0.5 μs, the compressive wave reaches the free surface. The velocity drop at 1.3 μs corresponds to the time when the tensile stress wave, which is produced by the interaction of rarefaction waves within the composite, arrives at the surface. When matrix damage is included, the aluminum matrix reaches the damage threshold stress and a spall signal is calculated. For this simulation, only normal debonding is predicted. Consequently, the effects of debonding are observed only after the composite is placed in tension. As in the one-element simulations, the effects of debonding are gradual. Decreasing the bond strength delays the arrival of the tensile wave and decreases the deceleration response. In the limit of very weak interfaces, the velocity profile does not become horizontal because the two matrix subcells, which are aligned with the direction of impact, continue to support a finite stress. When both matrix and interface damage are modeled, the velocity history is controlled initially by debonding in the tensile region. Ultimately, the matrix reaches the damage threshold, and a spall signal is again observed. The spall signal is delayed, however, because of the effects of interface damage.

The simulated response for spallation with a unidirectional fiber-reinforced composite oriented at 45° to the impact direction is shown in Fig. 8.10b. Simulations for the intact composite and including both matrix damage and debonding are provided. For this simulation, both quasi-longitudinal and quasi-transverse wave are generated in the composite. Furthermore, damage due to shear at the interface is evident in the compressive region of the response. No matrix damage is generated for this computation.

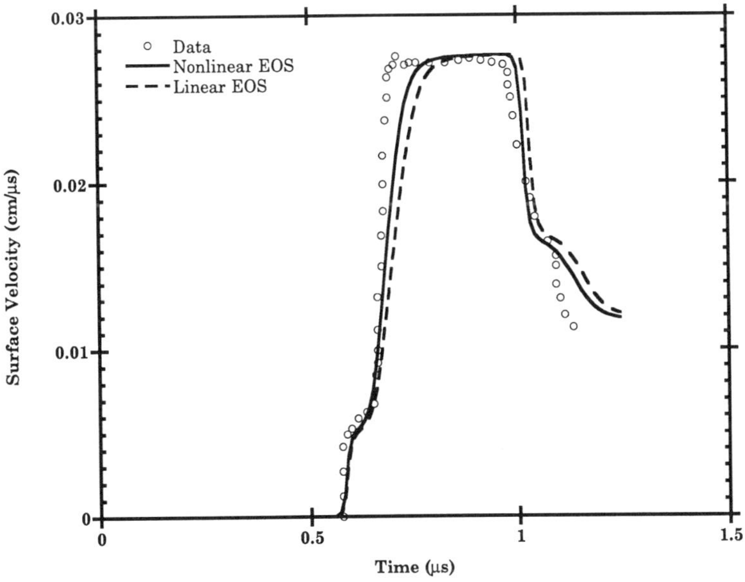

(a)

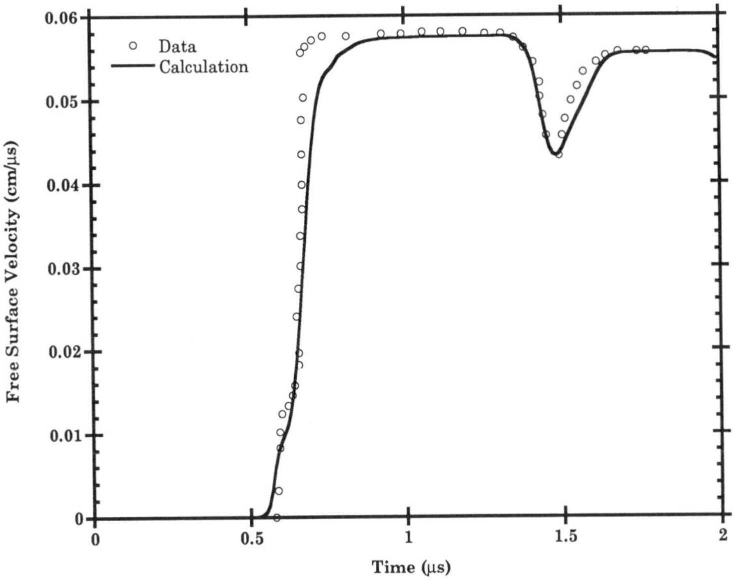

(b)

Figure 8.9. Pure aluminum impact simulations. (a) wave-profile experiment, and (b) spall experiment.

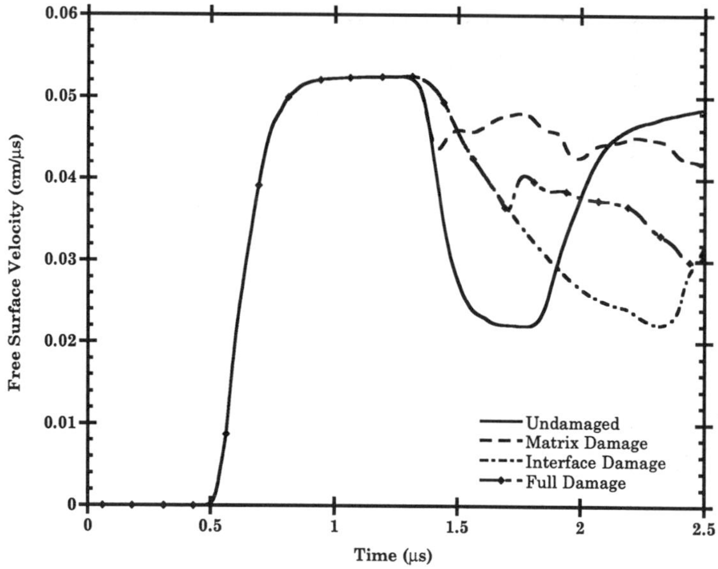

(a)

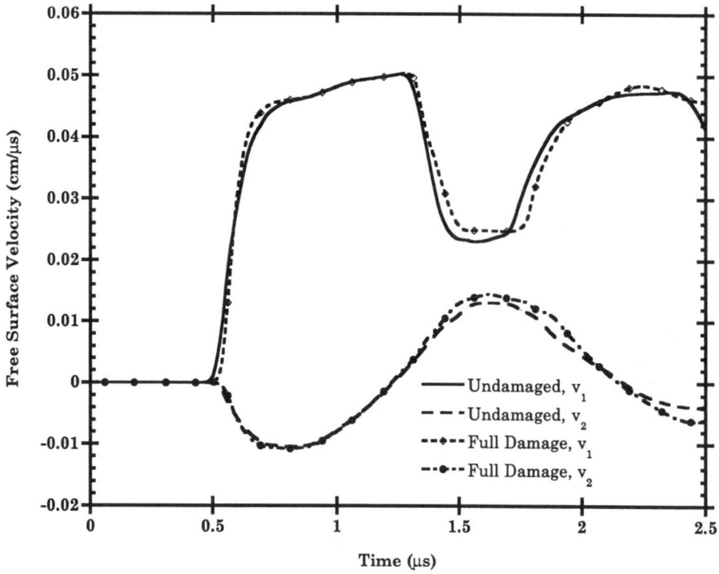

(b)

Figure 8.10. Unidirectional fiber-reinforced composite impact simulations. (a) 90° impact and (b) 45° impact.

8.6.2. Laminate Simulation

Spall simulations for the quasi-isotropic laminate are provided in Fig. 8.11. Again, only the fiber orientation varies among the layers. A stacking sequence of 0°/45°/90°/–45° is used. The impact direction is taken transverse to the direction of the layers. Only damage resulting from delamination is modeled. For this geometry, delamination resulting from the shear stress is absent. In Fig. 8.11a, the effect of reducing the bond strength of the layers is shown. Similar to the results for debonding in unidirectional composites, the release wave is delayed as the interfaces are weakened. For delamination, however, the damage response approaches a horizontal limit as the interface strength decreases; that is, complete interfacial separation is allowed. When this occurs, the rear portion of the sample is ejected and continues moving with the peak velocity attained by the free surface prior to delamination.The response for a spall simulation with the layers oriented at 45° to the impact direction is provided in Fig. 8.11b. Cases for both an intact laminate and for delamination are shown. As expected, both quasi-longitudinal and quasi-transverse waves are generated in the samples. Shear damage at the interfaces is observed now in the compressive region of the response. Extensive normal interfacial damage is calculated in the release wave.

8.7. Summary

Computational simulations of engineering structures require accurate material models. Phenomena including nonlinear elasticity and inelastic effects due to plasticity and damage are important for large-deformation, high-strain-rate problems. The additional complexity offered by the presence of interfaces also must be addressed for heterogeneous materials such as composites. Furthermore, viable material models must be computationally robust and numerically efficient. Consequently, a detailed response of the micromechanical structure of composites cannot be modeled using conventional continuum analyses. Macromechanical approaches to modeling composites, however, do not offer the versatility necessary to design engineering structures or interpret experimental results.

Homogenization techniques are not new [32,33]. The macromechanical response of the composite is obtained using an appropriate average of the responses of the microstructure in the heterogeneous material, using a homogenization approach. Therefore, both the response of the constituents and the interfaces are modeled. One homogenization technique, the Method of Cells, has demonstrated its versatility. This approach is applicable to general loading conditions,

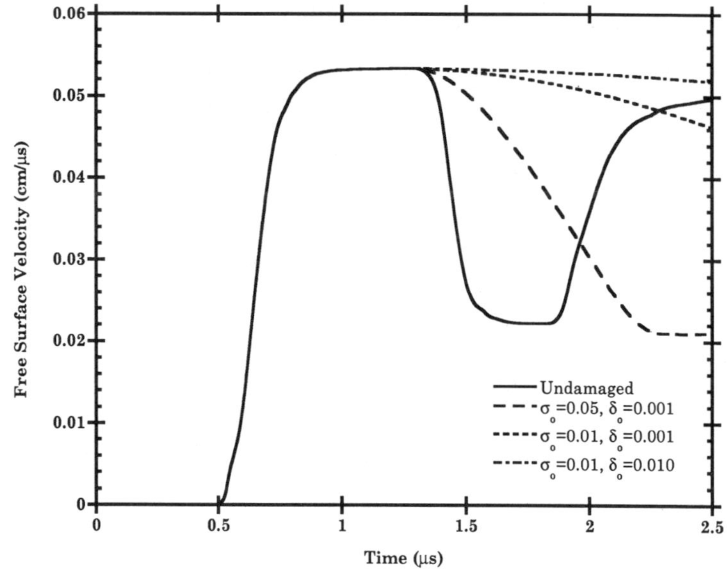

(a)

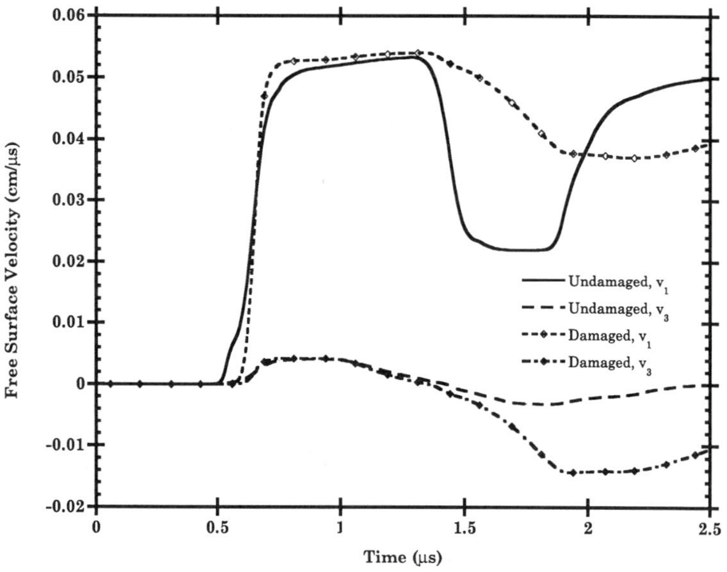

(b)

Figure 8.11. Laminated composite impact simulations. (a) 90° impact and (b) 45° impact.

compatible with a wide variety of constitutive models and easy to implement into computer analyses. This method has been extended to address high-strain-rate deformations. Anisotropic elasticity, matrix plasticity, and imperfect bonding models already have been implemented into the homogenization technique. Nonlinear elasticity and constituent damage have been included to pursue the response of composites under high-strain-rate, uniaxial-strain conditions. A computationally expedient approach to modeling laminates has been presented. The model has been implemented into a one-dimensional continuum computer code. Related investigations [34] have established the ability to implement this approach into multidimensional analyses. Furthermore, simulations have demonstrated the ability of this approach to pursue phenomena including constituent damage, debonding, and delamination within composite structures.

A homogenization approach to modeling composite materials does not provide a panacea. A price must be paid for the added details provided by a micromechanical approach both in computational storage and efficiency. Furthermore, assumptions such as the simplified geometries chosen to model a representative volume and the order of approximation used to define the microvariables must be made. Consequently, phenomena such as shear coupling are absent in a first-order theory. Questions regarding the ability to adequately model damage phenomena also must be considered. Despite these simplifications, a micromechanical approach to modeling composite materials has demonstrated numerous successes. However, a prudent approach to modeling composite structures should involve both micromechanical and macromechanical formulations. There also should be a strong coupling between analysis and experiment. Future efforts must be made to compare the predictive capability of micromechanical models for composites to data obtained from controlled experiments.

Acknowledgments

This research was performed at the Los Alamos National Laboratory. The financial support of the Joint DoD/DOE Munitions Technology Development Program is appreciated.

References

[1] G.L. Povirk, A. Needleman, and S.R. Nutt, *Mater. Sci. Eng.* **A125**, pp. 129–140 (1990).

[2] J. Aboudi, *Int. J. Engng. Sci.* **23**(7), pp. 773–787 (1985).

[3] J. Aboudi, *Int. J. Engng. Sci.* **20**, pp. 605–621 (1982).

[4] J. Aboudi, *Int. J. Engng. Sci.* **22**(4), pp. 439–449 (1984).

[5] M.J. Pindera and J. Aboudi, Int. *J. Plast.* **4**, pp. 195–214 (1988).

[6] J.M. McGlaun and P. Yarrington, in *High-Pressure Shock Compression of Solids* (ed. J.R. Asay and M. Shahinpoor), Springer-Verlag, New York (1992).

[7] D. Sulsky, Z. Chen, and H.L. Schreyer, *Comput. Meth. Appl. Engrg.* **118**, pp. 179–196 (1994).

[8] J.B. Aidun and F.L. Addessio, *J. Compos. Mater.* (in press).

[9] J. Aboudi, *Int. J. Solids Struct.* **17**, pp. 1005–1018 (1981).

[10] F.L. Addessio and J.N. Johnson, *J. Appl. Phys.* **74**(3), pp. 1640–1648 (1993).

[11] F.L. Addessio and J.N. Johnson, *J. Appl. Phys.* **67**(7), pp. 3275–3286 (1990).

[12] G.R. Johnson and W.H. Cook, *Eng. Fracture Mech.* **21**(1), pp. 31–48 (1985).

[13] P.S. Follansbee and U.F. Kocks, *Acta Metall.* **36**(1), pp. 81–93 (1988).

[14] F.J. Zerilli and R.W. Armstrong, *J. Appl. Phys.* **61**, pp. 1816–1825 (1987).

[15] S.R. Bodner and Y. Partom, *J. Appl Mech.* **4**, pp. 385–389 (1975).

[16] J.K. Dienes, *Mech. Mater.* **4**, pp. 325–335 (1985).

[17] A.L. Gurson, *J. Eng. Mater. Technol.* **99**, pp. 2–15 (1977).

[18] L.J. Sluys, *Wave Propagation, Localisation, and Dispersion in Softening Solids*, Ph.D. Thesis, Delft University of Technology, Netherlands (1992).

[19] N.R. Hansen, *Theories of Elastoplasticity Coupled with Continuum Damage Mechanics*, Sandia report SAND92-1436, Sandia National Laboratories, Albuquerque, New Mexico (1993).

[20] J.P. Jones, and J.S. Whittier, *J. Appl. Mech.* **34**, pp. 905–909 (1967).

[21] J. Aboudi, *Compos. Sci. Tech.* **28**, pp. 103–128 (1987).

[22] J. Aboudi, *Int. J. Plasticity* **4**, pp. 103–125 (1988).

[23] A. Needleman, *J. Appl. Mech.* **54**, pp. 525–531 (1987).

[24] A. Needleman, *J. Mech. Phys. Solids* **38** (3), pp. 289–324 (1990).

[25] A. Needleman, *Int. J. Fracture* **42**, pp. 21–40 (1990).

[26] M. Finot, Y.-L. Shen, A. Needleman, and S. Suresh, *Met. Mater. Trans. A* **25A**, pp. 2403–2420 (1994).

[27] J.D. McGee and C.T. Herakovich, *Micromechanics of Fiber/Matrix Debonding*, Center for Light Thermal Structures Interim Report AM-92-01, University of Virginia, Charlottesville, Virginia (1992).

[28] J.N. Johnson, R. S. Hixson, and G.T. Gray, *J. Appl. Phys.* **76**(10), pp. 5706–5718 (1994).

[29] J.K. Dienes, *Acta Mech.* **32**, pp. 217–232 (1979).

[30] D.P. Flanagan and L.M. Taylor, *Comput. Meth. Appl. Mech. Engr.* **62**, pp. 305–320 (1987).

[31] F.L. Addessio, D.E. Carroll, J.K. Dukowicz, F.H. Harlow, J.N. Johnson, B.A. Kashiwa, M.E. Maltrud, and H.M. Ruppel, *CAVEAT: A Computer Code for Fluid Dynamics Problems with Large Distortion and Internal Slip*, Los Alamos National Laboratory report LA-10613-MS, Los Alamos, New Mexico (1984).

[32] C.T. Sun, J.D. Achenbach, and G. Herrmann, *J. Appl. Mech.* **35**, pp. 467–475 (1968).

[33] R.A. Grot and J.D. Achenbach, *Acta Mech.* **9**, pp. 245–263 (1970).

[34] F.L. Addessio and J.B. Aidun, *Computational Modeling of Fiber-Reinforced Composites*, 14th IMACS World Congress Meeting, July 11–15, Atlanta (1994).

[35] J. Wackerle, J. *Appl. Phys.* **33**, pp. 922–937 (1962).

[36] L. Barker and R.E. Hollenbach, *J. Appl. Phys.* **41**, pp. 4208–4226 (1970).

CHAPTER 9

Attenuation of Longitudinal Elastoplastic Pulses

Lee Davison

9.1. Introduction

In this chapter, we discuss (in tutorial fashion) the attenuation of a uniaxial-strain pulse propagating in an elastoplastic solid. This is an old and familiar problem, but its solution seems not to have been presented in the detail required to properly describe the many and varied wave interactions that produce the propagated waveform and lead to attenuation of the pulse. These problems are usually solved by numerical means, often in the context of considerably more comprehensive theories than the one used here. However, some interesting phenomena are easily overlooked when examining the numerical results.

We shall use the theory of ideal elastoplastic response in the approximation of small deformation. It captures only the most basic aspect of the observed behavior of elastoplastic materials, but it is widely used and is well suited to illustrate the effects under discussion.

Numerous investigators have studied elastoplastic wave propagation in slender rods in the approximation of uniaxial stress [1–4], but that problem is quite different from the problem of uniaxial strain discussed here. In the case of wave propagation in rods, only the longitudinal component of principal stress is nonzero, so the yield criterion involves only a single stress-strain curve. In the case of waves of uniaxial strain, neither longitudinal nor transverse stress components vanish, so the yield criterion involves the stress-strain paths for both components. It is this difference that complicates the analysis of waves of uniaxial strain.

The problem of loading waves propagating into a half-space has been investigated in detail and Morland [5] has solved the problem discussed here, but in a more general version that leads to less explicit results.

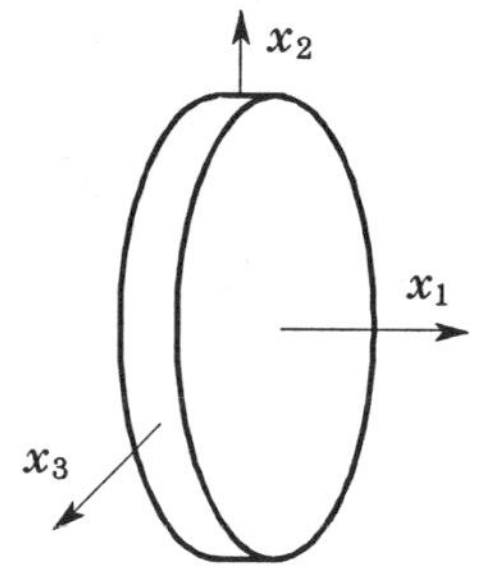

Figure 9.1. Thick slab, showing the coordinate frame used for the analysis.

9.2. Stress and Deformation Fields

Let us consider deformation of a homogeneous and isotropic material body in the Cartesian frame defined by coordinates x_i, for $i = 1, 2, 3$.

Attention is restricted to small deformations characterized by the displacement functions

$$u_i = u_i(x_1, x_2, x_3, t). \tag{9.1}$$

The strain of the material is given by the linear strain measure

$$E_{ij} = \frac{1}{2}\left[\frac{\partial u_i}{\partial x_j} + \frac{\partial u_j}{\partial x_i}\right] \tag{9.2}$$

and the particle velocity is given by

$$v_i = \frac{\partial u_i}{\partial t}. \tag{9.3}$$

Stresses are expressed in terms of components, t_{ij}, of the Cauchy stress tensor. For the simple case of a linear isotropic solid, the stress components are given by

$$t_{ij} = \lambda E_{kk} \delta_{ij} + 2\mu E_{ij}, \tag{9.4}$$

where the quantities λ and μ are material constants called Lamé parameters.

For the specific case of a slab (see Fig. 9.1) that is subject to uniform normal loading over its faces but is not permitted lateral movement, the deformation corresponds to the displacements

$$u_1 = u_1(x_1, t), \quad u_1 = u_2 = 0. \tag{9.5}$$

This displacement field leads to a nonvanishing value of E_{11} (positive for elongation) given by

$$E_{11} = \frac{\partial u_1(x_1, t)}{\partial x_1}, \tag{9.6}$$

with all other strain components vanishing. This defines the case called *uniaxial strain*. A longitudinal stress, t_{11} (positive for tension) must be imposed to produce the strain. The lateral components of stress are nonzero because a force must be applied to maintain the lateral constraint. Since the only coordinate appearing in the following analysis is x_1, we shall use the abbreviation $x_1 = x$.

9.3. Longitudinal Shocks

A *shock* is a propagating discontinuity in stress and particle velocity. In the special case of a plane longitudinal shock, the displacements are in the form given by Eqs. 9.5. The stress component $t_{11}(x,t)$, the strain component $E_{11}(x,t)$, and the particle velocity component $v_1(x,t)$ are all discontinuous at the shock, but the displacement component $u_1(x,t)$ is continuous at this surface.

The requirements for balance of mass and momentum at a shock are represented by the jump conditions

$$[\![-E_{11}]\!]\,U_S = [\![v_1]\!], \quad \rho[\![v_1]\!]\,U_S = [\![-t_{11}]\!], \tag{9.7}$$

where $[\![\varphi]\!] \equiv \varphi^+ - \varphi^-$ denotes the jump in the enclosed quantity from the initial value, φ^-, to the final value, φ^+, that occurs as the shock passes a material point. The quantity U_S is the (Lagrangean) shock velocity, taken positive for propagation in the $+x$ direction, and ρ is the density of the undeformed material. The principles of balance of energy and production of entropy must also be satisfied at a shock, but we shall restrict attention to weak shocks, in which case changes in the thermal variables can be neglected, leaving us with a purely mechanical theory.

When applied to a shock propagating into material in a known state $\mathcal{S}^- = \{t_{11}^-, E_{11}^-, v_1^-\}$, the two jump conditions involve the unknown quantities $\mathcal{S}^+ = \{t_{11}^+, E_{11}^+, v_1^+\}$ and U_S. One of the quantities comprising $\mathcal{S}^+$ is normally specified as a measure of the stimulus producing the shock.* The relevant material properties are characterized by equations, called *Hugoniot curves* or simply *Hugoniots*, relating any pair of the quantities $\mathcal{S}^+$ and depending parametrically on the known

* In the case of ideal elastoplastic materials in which the elastic response is linear, the shock velocity is excluded from a role in describing the material state because relations between the shock velocity and the other variables are not invertible.

initial state, $\mathcal{S}^-$ [6, p.15]. A Hugoniot curve corresponding to the initial state $\mathcal{S}^-$ (often described as *centered* on this state) is simply the locus of endstates that can be achieved by passing a shock through material in the given initial state. Hugoniot curves can be measured experimentally or calculated from an appropriate theory of material response.

For the analysis in this chapter, it is adequate to restrict attention to Hugoniots relating t_{11} and E_{11}, t_{22} and E_{11}, and t_{11} and v_1. These are called t_{11}–E_{11}, t_{22}–E_{11}, and t_{11}–v_1 Hugoniots, respectively, and each forms a curve through the state $\mathcal{S}^-$ in the plane of its variables. Schematic examples are shown in Fig. 9.2. The chord in each of these planes connecting the initial and final states associated with a given shock is called the *Rayleigh line* for the shock. The jump conditions yield relationships between the slope of the Rayleigh line and the shock velocity. As we shall draw these curves, the slope of the Rayleigh line in the t_{11}–E_{11} plane is always positive, but its slope in the t_{11}–v_1 plane changes sign according to the direction of shock propagation. This slope is positive for shocks propagating in the $+x$ direction and negative for shocks propagating in the $-x$ direction. The t_{11}–v_1 Hugoniots (and associated Rayleigh lines) for left- and right-propagating shocks are reflections of one another in a vertical line through $\mathcal{S}^-$.

The shock interactions discussed in this chapter occur when one shock overtakes another, when shocks propagating toward one another collide, and when shocks encounter an interface at which the material state changes. Usually, but not inevitably, interaction pro-

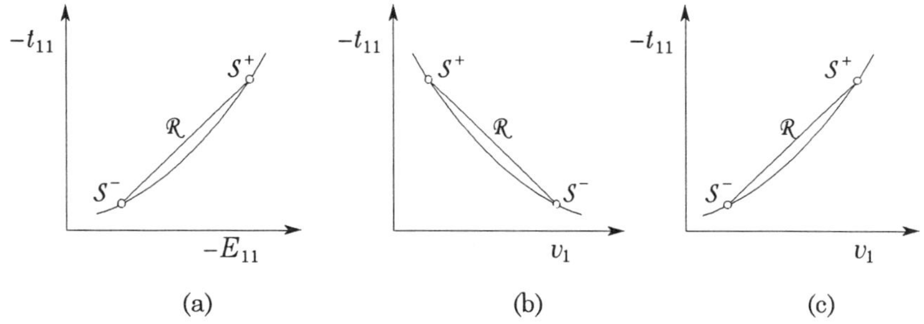

Figure 9.2. Generic Hugoniot curves for a shock transition from state $\mathcal{S}^-$ to state $\mathcal{S}^+$. The secant line designated $\mathcal{R}$ is the Rayleigh line for the shock. (a) stress–strain Hugoniot; (b) stress–particle velocity Hugoniot for a left-propagating shock; (c) stress–particle velocity Hugoniot for a right-propagating shock. The slope of the Rayleigh line in the stress–strain plane is ρU_S^2 and the slope in the stress–particle velocity plane is ρU_S, with $U_S > 0$ for a right-propagating shock and $U_S < 0$ for a left-propagating shock.

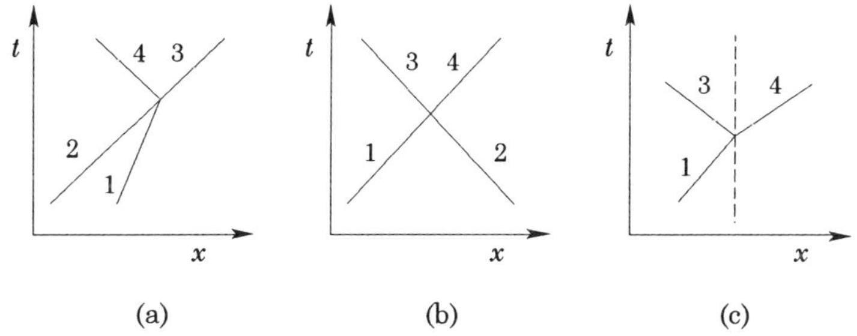

Figure 9.3. Shock interactions shown in the x–t plane. (a) Shock 2 overtaking shock 1 producing an interaction in which shocks 3 and 4 are formed; (b) shocks 1 and 2 colliding to form shocks 3 and 4; (c) shock 1 encountering a material interface, producing shocks 3 and 4.

duces a shock propagating in each direction from the plane at which the interaction takes place. These situations can be visualized using a plot of the position of shocks as they evolve in time, an x–t diagram, as shown in Fig. 9.3.

The t_{11}–v_1 Hugoniot plays an essential role in analysis of shock interactions. Since the longitudinal stress and particle velocity fields in the material between the two separating shocks produced by an interaction must be continuous, the jumps in these variables at each of the shocks must result in the same endstate values t_{11}^+ and v_1^+. This means that these field values are defined by the intersection of the t_{11}–v_1 Hugoniots centered on the state ahead of each of the interacting shocks and having the slope appropriate to the direction of propagation of the shock in question.

9.4. Material Response Model: Ideal Elastoplasticity at Small Strain

We must now determine shock Hugoniots for the elastoplastic material that is the subject of this chapter. As mentioned in the previous section, neglect of thermal effects means that the response of a material to shock-induced deformation is the same as its response when the deformation is effected quasi-statically; i.e., the t_{11}–E_{11} Hugoniot and the static stress-strain curve are the same.

The setting in which elastoplastic response is most easily observed is that in which a ductile metal rod is subjected to tension along its axis. When the applied stress is plotted against the associated strain, a curve such as that of Fig. 9.4b is obtained. If the maximum strain imposed is sufficiently small, the elongation process is reversible (i.e., the curve is retraced upon unloading) and is described by elasticity theory. When the rod is stretched beyond a limit

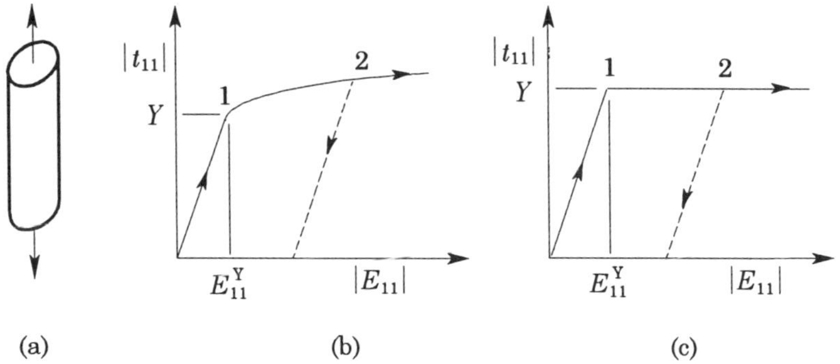

Figure 9.4. Stress-strain response of a rod in states of uniaxial stress. (a) Test configuration; (b) typical observed stress-strain curve; (c) idealized stress-strain curve. The broken lines are unloading paths from state 2.

called the *elastic limit* or *yield point,* much smaller stress increments suffice to produce a given strain increment and the loading curve is no longer retraced when the applied tension is reduced. When the applied tension is completely removed, some residual elongation, called *plastic deformation,* is observed, as shown in the figure. The unloading is accompanied by an axial contraction of the rod that is described by elasticity theory based on strain measured relative to the residual configuration. The simplest theory of the phenomena described, that of an ideal elastoplastic material, corresponds to the stress-strain curve of Fig. 9.4c; i.e., once the bar has been deformed beyond the yield point of the material, it can be further extended without application of additional stress.

Elastoplasticity theory is based on the postulate that yielding (i.e., onset of plastic deformation) occurs when some critical state of stress is reached. For example, the rod yields at a critical axial stress $t_{11} = Y$. Although we have been discussing plasticity in terms of tensile response, it is also observed in compression. Often the magnitude of the stress is the same for yielding in compression as in tension, and we shall limit ourselves to this case. Two functions that have been proposed as criteria for yield of material subject to general three-dimensional stress fields are the *Tresca criterion* and the *von Mises criterion.* According to the Tresca criterion, yielding at a material point occurs when the maximum shear stress at that point reaches a critical value. The von Mises criterion is a simple function of the invariants of the stress tensor that closely approximates the Tresca condition. These criteria do not predict the onset of yielding at the same stress under all conditions, but they do agree that yielding of the rod occurs when

$$|t_{11}| = Y \tag{9.8}$$

and yielding of the slab (in uniaxial strain) occurs when

$$|t_{11} - t_{22}| = Y. \tag{9.9}$$

The *yield stress* or *elastic limit* Y is a material constant that can be determined from the measured stress-strain response of a rod, as suggested in Fig. 9.4.

Our consideration of small elastoplastic deformations begins with the assumption that the strain components, E_{ij}, can be additively decomposed into parts E^{e}_{ij} and E^{p}_{ij} corresponding to elastic and plastic contributions, respectively, to the deformation:

$$E_{ij} = E^{e}_{ij} + E^{p}_{ij}. \tag{9.10}$$

Both experimental observation and microscopic models of plastic deformation indicate that no volume change results from the plastic part of the deformation, so the constraint

$$E^{p}_{kk} = 0 \tag{9.11}$$

is incorporated into the theory.

The stress is determined solely by the elastic part of the strain and is obtained by substituting E^{e}_{ij} into Eq. 9.4:

$$t_{ij} = \lambda E^{e}_{kk} \delta_{ij} + 2\mu E^{e}_{ij}. \tag{9.12}$$

Let us apply this theory to analysis of uniaxial quasi-static compression* of a slab.

Elastic Range

We begin by obtaining the elastic solution. As noted, we have

$$E_{22} = E_{33} = 0. \tag{9.13}$$

According to Eq. 9.12, the nonvanishing stress components are

$$t_{11} = (\lambda + 2\mu)E^{e}_{11}, \quad t_{22} = t_{33} = \lambda E^{e}_{11} = \frac{\lambda}{\lambda + 2\mu} t_{11}, \tag{9.14}$$

so the pressure is

---

* The analysis for tensile loading follows the same procedure and leads to similar results, but treatment of the two cases simultaneously is complicated by sign changes that occur due to the absolute values that appear in the yield condition. In this chapter, we shall restrict attention to the compressive case because it is the one normally encountered in experimental investigations.

$$p \equiv -\frac{1}{3}t_{kk} = -\left[\lambda + \frac{2}{3}\mu\right]\lambda + \frac{2}{3}\mu E^{\text{e}}_{11} = -\frac{3\lambda + 2\mu}{3(\lambda + 2\mu)}t_{11}. \tag{9.15}$$

The criterion of Eq. 9.9 gives the stress at the onset of compressive yielding as

$$-t_{11} = \frac{\lambda + 2\mu}{2\mu}Y, \quad -t_{22} = \frac{\lambda}{2\mu}Y. \tag{9.16}$$

We shall call these yield stresses *Hugoniot elastic limits*, in anticipation of their appearance in analyses of shock propagation phenomena:

$$t^{\text{HEL}}_{11} \equiv \frac{\lambda + 2\mu}{2\mu}Y, \quad t^{\text{HEL}}_{22} \equiv \frac{\lambda}{2\mu}Y. \tag{9.17}$$

Plastic Range

In the plastic regime, we have the usual elastic stress relations, but, in contrast to the analysis of the elastic regime, E^{e}_{ij} is no longer the total strain. Since the two transverse components of each of the variables are equal, we need only give results for the x_2 components. Specialization of the stress relations of Eq. 9.4 to the case at hand gives

$$t_{11} = (\lambda + 2\mu)E^{\text{e}}_{11} + 2\lambda E^{\text{e}}_{22}, \quad t_{22} = \lambda E^{\text{e}}_{11} + 2(\lambda + \mu)E^{\text{e}}_{22}, \tag{9.18}$$

which we use in the equivalent form

$$E^{\text{e}}_{11} = \frac{(\lambda + \mu)t_{11} - \lambda t_{22}}{\mu(3\lambda + 2\mu)}, \quad E^{\text{e}}_{22} = \frac{-\lambda t_{11} + (\lambda + 2\mu)t_{22}}{2\mu(3\lambda + 2\mu)}. \tag{9.19}$$

The axial and lateral strains are each made up of both elastic and plastic parts, as given by Eq. 9.10. The lateral constraint takes the form

$$E_{22} = E^{\text{e}}_{22} + E^{\text{p}}_{22} = 0, \tag{9.20}$$

and the condition that the plastic part of the deformation be isochoric is expressed by the equation

$$E^{\text{p}}_{11} + 2E^{\text{p}}_{22} = 0. \tag{9.21}$$

Finally, the yield condition expressed by Eq. 9.9 must be satisfied as the material is loaded into the plastic regime. To determine the sign of the quantity $t_{11} - t_{22}$ appearing in this criterion, it is only necessary to evaluate it at the yield point, where $E^{\text{e}}_{22} = 0$. At yield, we have

$$t_{11} = (\lambda + 2\mu)E^{\text{e}}_{11}, \quad t_{22} = \lambda E^{\text{e}}_{11}. \tag{9.22}$$

As the slab is compressed, $E_{11}^{e} < 0$ and we have $t_{11} < t_{22} < 0$, so that $t_{11} - t_{22} < 0$ and the yield condition can be written

$$t_{11} - t_{22} = -Y. \tag{9.23}$$

When this condition is satisfied, the material of the slab is said to be in a state of *forward yield.*

The five equations 9.19–9.21 and 9.23 can be solved for any five of the unknown quantities t_{11}, t_{22}, E_{11}^{e}, E_{11}^{p}, E_{22}^{e}, and E_{22}^{p} once the value of the remaining unknown is given as a measure of the stimulus producing the deformation. Let us assume that t_{11} is given.

From Eqs. 9.19$_2$, 9.20, and 9.21 we obtain

$$E_{11}^{p} = \frac{-\lambda t_{11} + (\lambda + 2\mu)t_{22}}{\mu(3\lambda + 2\mu)}, \quad E_{22}^{p} = \frac{\lambda t_{11} - (\lambda + 2\mu)t_{22}}{2\mu(3\lambda + 2\mu)}. \tag{9.24}$$

We can now use the yield condition 9.23 to express t_{22} in terms of t_{11}. We obtain, from Eqs. 9.19 and 9.24,

$$\begin{aligned} E_{11}^{e} &= \frac{\mu t_{11} - \lambda Y}{\mu(3\lambda + 2\mu)}, & E_{22}^{e} &= \frac{2\mu t_{11} + (\lambda + 2\mu)Y}{2\mu(3\lambda + 2\mu)} \\ E_{11}^{p} &= \frac{2\mu t_{11} + (\lambda + 2\mu)Y}{\mu(3\lambda + 2\mu)}, & E_{22}^{p} &= \frac{-2\mu t_{11} - (\lambda + 2\mu)Y}{2\mu(3\lambda + 2\mu)}. \end{aligned} \tag{9.25}$$

The total longitudinal strain is

$$-E_{11} = \frac{1}{B}\left(-t_{11} - \frac{2}{3}Y\right), \tag{9.26}$$

where

$$B = \lambda + \frac{2}{3}\mu \tag{9.27}$$

is called the bulk modulus of elasticity. Equation 9.26 can be solved for t_{11} to yield

$$-t_{11} = -BE_{11} + \frac{2}{3}Y. \tag{9.28}$$

We recall that all of these relationships are valid only in the plastic range, which (in compression) is defined by either of the equivalent relations

$$-E_{11} > E_{11}^{\text{HEL}} \text{ or } -t_{11} > t_{11}^{\text{HEL}}, \tag{9.29}$$

where we have introduced the addition

$$E_{11}^{\text{HEL}} \equiv Y/(2\mu) \tag{9.30}$$

to the definitions of Eqs. 9.17.

Using Eqs. 9.6, 9.13, 9.15, and 9.16, we can show that

$$t_{22} = \frac{1}{2}\left[(3\lambda + 2\mu)E_{11} - t_{11}\right], \tag{9.31}$$

a result that holds in both the elastic and plastic ranges.

Stress-Strain Paths

Figure 9.5 shows stress-strain curves for compressive loading of the slab. The line $t_{11} = [\lambda + (2\mu/3)]E_{11}$ represents the contribution of the pressure to the total stress. The lines $t_{11} = (\lambda + 2\mu)E_{11}$ and $t_{22} = \lambda E_{11}$ represent the elastic response below the yield point, and the segments through the yield point and extending parallel to the pressure response represent the total stress as the slab is loaded beyond the yield point. When loaded beyond yield in this way, the material is said to be in a state of *forward yield*. At all points beyond the forward yield point, the longitudinal compressive stress exceeds the pressure by $2Y/3$ and the lateral compressive stress is less than the pressure by $Y/3$.

It is also important to determine the stress-strain paths followed on decompression, recompression, etc. All of these responses play a role in analysis of elastoplastic wave propagation. Let us consider decompression from a stress state $t_{11}^{(2)}$ to the state $t_{11} = 0$. The initial part of the decompression path is an elastic load release. This process

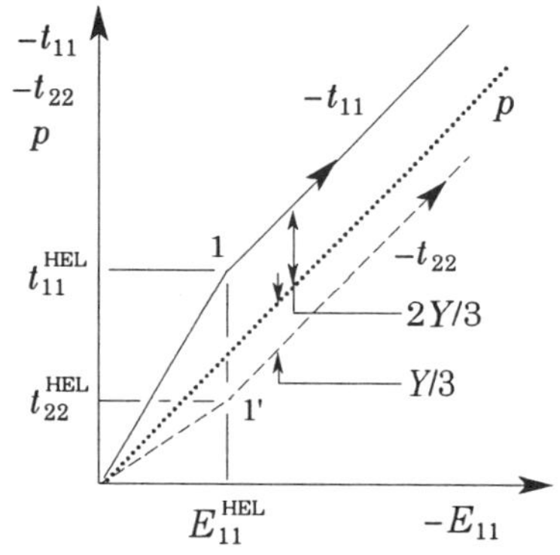

Figure 9.5. Stress–strain paths for loading of a slab that is laterally restrained and subjected to uniform compression over its faces. The upper path is the longitudinal stress. In the elastic range, it has the equation $t_{11} = (\lambda + 2\mu)E_{11}$. In the plastic range, the equation is $t_{11} = BE_{11} - (2Y/3)$. The middle line is the pressure response, given by the equation $p = -BE_{11}$. Finally, the lowest path represents the lateral component of stress. In the elastic range, it has the equation $t_{22} = \lambda E_{11}$. In the plastic range, its equation is $t_{22} = BE_{11} + (Y/3)$.

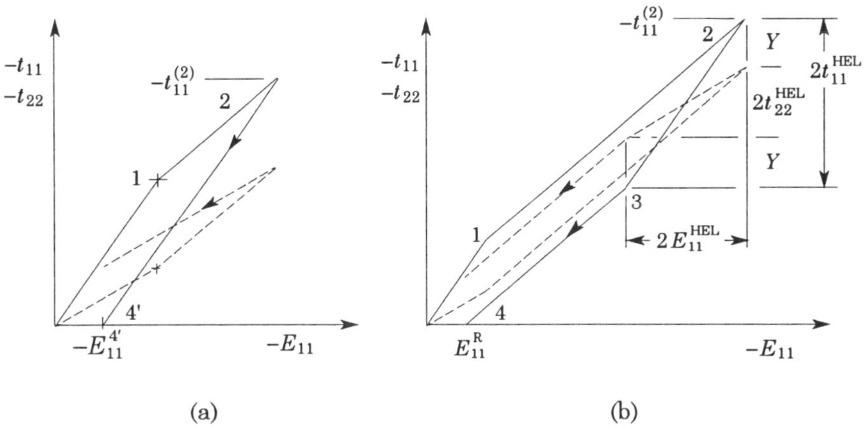

Figure 9.6. Stress-strain paths for decompression (from state 2) of a laterally restrained slab that is subject to a uniform normal compression. Part (a) is for the range, $t_{11}^{HEL} < -t_{11}^{(2)} \le 2t_{11}^{HEL}$, in which the decompression is entirely elastic. Part (b) is for $-t_{11}^{(2)} > 2t_{11}^{HEL}$, the range in which decompression involves both elastic and plastic strains. Solid lines refer to $-t_{11}$ and broken lines refer to $-t_{22}$. The quantity $E_{11}^{R} = 2Y/(3B)$ is a material property but $E_{11}^{(4')}$ is a function of $t_{11}^{(2)}$.

is described by the elasticity theory that we have been using, with all of the decompression being associated with the elastic part of the strain. In this simple one-dimensional example, it is easiest to follow the process graphically. The state point defined by $-t_{11}$ moves to lower stresses and strains along a line having the same slope, $\lambda + 2\mu$, as the elastic compression process. As the state point crosses the pressure line there is a reversal of shear-stress direction, but the point continues to move along the same path until this reversed shear stress reaches the yield value. At this transition, the state point moves along the yield surface, i.e., along a line having slope $\lambda + (2\mu/3)$ in the stress-strain plane. This line intercepts the $-E_{11}$ axis at the point $-E_{11} = 2Y/(3B) \equiv E_{11}^{R}$ defining the state of residual plastic strain. The stress component t_{22} also changes during the decompression process, following the path indicated by arrow on the broken line in Fig. 9.6. It is important to note that t_{22} does not vanish when the unloading of the faces of the slab is completed. A residual in-plane stress is required to maintain the constraint of uniaxial deformation.

We shall call a yield condition reached by compression a *forward yield point* and the material states reached by continuation of the deformation beyond this point *states of forward yield*. A yield point reached by a decompression process is called a *reverse yield point* and states achieved by continuation of the deformation beyond this

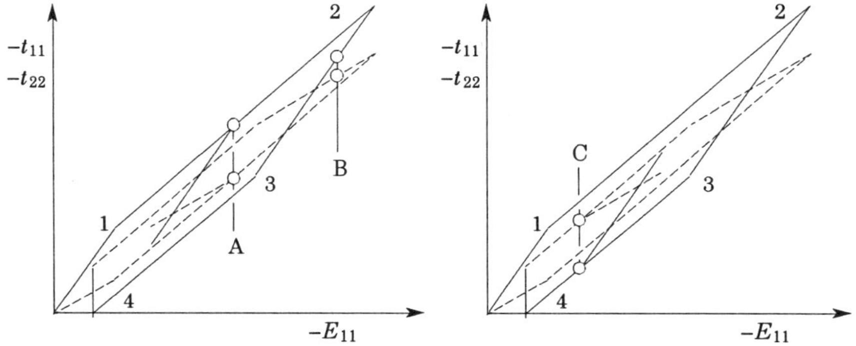

Figure 9.7. Stress-strain curves illustrating unloading and reloading processes. State A is one of forward yield, state B is in the elastic range, and state C is one of reverse yield.

point *states of reverse yield.* In all cases, a compression or decompression process that has been interrupted can be resumed (loading continued in the same direction as before the interruption) without affecting the stress-strain path followed. Decompression from any state of forward yield (such as state A in Fig. 9.7a) begins with an elastic deformation and, if it proceeds beyond the reverse yield point (defined by the criterion $t_{11} - t_{22} = +Y$, cf. Eq. 9.23), continues as a process involving both elastic and plastic deformation, as discussed previously. Both compression and decompression of material in an elastic state (such as state B in Fig. 9.7a) begins as an elastic process and, if it proceeds beyond a yield point, continues as a process involving both elastic and plastic deformation. When material is in a state of reverse yield, as illustrated by state C in Fig. 9.7b, further decompression is simply a continuation of the first part of the decompression. Recompression is an elastic process until it reaches the forward yield point, after which it proceeds as with the initial compression.

9.5. Shock Propagation in a Slab

In the context of the theory of ideal elastoplastic response of a slab, all responses to instantaneous load application or release take the form of propagating shocks. Bilinear Hugoniots such as are shown in Fig. 9.6 correspond to shocks that are unstable (for both compression and decompression) when their amplitude is such as to include the slope discontinuity in the path [6, p. 18].

Several quantitative examples are given in the remaining sections of this chapter. They are all calculated for a steel characterized by the elastic moduli $\lambda = 123.1$ GPa and $\mu = 79.29$ GPa (corresponding to a Young's modulus of 30×10^6 psi and a shear modulus of 11.5×10^6 psi), a density $\rho = 7870\ \text{kg/m}^3$, and a yield stress $Y = 1.00$ GPa (145000 psi).

These parameters lead to the elastic wavespeeds (defined later) $C_0 = 5.982$ km/s and $C_B = 4.729$ km/s. In all cases where a result is said to be for "steel," these are the material constants that were used in the calculation.

9.5.1. Compression Shocks

Let us begin by analysis of shocks produced by sudden application of a uniform compressive stress to the boundary $x = 0$ of the slab.

Examination of the Hugoniot shown in Fig. 9.8a shows that waves propagating into undeformed material and having compressive amplitude less than the Hugoniot elastic limit involve only elastic response. Shocks of greater amplitude are unstable, causing them to separate into two shocks. The leading shock, called the *elastic precursor*, is of amplitude $t_{11} = -t_{11}^{\text{HEL}}$ and propagates at the longitudinal elastic wavespeed

$$C_0 = \sqrt{(\lambda + 2\mu)/\rho}\,. \qquad (9.32)$$

The second shock, called the *plastic shock* or *plastic wave*, takes the material to the state imposed on the boundary and propagates at the bulk wavespeed

$$C_B = \sqrt{[\lambda + (2/3)\mu]/\rho} = \sqrt{B/\rho}\,. \qquad (9.33)$$

(As is obvious from these expressions, $C_0 > C_B$.) This waveform is shown in Fig. 9.8b.

The elastic and plastic shocks, and their interactions with each other, material interfaces, etc., can be analyzed using the jump conditions 9.7. The shock velocities to be used in these equations are $U_S = \pm C_0$ for the elastic shock and $U_S = \pm C_B$ for the plastic shock.

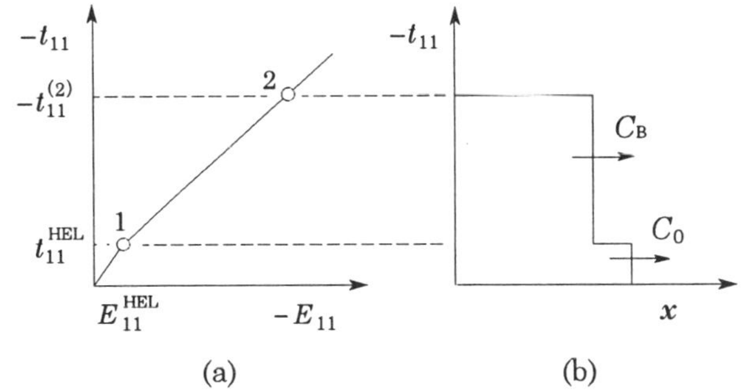

Figure 9.8. (a) Uniaxial-strain Hugoniot and (b) elastic–plastic waveform of stress amplitude $t_{11}^{(2)}$ propagating into undeformed material (steel). The slope of the elastic segment of the Hugoniot is ρC_0^2 and that of the plastic segment is ρC_B^2.

Applying these equations to shocks in the elastic range that are propagating in the $+x$ direction into undeformed material at rest gives

$$\begin{aligned} t_{11} &= \rho C_0^2 E_{11}, \quad 0 \le -E_{11} \le E_{11}^{\mathrm{HEL}} \\ v_1 &= \frac{-1}{\rho C_0} t_{11}, \quad 0 \le -t_{11} \le t_{11}^{\mathrm{HEL}}. \end{aligned} \tag{9.34}$$

Since the amplitude of the elastic precursor is $-t_{11}^{\mathrm{HEL}}$, the particle velocity and strain of the material behind it are given by

$$v_1^{(1)} = v_1^{\mathrm{HEL}} = \frac{1}{\rho C_0} t_{11}^{\mathrm{HEL}} = C_0 E_{11}^{\mathrm{HEL}} \tag{9.35}$$

$$E_{11}^{(1)} = -E_{11}^{\mathrm{HEL}} = \frac{-1}{C_0} v_1^{\mathrm{HEL}} = \frac{-1}{\rho C_0^2} t_{11}^{\mathrm{HEL}}, \tag{9.36}$$

and the transverse stress component is

$$t_{22}^{(1)} = -t_{22}^{\mathrm{HEL}} = -\frac{3C_{\mathrm{B}}^2 - C_0^2}{2C_0^2} t_{11}^{\mathrm{HEL}}. \tag{9.37}$$

In writing these equations, we have regarded E_{11}^{HEL}, v_1^{HEL}, t_{11}^{HEL}, and t_{22}^{HEL} as material properties given by positive numbers. These quantities can be expressed in other forms, as listed in Table 9.1.

Suppose the state behind the plastic wave is characterized by the constant field values $\mathcal{S}^{(2)} = \{t_{11}^{(2)}, E_{11}^{(2)}, v_1^{(2)}\}$. Equations 9.7, applied to the plastic wave, yield the two relations

$$\begin{aligned} C_{\mathrm{B}}(-E_{11}^{(2)} - E_{11}^{\mathrm{HEL}}) &= v_1^{(2)} - v_1^{\mathrm{HEL}} \\ \rho C_{\mathrm{B}}(v_1^{(2)} - v_1^{\mathrm{HEL}}) &= -t_{11}^{(2)} - t_{11}^{\mathrm{HEL}} \end{aligned} \tag{9.38}$$

among the variables defining the state $\mathcal{S}^{(2)}$. If $v_1^{(2)}$ is given as a boundary condition, it is convenient to write these equations in the form

$$\left.\begin{aligned} E_{11}^{(2)} &= -\frac{1}{C_{\mathrm{B}}}\left\{v_1^{(2)} - \left(\frac{C_0 - C_{\mathrm{B}}}{C_0}\right) v_1^{\mathrm{HEL}}\right\} \\ t_{11}^{(2)} &= -\rho C_{\mathrm{B}} v_1^{(2)} - \rho(C_0 - C_{\mathrm{B}}) v_1^{\mathrm{HEL}} \end{aligned}\right\}, \quad v_1^{(2)} \ge v_1^{\mathrm{HEL}}. \tag{9.39}$$

Similarly, if $t_{11}^{(2)}$ is given, we have

$$\left.\begin{aligned} E_{11}^{(2)} &= \frac{1}{\rho C_{\mathrm{B}}^2}\left\{t_{11}^{(2)} + \left(\frac{C_0^2 - C_{\mathrm{B}}^2}{C_0^2}\right) t_{11}^{\mathrm{HEL}}\right\} \\ v_1^{(2)} &= \frac{1}{\rho C_{\mathrm{B}}}\left\{-t_{11}^{(2)} - \left(\frac{C_0 - C_{\mathrm{B}}}{C_0}\right) t_{11}^{\mathrm{HEL}}\right\} \end{aligned}\right\}, \quad -t_{11}^{(2)} \ge t_{11}^{\mathrm{HEL}}. \tag{9.40}$$

Table 9.1. Expressions for the Hugoniot elastic limit

	t_{11}^{HEL}	t_{22}^{HEL}	E_{11}^{HEL}	v_1^{HEL}
λ,μ	$\dfrac{\lambda+2\mu}{2\mu}Y$	$\dfrac{\lambda}{2\mu}Y$	$\dfrac{1}{2\mu}Y$	$\left[\dfrac{\lambda+2\mu}{4\rho_R\,\mu^2}\right]^{1/2}Y$
B,μ	$\left[\dfrac{B}{2\mu}+\dfrac{2}{3}\right]Y$	$\left[\dfrac{B}{2\mu}-\dfrac{1}{3}\right]Y$	$\dfrac{1}{2\mu}Y$	$\left[\dfrac{B+\frac{4}{3}\mu}{4\rho_R\,\mu^2}\right]^{1/2}Y$
C_0,C_B	$\dfrac{2C_0^2Y}{3(C_0^2-C_B^2)}$	$\dfrac{3C_B^2-C_0^2}{3(C_0^2-C_B^2)}Y$	$\dfrac{2Y}{3\rho_R(C_0^2-C_B^2)}$	$\dfrac{2C_0Y}{3\rho_R(C_0^2-C_B^2)}$
t_{11}^{HEL}	t_{11}^{HEL}	$\dfrac{3C_B^2-C_0^2}{2C_0^2}t_{11}^{\mathrm{HEL}}$	$\dfrac{1}{\rho_R C_0^2}t_{11}^{\mathrm{HEL}}$	$\dfrac{1}{\rho_R C_0}t_{11}^{\mathrm{HEL}}$
t_{22}^{HEL}	$\dfrac{2C_0^2}{3C_B^2-C_0^2}t_{22}^{\mathrm{HEL}}$	t_{22}^{HEL}	$\dfrac{2}{\rho(3C_B^2-C_0^2)}t_{22}^{\mathrm{HEL}}$	$\dfrac{2C_0\,t_{22}^{\mathrm{HEL}}}{\rho(3C_B^2-C_0^2)}$
E_{11}^{HEL}	$\rho C_0^2E_{11}^{\mathrm{HEL}}$	$\dfrac{\rho}{2}(3C_B^2-C_0^2)E_{11}^{\mathrm{HEL}}$	E_{11}^{HEL}	$C_0E_{11}^{\mathrm{HEL}}$
v_1^{HEL}	$\rho C_0 v_1^{\mathrm{HEL}}$	$\dfrac{\rho(3C_B^2-C_0^2)}{2C_0}v_1^{\mathrm{HEL}}$	$\dfrac{v_1^{\mathrm{HEL}}}{C_0}$	v_1^{HEL}

We can use these equations to plot the t_{11}–v_1 Hugoniot curve corresponding the Hugoniot relations of Fig. 9.6. In the elastic range, Eq. 9.34 gives

$$-t_{11}=\rho C_0 v_1 \quad \text{for } v_1\le v_1^{\mathrm{HEL}} \text{ or } -t_{11}\le t_{11}^{\mathrm{HEL}} \tag{9.41}$$

and in the plastic range, Eq. 9.39_2 can be written

$$-t_{11}=\rho C_B v_1+\left(\frac{C_0-C_B}{C_0}\right)t_{11}^{\mathrm{HEL}} \quad \text{for } v_1\ge v_1^{\mathrm{HEL}} \text{ or } -t_{11}\ge t_{11}^{\mathrm{HEL}}. \tag{9.42}$$

The result is shown as the compression branch on each of the diagrams in Fig. 9.9b.

Finally, using Eq. 9.31, we can write the transverse stress components in either of the equivalent forms

$$t_{22}^{(2)}=t_{11}^{(2)}+\frac{3(C_0^2-C_B^2)}{2C_0^2}t_{11}^{\mathrm{HEL}}=t_{11}^{(2)}+Y \tag{9.43}$$

and

$$t_{22}^{(2)}=-\rho C_B v_1^{(2)}+\frac{\rho(C_0+3C_B)(C_0-C_B)}{2C_0}v_1^{\mathrm{HEL}}. \tag{9.44}$$

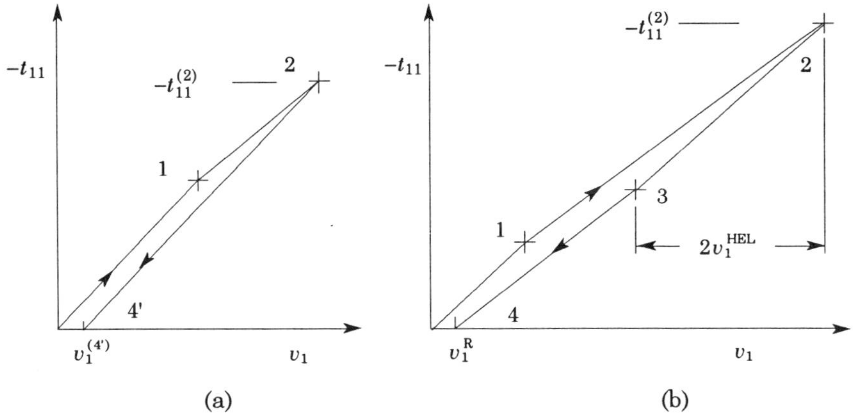

Figure 9.9. Stress–particle velocity Hugoniot curves corresponding to the elastoplastic stress relations of Fig. 9.6 (scaled for steel).

9.5.2. Decompression Shocks

Let us now consider the case in which the applied stress producing the waveform of Fig. 9.8b is suddenly removed from the boundary of the slab after a time interval t^*. The three cases that must be considered are i) $-t_{11}^{(2)} \leq t_{11}^{HEL}$, ii) $t_{11}^{HEL} < -t_{11}^{(2)} \leq 2t_{11}^{HEL}$, and iii) $2t_{11}^{HEL} < -t_{11}^{(2)}$. The first case involves only linear elastic response. The propagating disturbance takes the form of a nonattenuating flat-topped pulse advancing at the velocity C_0. We shall not discuss this case further. In the second case, decompression to $t_{11} = 0$ occurs by means of a single, elastic decompression shock. In the third case, the decompression shock is unstable, separating into elastic and plastic shocks. Since both compression and decompression shocks are (marginally) stable, discontinuous loading and unloading problems can all be solved using the shock jump conditions. This characteristic of elastoplastic wave propagation is attributable to the bilinear hysteretic response of these materials and does not occur in most other materials.

Elastic Decompression

We begin our consideration of this low-stress case of the attenuation problem by determining the state behind a decompression shock propagating into compressed material in the state discussed in Sec. 9.5.1. The case is defined by the requirement that the peak stresses in the compressed state (called state 2) are restricted to the range $t_{11}^{HEL} < -t_{11}^{(2)} \leq 2t_{11}^{HEL}$. Examination of the Hugoniot curves for this case, as illustrated in Figs. 9.10a and 9.10b, shows that decompression to zero longitudinal stress can be accomplished by a single elastic shock taking the material to a state 4′ that is a function of state 2.

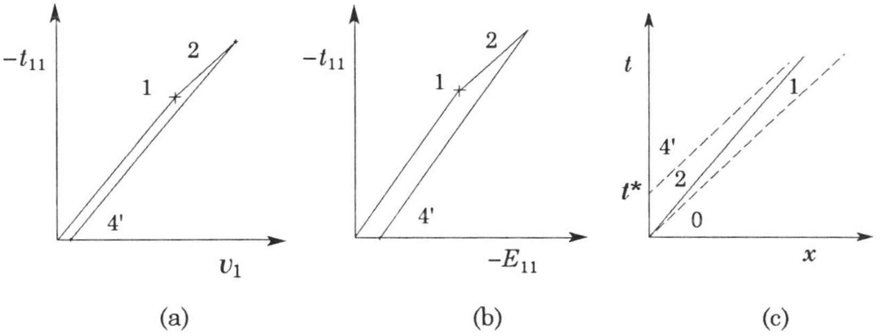

Figure 9.10. Hugoniot curves and an x–t diagram for the case in which the peak stress satisfies $t_{11}^{\mathrm{HEL}} < -t_{11}^{(2)} \leq 2t_{11}^{\mathrm{HEL}}$. In this and all following x–t diagrams, the solid lines designate plastic waves and the broken lines designate elastic waves.

Application of the jump conditions 9.7 to this shock gives the particle velocity, strain, and transverse stress in the decompressed state (state 4′) as

$$v_1^{(4')} = -\frac{C_0 - C_{\mathrm{B}}}{\rho C_0 C_{\mathrm{B}}}\left[t_{11}^{(2)} + t_{11}^{\mathrm{HEL}}\right]$$

$$E_{11}^{(4')} = \frac{C_0^2 - C_{\mathrm{B}}^2}{\rho C_0^2 C_{\mathrm{B}}^2}\left[t_{11}^{(2)} + t_{11}^{\mathrm{HEL}}\right] \tag{9.45}$$

$$t_{22}^{(4')} = \frac{3(C_0^2 - C_{\mathrm{B}}^2)}{2\rho C_{\mathrm{B}}^2}\left[t_{11}^{(2)} + t_{11}^{\mathrm{HEL}}\right].$$

Examination of the x–t diagram of Fig. 9.10c indicates that the elastic decompression wave will overtake the plastic part of the compression wave at

$$t_{\mathrm{a}} = \frac{C_0}{C_0 - C_{\mathrm{B}}}t^*, \quad x_{\mathrm{a}} = \frac{C_0 C_{\mathrm{B}}}{C_0 - C_{\mathrm{B}}}t^*, \tag{9.46}$$

where t^* is the duration of load application. The waveform produced in this case is shown in the first column of Fig. 9.12.

Elastic–Plastic Decompression

Stress-strain paths for unloading from a state $t_{11}^{(2)}$–$E_{11}^{(2)}$ have been given in Fig. 9.6. The associated t_{11}–v_1 Hugoniot of Eqs. 9.41 and 9.42 is plotted in Fig. 9.9b, along with an unloading Hugoniot for right-propagating waves, which can be found using the jump conditions as above, or by graphical means. The residual strain, particle velocity, and lateral component of stress (corresponding to state 4 in the figures) are given by

$$E_{11}^{(4)} = -E_{11}^{\mathrm{R}} = \frac{-2Y}{3B}$$
$$v_1^{(4)} = v_1^{\mathrm{R}} = \frac{C_0 - C_{\mathrm{B}}}{\rho C_0 C_{\mathrm{B}}} t_{11}^{\mathrm{HEL}} \tag{9.47}$$
$$t_{22}^{(4)} = -t_{22}^{\mathrm{R}} = -Y.$$

Several other expressions for these quantities are

$$E_{11}^{\mathrm{R}} = \left(\frac{C_0^2 - C_{\mathrm{B}}^2}{\rho C_0^2 C_{\mathrm{B}}^2}\right) t_{11}^{\mathrm{HEL}} = \left(\frac{C_0^2 - C_{\mathrm{B}}^2}{C_0 C_{\mathrm{B}}^2}\right) v_1^{\mathrm{HEL}} = \left(\frac{C_0^2 - C_{\mathrm{B}}^2}{C_{\mathrm{B}}^2}\right) E_{11}^{\mathrm{HEL}}$$
$$v_1^{\mathrm{R}} = \left(\frac{C_0 - C_{\mathrm{B}}}{\rho C_0 C_{\mathrm{B}}}\right) t_{11}^{\mathrm{HEL}} = \left(\frac{C_0 - C_{\mathrm{B}}}{C_{\mathrm{B}}}\right) v_1^{\mathrm{HEL}} = \frac{C_0}{C_{\mathrm{B}}} (C_0 - C_{\mathrm{B}}) E_{11}^{\mathrm{HEL}}. \tag{9.48}$$

Note that, in contrast to the previous case of elastic decompression where the decompressed state varied with varying peak stress, the state 4 obtained upon release of material compressed into the plastic region can be completely described in terms of material properties and is independent of peak stress.

Let us suppose that a stress $-t_{11}^{(2)} > 2t_{11}^{\mathrm{HEL}}$ is applied at the boundary at the time $t = 0$ and removed at the time $t = t^*$. The t_{11}–v_1 Hugoniot and the x–t diagram for this problem are given in Fig. 9.11 and two examples of the waveform as it evolves in time are shown in Figs. 9.12b and 9.12c. It is apparent from the x–t diagram that the first step of the unloading wave will overtake the second step of the loading wave at the time and place given by Eqs. 9.46.

The field values in region 1 of Fig. 9.11b are given by Eqs. 9.35 and 9.36, and in region 2, we have fields given by Eqs. 9.39 or 9.40.

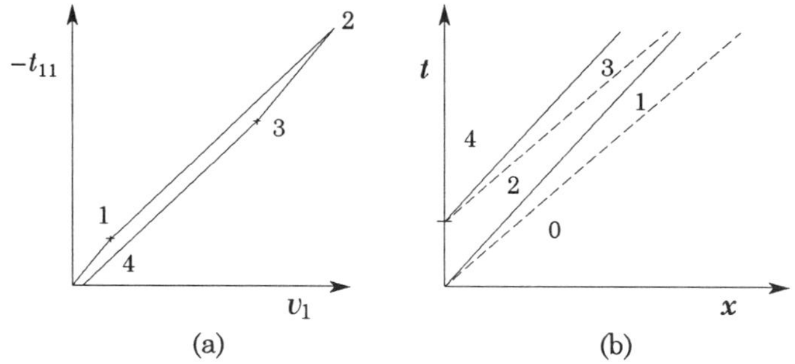

Figure 9.11. Stress–particle velocity Hugoniot and x–t diagram for a compression–decompression process. Hugoniot states and the regions of x–t space in which they prevail are matched by the numbers.

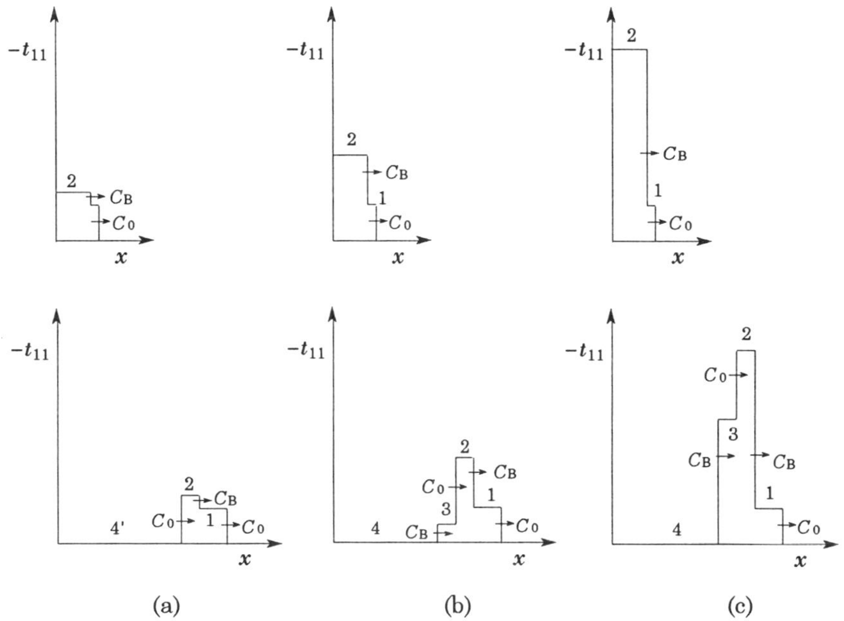

Figure 9.12. Decompression wave overtaking a compression wave. This figure is drawn for waves of amplitude (a) $t_{11}^{\mathrm{HEL}} < -t_{11}^{(2)} < 2t_{11}^{\mathrm{HEL}}$ (–2.5 GPa), (b) $2t_{11}^{\mathrm{HEL}} < -t_{11}^{(2)} < 3t_{11}^{\mathrm{HEL}}$ (–4.5 GPa), and (c) $-t_{11}^{(2)} > 3t_{11}^{\mathrm{HEL}}$ (–10 GPa) propagating in steel. Waveforms in the top row correspond to a time of $t^* = 1\,\mu\mathrm{s}$ (at which time the two waves have advanced 5.975 and 4.729 mm, respectively), and waveforms in the lower row are plotted for $t = 3\ \mu s$. The elastic part of the unloading wave will overtake the plastic part of the loading wave at $t_a = 4.8\,\mu\mathrm{s}$, $x_a = 22.6$ mm.

In region 3, the fields are determined by application of Eqs. 9.7 to a right-propagating elastic wave ($U_{\mathrm{S}} = +C_0$) taking material from the state $t_{11}^{(2)} - v^{(2)}$ to a state in which the stress is $t_{11}^{(2)} + 2t_{11}^{\mathrm{HEL}}$. The result is

$$
\begin{aligned}
t_{11}^{(3)} &= t_{11}^{(2)} + 2t_{11}^{\mathrm{HEL}} \\
v_1^{(3)} &= v_1^{(2)} - 2v_1^{\mathrm{HEL}} = \frac{-1}{\rho C_{\mathrm{B}}}\left[t_{11}^{(2)} + \left(\frac{C_0 + C_{\mathrm{B}}}{C_0}\right)t_{11}^{\mathrm{HEL}}\right] \qquad (9.49) \\
E_{11}^{(3)} &= E_{11}^{(2)} + 2E_{11}^{\mathrm{HEL}} = \frac{1}{\rho C_{\mathrm{B}}^2}\left[t_{11}^{(2)} + \left(\frac{C_0^2 + C_{\mathrm{B}}^2}{C_0^2}\right)t_{11}^{\mathrm{HEL}}\right].
\end{aligned}
$$

In writing these results, we have expressed the strength of the loading disturbance in terms of its stress amplitude $t_{11}^{(2)}$ and have expressed all of the measures of the HEL in terms of t_{11}^{HEL}. One could, of course, express the same results in terms of other measures of the applied load and material yield point.

It is important to note that the yield condition, which is always given by Eq. 9.9, changes from the expression $t_{11} - t_{22} = -Y$ used for the compression processes to the expression $t_{11} - t_{22} = +Y$ for the decompression processes. The point defined by this latter equation is often called the *reverse yield point*. Using this yield condition and the previous results, we obtain the transverse stress in state 3 as

$$t_{22}^{(3)} = t_{22}^{(2)} + 2t_{22}^{\mathrm{HEL}} = t_{11}^{(3)} - Y = t_{11}^{(2)} + \frac{C_0^2 + 3C_\mathrm{B}^2}{2C_0^2} t_{11}^{\mathrm{HEL}}. \tag{9.50}$$

As we shall see, it is always necessary to keep track of the transverse stress in solving elastoplastic wave propagation problems because it is essential to determining whether a wave interaction produces an elastic or plastic response.

9.6. Elastoplastic Pulse Attenuation

Attenuation of an elastoplastic pulse such as we have discussed in the previous section occurs in discrete steps when elastic decompression shocks overtake plastic shocks. Solution of the overtake problem is complicated by the need for sequential analysis of numerous shock interactions, with analysis of a given case being concluded only when it has been carried to a time after which no further interactions are possible. Elastoplastic pulses do not attenuate to zero strength, but only to the amplitude of the elastic precursor.

The details of the shock interactions change as $t_{11}^{(2)}$ occupies various ranges that depend on t_{11}^{HEL} and the wavespeeds. The lowest stress range, $0 < -t_{11}^{(2)} \le t_{11}^{\mathrm{HEL}}$, is one of elastic response. Since no attenuation occurs in this range, we shall ignore it. In the next higher range, $t_{11}^{\mathrm{HEL}} < -t_{11}^{(2)} \le 2t_{11}^{\mathrm{HEL}}$, complete decompression is achieved by a single elastic shock and a simple, direct analysis applies throughout the range. When the peak compressive stress exceeds $2t_{11}^{\mathrm{HEL}}$, the decompression process involves both elastic and plastic transitions and presents a sequence of cases distinguished by qualitative changes in the shock interactions that occur. The cases are defined in terms of a range of the peak stress, $-t_{11}^{(2)}$, throughout which the interactions are qualitatively the same so that a single analytical solution is valid. Several members of this sequence of cases are listed in Table 9.2 and discussed in the following sections of the chapter. The analysis is accompanied by quantitative calculation of field values, waveforms, etc. for a pulse of 1 μs duration introduced into the steel described earlier.

Table 9.2. Similarity ranges for pulse attenuation

Stress range designation	Peak stress limits	Stress limits in steel (−GPa)
A	$0 < -t_{11}^{(2)} \le t_{11}^{\mathrm{HEL}}$	0–1.78
B	$t_{11}^{\mathrm{HEL}} < -t_{11}^{(2)} \le 2t_{11}^{\mathrm{HEL}}$	1.78–3.55
C	$2t_{11}^{\mathrm{HEL}} < -t_{11}^{(2)} \le \dfrac{C_0 + 3C_{\mathrm{B}}}{2C_{\mathrm{B}}} t_{11}^{\mathrm{HEL}}$	3.55–3.79
D	$\dfrac{C_0 + 3C_{\mathrm{B}}}{2C_0} t_{11}^{\mathrm{HEL}} < -t_{11}^{(2)} \le \dfrac{C_0 + 5C_{\mathrm{B}}}{C_0 + C_{\mathrm{B}}} t_{11}^{\mathrm{HEL}}$	3.79–4.91
E	$\dfrac{C_0 + 5C_{\mathrm{B}}}{C_0 + C_{\mathrm{B}}} t_{11}^{\mathrm{HEL}} < -t_{11}^{(2)} \le \dfrac{C_0^2 + 10C_0 C_{\mathrm{B}} + C_{\mathrm{B}}^2}{(C_0 + C_{\mathrm{B}})^2} t_{11}^{\mathrm{HEL}}$	4.91–5.28
F	$\dfrac{C_0^2 + 10C_0 C_{\mathrm{B}} + C_{\mathrm{B}}^2}{(C_0 + C_{\mathrm{B}})^2} t_{11}^{\mathrm{HEL}} < -t_{11}^{(2)} \le \dfrac{2(C_0 + 3C_{\mathrm{B}})}{C_0 + C_{\mathrm{B}}} t_{11}^{\mathrm{HEL}}$	5.28–6.69

9.6.1. Attenuation of a Pulse of Amplitude in Stress Range B

Stress range B is defined by the inequality $t_{11}^{\mathrm{HEL}} < -t_{11}^{(2)} \le 2t_{11}^{\mathrm{HEL}}$, representing the condition that the decompression shock involve only elastic deformation. The field values in the state of peak compression (called state 2) are as given in Sec. 9.5.1. In the state behind the decompression shock (called state 4') the field values are given by Eqs. 9.45.

Examination of the x–t diagram of Fig. 9.13c indicates that the elastic decompression wave will overtake the plastic part of the compression wave at the time and place given by Eqs. 9.46. This interaction produces a right-propagating shock moving into the material in state 1 and a left-propagating shock moving into material in state 4'. The t_{11}–v_1 Hugoniots appropriate to this problem are shown in Fig. 9.13a. State 5 is the new state of the material in the region between the waves propagating away from the plane of interaction (see Fig. 9.13c). This state must lie at the intersection of the t_{11}–v_1 Hugoniot for left-propagating shocks centered on state 4' and the Hugoniot for right-propagating shocks centered on state 1. By examination of the stress-strain Hugoniots of Fig. 9.13b, we see that state 1 is

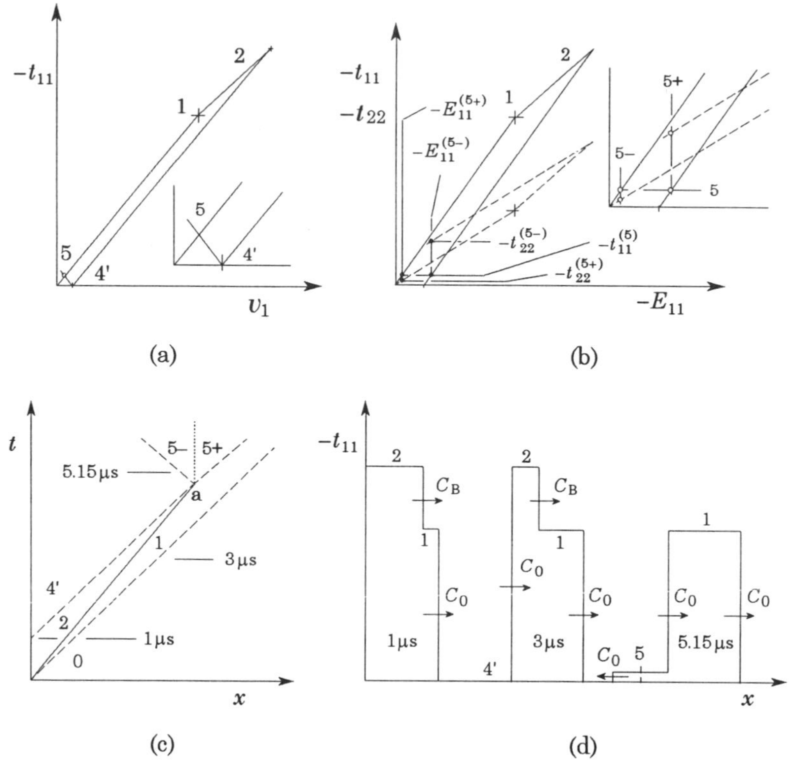

Figure 9.13. Analysis of attenuation of a pulse of amplitude $-t_{11}^{(2)}$ falling in the stress range B. (a) The stress–particle velocity Hugoniot curve, (b) stress-strain Hugoniot curves, (c) an x–t plot on which the solid lines designate plastic shocks propagating at the speed C_B, the broken lines designate elastic shocks propagating at the speed C_0 and the dotted vertical line designates a contact surface, and (d) selected waveforms. The example is drawn for a stress pulse of 1 μs duration and amplitude −2.5 GPa propagating in steel.

a condition of forward yield so that a decompression shock centered on this state produces elastic deformation. A compression shock centered on state 4' is elastic so long as it does not produce stresses in excess of $-t_{11}^{(2)}$. The intersection of these Hugoniots defining the t_{11}–v_1 state 5 is shown in Fig. 9.13a.

An important result of this interaction is the formation of a nonpropagating *contact surface* in the material at the plane of the interaction. The longitudinal stress and particle velocity are continuous at this surface, in accordance with the requirements of continuity and equilibrium, but the hysteretic aspects of elastoplastic response lead to discontinuities in the plastic part of the longitudinal strain and in the lateral components of stress and strain. The contact surface is shown on the x–t diagram of Fig. 9.13c, with the regions occupied by material in the two substates of state 5 being designated 5^+ and 5^-. Examination of the stress-strain Hugoniot segments connecting states 4' and 5 and those connecting states 1 and 5 shows that each transition can be effected by a single shock.

Quantitative analysis of the interaction is achieved by application of the jump conditions 9.7 to each of these waves. This gives

$$\begin{aligned} C_0(-E_{11}^{(5+)} + E_{11}^{(1)}) &= v_1^{(5)} - v_1^{(1)} \\ \rho C_0(v_1^{(5)} - v_1^{(1)}) &= -t_{11}^{(5)} + t_{11}^{(1)} \end{aligned} \tag{9.51}$$

for the right-propagating shock ($U_S = C_0$) and

$$\begin{aligned} -C_0(-E_{11}^{(5-)} + E_{11}^{(4')}) &= v_1^{(5)} - v_1^{(4')} \\ -\rho C_0(v_1^{(5)} - v_1^{(4')}) &= -t_{11}^{(5)} + t_{11}^{(4')} \end{aligned} \tag{9.52}$$

for the left-propagating shock ($U_S = -C_0$). In writing these equations, we have recognized the fact that t_{11} and v_1 must be continuous in region 5 of the x–t plane, but have allowed for the possible formation of a contact surface at the time and place of the interaction. When we substitute HEL values for the state 1 variables and solve these equations, we get

$$\begin{aligned} t_{11}^{(5)} &= \frac{C_0 - C_B}{2C_B}\left[t_{11}^{(2)} + t_{11}^{\mathrm{HEL}}\right] \\ v_1^{(5)} &= -\frac{C_0 - C_B}{2\rho C_0 C_B}\left[t_{11}^{(2)} + t_{11}^{\mathrm{HEL}}\right] \\ E_{11}^{(5-)} &= \frac{2C_0^2 + C_0 C_B - 3C_B^2}{2\rho C_0^2 C_B^2}\left[t_{11}^{(2)} + t_{11}^{\mathrm{HEL}}\right] \\ E_{11}^{(5+)} &= \frac{C_0 - C_B}{2\rho C_0^2 C_B}\left[t_{11}^{(2)} + t_{11}^{\mathrm{HEL}}\right]. \end{aligned} \tag{9.53}$$

Since $E_{11}^{(5-)}$ differs from $E_{11}^{(5+)}$, we see that the potential for formation of a contact surface is actually realized. The transverse stresses, as obtained from Eq. 9.31, are

$$\begin{aligned} t_{22}^{(5-)} &= \frac{-C_0^3 + 7C_0^2 C_B + 3C_0 C_B^2 - 9C_B^3}{4C_0^2 C_B}\left[t_{11}^{(2)} + t_{11}^{\mathrm{HEL}}\right] \\ t_{22}^{(5+)} &= \frac{-C_0^3 + C_0^2 C_B + 3C_0 C_B^2 - 3C_B^3}{4C_0^2 C_B}\left[t_{11}^{(2)} + t_{11}^{\mathrm{HEL}}\right]. \end{aligned} \tag{9.54}$$

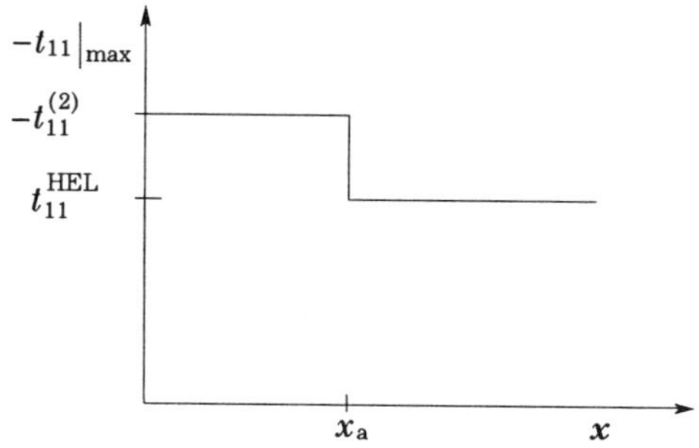

Figure 9.14. Attenuation curve for a pulse in stress range B. The point x_a is defined in Fig. 9.13c (steel).

As shown on the x–t diagram, all of the shocks present after the interaction propagate at the elastic wavespeed and are moving away from the contact surface. This means that no further wave interactions can occur, so the solution given persists indefinitely. The waveform arising in this problem is plotted in Fig. 9.13d at the three instants in time marked on the x–t diagram.

The end result of a pulse attenuation calculation is often presented in the form of an *attenuation curve* giving the maximum stress reached at each point x, without regard to the time or duration of its application. Examination of Fig. 9.13c shows that, for any point to the left of the interaction point x_a, the stress is successively 0, $-t_{11}^{(1)}$, $-t_{11}^{(2)}$, and 0. Beyond the interaction point, the succession of stress values is 0, $-t_{11}^{(1)}$, and $-t_{11}^{(5)}$. Since $0 < -t_{11}^{(5)} < -t_{11}^{(1)} < -t_{11}^{(2)}$, the attenuation curve is as given in Fig. 9.14. Note that the pulse amplitude never falls below $-t_{11}^{(1)} = t_{11}^{HEL}$ since no attenuating disturbance can overtake the leading elastic wave.

9.6.2. Attenuation of a Pulse of Amplitude in the Stress Range C

Stress range C is the lowest-stress member of the sequence of attenuation ranges listed in Table 9.2 in which an elastic–plastic decompression wave appears. The stress pulse to be analyzed is qualitatively like that shown in Fig. 9.12b; i.e., the elastic decompression takes the material to a state of lower compression than that behind the elastic precursor, but not to zero longitudinal stress.

The x–t diagram for this problem is shown in Fig. 9.15. (This and other diagrams associated with each solution must be drawn in stages as the wave interactions are analyzed; one cannot, for example, know in advance whether an interaction produces an elastic or plastic wave.) The field variables in regions 1–4 are as given in Secs. 9.5.1 and 9.5.2 and summarized in Appendix 9.A.

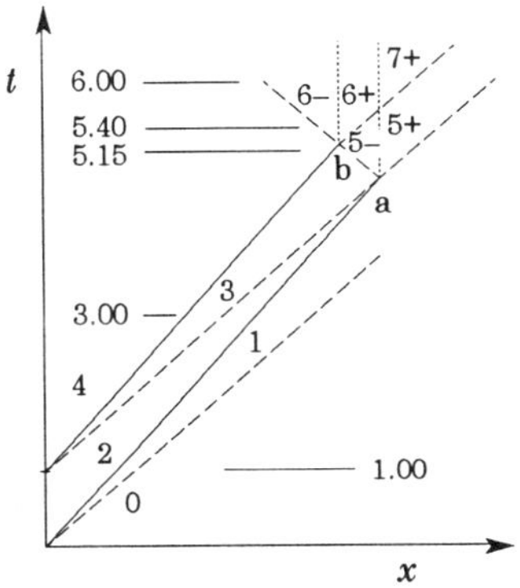

Figure 9.15. Space–time diagram for pulse propagation in stress range C ($2t_{11}^{\text{HEL}} < -t_{11}^{(2)} \le [(C_0 + 3C_{\text{B}})/(2C_{\text{B}})]t_{11}^{\text{HEL}}$). The figure is scaled for steel; times marked along the ordinate are in microseconds and correspond to plots of waveforms appearing in Fig. 9.17.

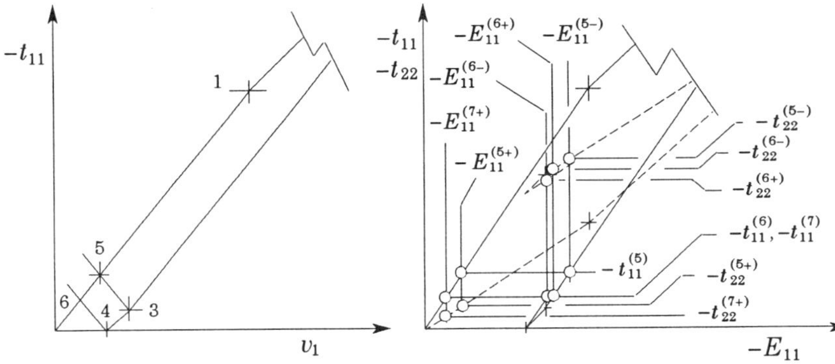

(a) (b)

Figure 9.16. Low-stress portions of the (a) stress–particle velocity Hugoniot curve and (b) the stress-strain Hugoniot curves for a pulse of amplitude falling in stress range C. The example is drawn for a stress pulse of amplitude −3.7 GPa propagating in steel.

As is always the case, the attenuation process begins when the elastic part of the decompression wave overtakes the plastic part of the compression wave (point "a" in Fig. 9.15). This interaction produces a right-propagating shock moving into the material in state 1 and a left-propagating shock moving into material in state 3. The Hugoniots appropriate to this analysis are shown in Fig. 9.16. State 5 (which differs from state 5 of Sec. 9.6.1) is the state of the material in the region between the newly created waves (see Fig. 9.15). By examination of the stress-strain Hugoniots of Fig. 9.16b, we see that a decompression shock centered on state 1 produces elastic deformation as does a compression shock centered on state 3 (so long as it does not produce stresses in excess of $-t_{11}^{(2)}$). The intersection of these Hugoniots defines the $t_{11}-v_1$ state 5.

The intersection point and the associated strain values are determined by application of the jump conditions 9.7 to each of the transitions. For the right-propagating elastic wave ($U_S = C_0$), we have

$$\begin{aligned} C_0(-E_{11}^{(5+)} + E_{11}^{(1)}) &= v_1^{(5)} - v_1^{(1)} \\ \rho C_0(v_1^{(5)} - v_1^{(1)}) &= -t_{11}^{(5)} + t_{11}^{(1)}, \end{aligned} \tag{9.55}$$

and for the left-propagating wave ($U_S = -C_0$), we have

$$\begin{aligned} -C_0(-E_{11}^{(5-)} + E_{11}^{(3)}) &= v_1^{(5)} - v_1^{(3)} \\ -\rho C_0(v_1^{(5)} - v_1^{(3)}) &= -t_{11}^{(5)} + t_{11}^{(3)}. \end{aligned} \tag{9.56}$$

In writing these equations, we have recognized the continuity of stress and particle velocity, but have allowed for formation of a contact surface at the plane of interaction.

Equations 9.55 and 9.56 can be solved for the quantities $t_{11}^{(5)}$, $v_1^{(5)}$, $E_{11}^{(5+)}$, and $E_{11}^{(5-)}$ characterizing state 5 in terms of known quantities ahead of each of the waves, as given by Eqs. 9.35, 9.36, and 9.49. To complete the analysis of this interaction, we must calculate the transverse stresses in region 5. This is most easily accomplished using Eq. 9.31, yielding

$$t_{22}^{(5\pm)} = \frac{1}{2}\left[3\rho C_B^2 E_{11}^{(5\pm)} - t_{11}^{(5)}\right].$$

The rather lengthy result of these calculations is given in Appendix 9.A. When necessary, the strain components can be decomposed into their elastic and plastic parts using Eqs. 9.10, 9.19_2, and 9.21.

A diagram of the state points realized for this case is given in Fig. 9.16. It is easy to show that the material in both regions 5^+ and 5^- is in the elastic state. This solution can be extended to the next wave interaction by simply advancing each of the waves at its propagation velocity. Further extension of the solution involves calculation of additional shock interactions.

The next interaction (see Fig. 9.15) is the collision of the left-propagating shock taking the material from state 3 to state 5^- with the right-propagating shock taking material from state 3 to state 4. The state produced by this interaction (let us call it state 6) lies at the intersection of the Hugoniot centered on state 4 with that centered on 5^-, as shown in Fig. 9.16a. We see from Fig. 9.16b that both the transition from state 4 to state 6 and the transition from state 5^- to state 6 are elastic shocks. The analysis of region 6 proceeds in the same way as that of region 5. It shows that this interaction also produces a contact surface, so state 6 comprises two substates which we designate 6^+ and 6^-. The field quantities in regions 6^- and 6^+ are reported in Appendix 9.A and plotted on the stress-strain plane of Fig. 9.16b. We find that states 6^+ and 6^- both lie inside the yield surface, confirming the assumption that the transitions to these states are both elastic shocks.

It is important to note that if the analysis were being carried out for a slightly larger value of $-t_{11}^{(2)}$, $-t_{11}^{(6)}$ would lie below $-t_{11}^{(3)}$ and the transition from state 5^- to state 6 would pass over the yield point, altering the qualitative nature of the solution. It is the need to exclude this change in the nature of the solution that establishes the upper limit on $-t_{11}^{(2)}$ defining stress range C.

Completion of the solution requires analysis of one more interaction, as indicated in Fig. 9.15. State 7 is formed when the right-propagating transition from state 5^- to state 6^+ encounters the contact surface separating regions 5^- and 5^+. We expect formation of a left-propagating transition from state 6^+ to state 7^- and a right-propagating transition from state 5^+ to state 7^+. We see from Fig. 9.16 that states 5^+ and 6^+ are both in the elastic range, so compression and decompression from these states begins with an elastic deformation. The t_{11}–v_1 Hugoniots for a right-propagating shock centered on state 5^+ and a left-propagating shock centered on state 6 intersect at the point corresponding to state 6, so we have the solution

$$t_{11}^{(7)} = t_{11}^{(6)}, \quad v_1^{(7)} = v_1^{(6)}. \tag{9.57}$$

The left-propagating shock is of vanishing strength since there is no jump in either t_{11} or v_1. In effect, there is no distinct region 7^-; region 6^+ is simply extended to the contact surface. Application of the jump condition 9.7_1 to the transition from region 5^+ to region 7^+ yields $E_{11}^{(7+)}$ as given in Appendix 9.A. Waveforms at several times noted on the x–t diagram are shown in Fig. 9.17.

It is easy to verify that the attenuation curve for this problem is as given in Fig. 9.14.

9.6.3. Attenuation of a Pulse of Amplitude in the Stress Range D

Stress range D is the second-lowest characteristic-response range in which an elastic–plastic decompression wave appears. As with the case of Sec. 9.6.2, the nonattenuated pulse is qualitatively like that shown in Fig. 9.12b. The x–t diagram for this problem is given in Fig. 9.18, and the solution in regions 1–4 is as given in Secs. 9.5.1 and 9.5.2.

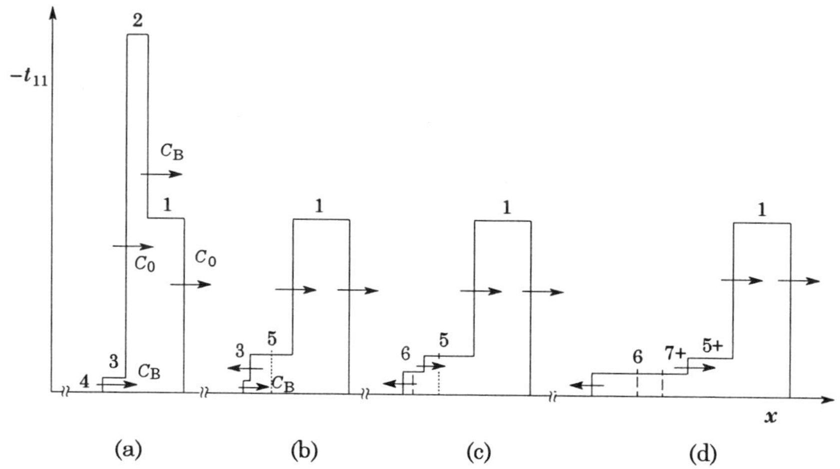

Figure 9.17. Unattenuated and three attenuated waveforms for a 1-μs duration pulse of −3.7 GPa amplitude propagating in steel: (a) at 3 μs, (b) at 5.15 μs, (c) at 5.40 μs, and (d) at 6.00 μs. Arrows designate the direction of wave propagation. Elastic shocks often carry no notation, but plastic waves are all marked with their wavespeed, C_B. Numerals relate portions of the waveform to the corresponding region of the x–t diagram.

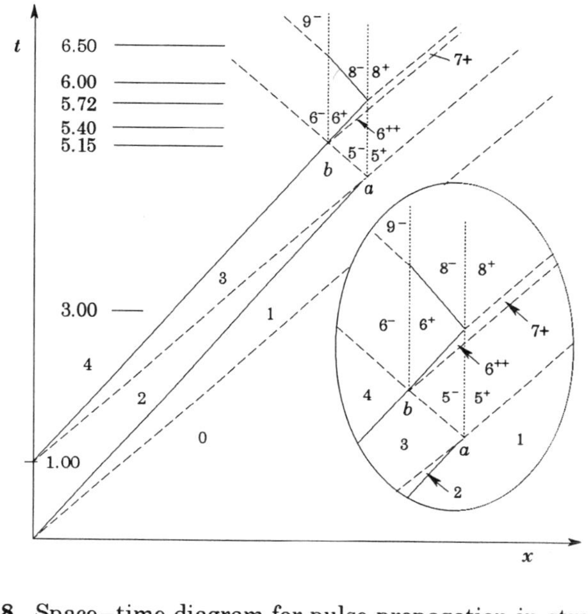

Figure 9.18. Space–time diagram for pulse propagation in stress range D, $[(C_0 + 3C_B)/(2C_B)]t_{11}^{\text{HEL}} < -t_{11}^{(2)} \le [(C_0 + 5C_B)/(C_0 + C_B)]t_{11}^{\text{HEL}}$. The notational conventions are the same as those of Fig. 9.13 and the scale is for steel.

The attenuation process begins when the elastic part of the decompression wave overtakes the plastic part of the compression wave (see Figs. 9.12b and 9.18), an interaction producing a right-propagating shock moving into the material in state 1 and a left-propagating shock moving into material in state 3. The analysis of this interaction proceeds exactly as described in Sec. 9.6.2 and leads to the same solution. The waveform at times slightly before and after the formation of region 5 is plotted in Fig. 9.19.

Note that, if $-t_{11}^{(2)}$ were somewhat larger, $-t_{11}^{(5)}$ would exceed t_{11}^{HEL} and the Hugoniot connecting state 1 to state 5 would encounter the plastic rather than the elastic segment of the compression Hugoniot. The need to separate response ranges at this point determines the upper limit on peak compressive stress defining response range D.

The next interaction (see Fig. 9.18) is the collision of the left-propagating shock taking the material from state 3 to state 5^- with the right-propagating shock taking material from state 3 to state 4. To analyze this interaction, we need the $t_{11}-v$ Hugoniots for right-propagating shocks centered on state 5^- and for left-propagating shocks centered on state 4. Determination of these Hugoniots requires examination of the $t_{11}-E_{11}$ and $t_{22}-E_{11}$ Hugoniots in order to establish whether the transition process at issue is elastic or plastic. State 4 is at the reverse yield point, so the Hugoniot begins with an elastic segment that extends to the forward yield point. Further loading produces plastic deformation. Since state 5^- is not at yield (see Fig. 9.19b), the Hugoniot centered on it begins with an elastic segment taking the material to the reverse yield point and continues with a plastic segment. Let us call the state at the intersection of these Hugoniots state 6 (which we shall assume may contain a contact surface separating substates 6^- and 6^+). Examination of Fig. 9.20 indicates that the intersection of these Hugoniots is such that the segment of the Hugoniot connecting state 5^- to state 6^+ passes over

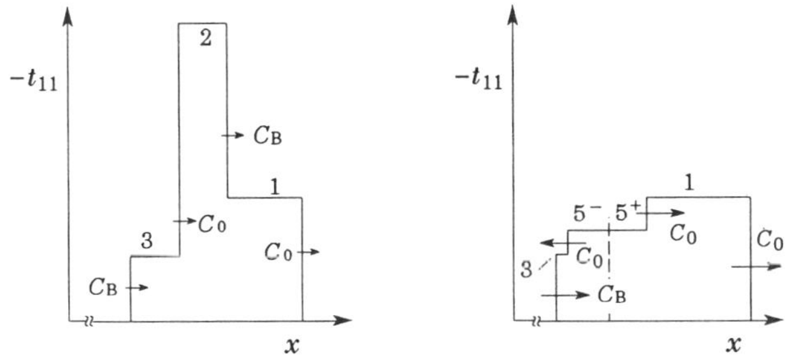

Figure 9.19. Waveform at $t = 3.00$ and 5.15 μs (slightly before and after the first wave interaction) for a pulse of amplitude in stress range D. The example refers to a –4.5-GPa pulse of 1 μs duration propagating in steel.

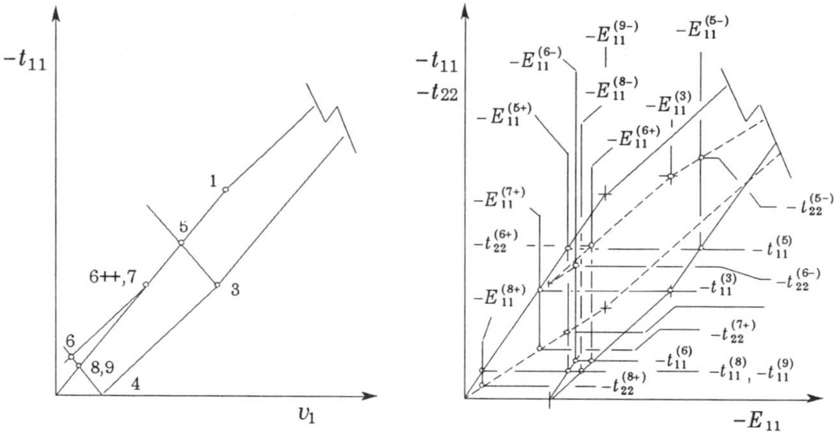

(a) (b)

Figure 9.20. (a) Stress–particle velocity and (b) stress–strain Hugoniots for attenuation of a pulse in stress range D. The example is drawn for a stress pulse of amplitude –4.5 GPa in steel.

the reverse yield point. This means that the decompression shock will be unstable, separating into an elastic decompression from state 5^- to the yield state (which we call state 6^{++}) followed by a plastic decompression from state 6^{++} to state 6^+. The transtion from state 4 to state 6^- is an elastic compression.

Applying the jump conditions 9.7 to the right-propagating elastic shock transition from state 5^- to state 6^{++} yields the values for the field variables in the latter region. The analysis of the interaction is now completed by applying the jump conditions to the right-propagating plastic shock, taking the material from state 6^{++} to the state 6^+, and the left-propagating elastic shock, taking the material from state 4 to state 6^-. The result is recorded in Appendix 9.B and plotted in Fig. 9.20. It is easy to verify that state 6^{++} lies on the yield surface and that the stresses and strains are the same as in state 3, although the particle velocity is lower. We find that state 6^+ lies below the HEL and is on the yield surface. State 6^- involves even lower stresses and lies inside the yield surface.

Continuation of the solution requires analysis of several more interactions, as shown in Fig. 9.18. Some of these interactions are between shocks and some are of shocks with contact interfaces. In the following paragraphs, we briefly sketch the results of the analysis of the remaining regions shown on Fig. 9.18. The state points associated with each of these regions are plotted on the Hugoniots of Fig. 9.20. The field values in each of the regions shown in the figure are given in Appendix 9.B and illustrated in the stress–strain and stress–particle velocity planes and as stress profiles in Fig. 9.21.

State 7 is formed when the right-propagating transition from state 5^- to state 6^{++} encounters the contact surface separating regions 5^- and 5^+. We expect formation of a left-propagating transition from state 6^{++} to state 7^- and a right-propagating transition from state 5^+ to state 7^+. Since state 5^+ is in the elastic range, both compression and decompression from this state begin with an elastic deformation. State 6^{++} is at the reverse yield point, so compression is elastic and decompression is plastic. The t_{11}–v_1 Hugoniots for a right-propagating shock centered on state 5^+ and a left-propagating shock centered on state 6^{++} intersect at the point corresponding to state 6^{++} so we have the solution

$$t_{11}^{(7)} = t_{11}^{(6++)}, \quad v_1^{(7)} = v_1^{(6++)}. \tag{9.58}$$

The left-propagating shock is of vanishing strength since there is no jump in either t_{11} or v_1. From the jump condition 9.7_1, we have

$$E_{11}^{(7-)} = E_{11}^{(6++)} \tag{9.59}$$

and we find from Eq. 9.31 that

$$t_{22}^{(7-)} = t_{22}^{(6++)}. \tag{9.60}$$

In effect, there is no distinct region 7^-; region 6^{++} is simply extended to the contact surface. Application of the jump condition 9.7_1 to the transition from region 5^+ to region 7^+ yields $E_{11}^{(7+)}$ as given in Appendix 9.B. Waveforms at various times after formation of region 6 are shown in Fig. 9.21.

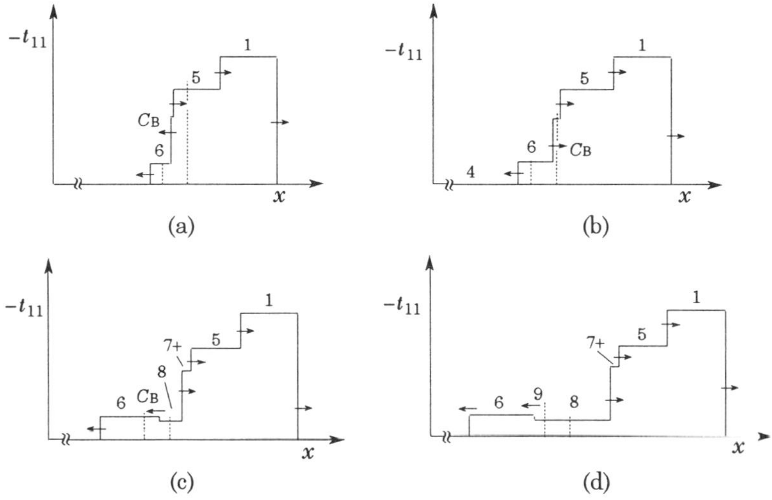

Figure 9.21. This is a continuation of the sequence of waveforms begun in Figs. 9.12b and 9.19, but with an enlargement of both horizontal and vertical scale. Waveforms are for the times (a) $t = 5.40$, (b) $t = 5.72$, (c) $t = 6.00$, and (d) $t = 6.50$ μs. Notational conventions are as before.

State 8 is formed as a result of the encounter of the plastic decompression wave with the contact surface through region 5. We expect formation of a left-propagating transition from state 6^+ to state 8 and a right-propagating transition from state 7^+ to state 8. State 6^+ is at reverse yield, so compression is an elastic process and decompression is a plastic process. State 7^+ is elastic: a small range of elastic compression and a larger range of elastic decompression is available before an elastic–plastic transition point is reached and deformation continues as a plastic process. The intersection of the two Hugoniots of interest falls on the main compression branch of the $t_{11}-v_1$ and $t_{11}-E_{11}$ paths at a point well below state 6^{++}. Application of the jump conditions to these shocks leads to the solution given in Appendix 9.B.

State 9 is analyzed in the same way as the regions discussed previously. Like the interaction producing region 7, this interaction produces only one nontrivial shock, a left-propagating elastic shock. At the completion of this interaction, all remaining shocks are elastic and are propagating away from the contact interfaces, so there can be no further interactions.

Again, it is easy to verify that the attenuation curve for this case is similar to the one plotted in Fig. 9.14. Only the value of $t_{11}^{(2)}$ is changed. It is worth pointing out that the attenuation process, as represented by the peak stress criterion used in drawing the attenuation curve, is completed with formation of region 5. The several subsequent interactions that must be analyzed to complete the description of the entire waveform cannot cause reduction of the peak compressive stress below t_{11}^{HEL}.

9.6.4. Attenuation of a Pulse of Amplitude in the Stress Range E

Stress range E of Table 9.2 is the lowest-stress range in which the longitudinal stress resulting from the first wave interaction lies above the Hugoniot elastic limit. As with the previous case, the non-attenuated pulse is qualitatively like that shown in Fig. 9.12b.

The analysis of this case proceeds in the same way as the cases already discussed. The results of the analysis are shown in Figs. 9.22–9.26 and given in Appendix 9.C. The upper limit on stress range E is set by the requirement that $-t_{11}^{(9)} \leq t_{11}^{\mathrm{HEL}}$ (see Fig. 9.23). Several very weak interactions occur beginning with the encounter of the left-propagating shock bounding region 8 with the contact surface b. These interactions do not significantly affect the waveform, do not affect the attenuation curve, and have not been analyzed.

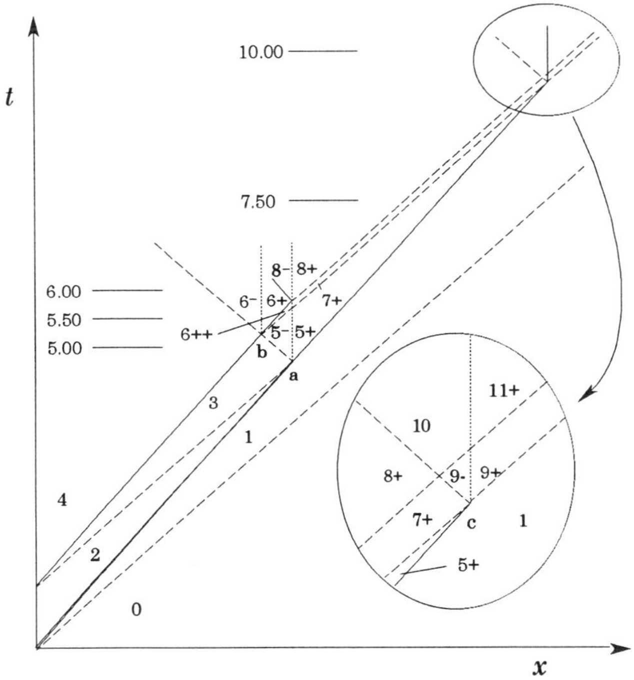

Figure 9.22. Space–time diagram for pulse propagation in the stress range E. The figure is scaled for steel; times marked along the ordinate are in microseconds and correspond to plots of waveforms appearing in Fig. 9.25.

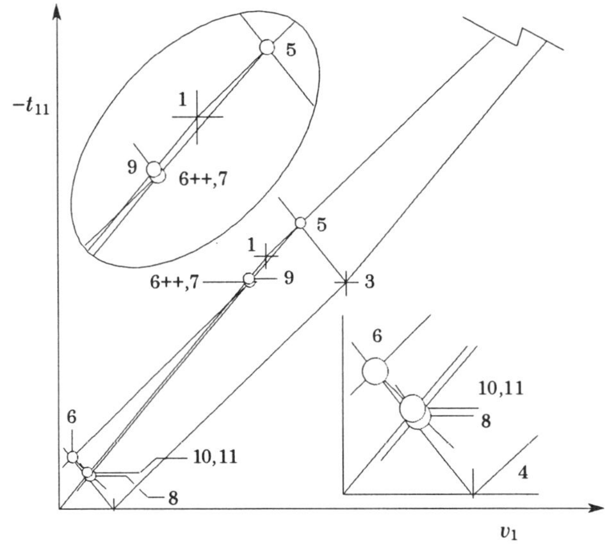

Figure 9.23. Low-stress portions of the stress–particle velocity Hugoniot for a pulse of amplitude $-t_{11}^{(2)}$ falling in the stress range E. The example is drawn for a −5.15-GPa pulse propagating in steel.

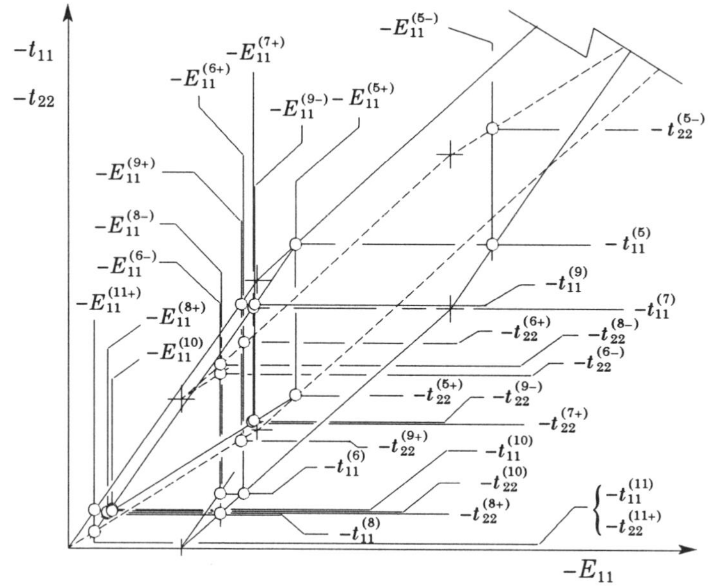

Figure 9.24. Low-stress portions of the stress-strain Hugoniot for a pulse falling in the stress range E. The example is a –5.15-GPa pulse in steel.

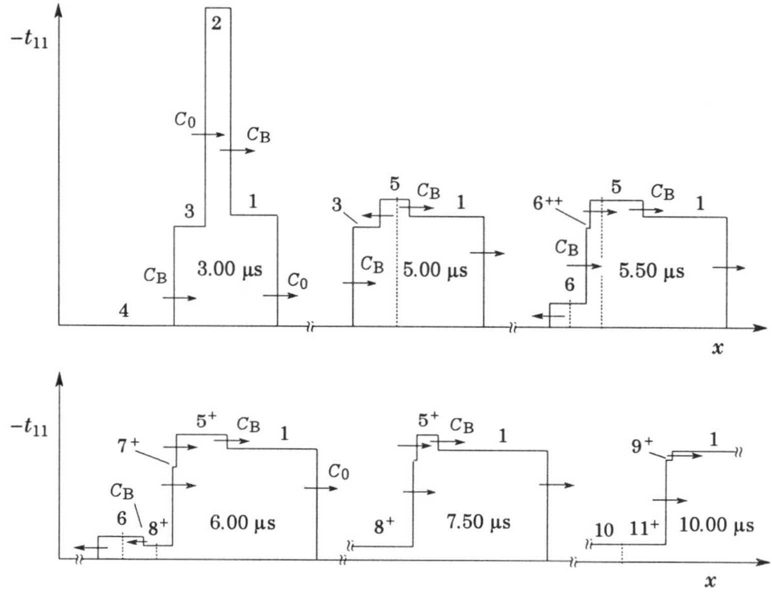

Figure 9.25. Waveforms for a pulse in stress range E. The example is a 1-μs-duration, –5.15-GPa-amplitude pulse in steel. Numerals relate portions of the waveform to the corresponding region of the x–t diagram of Fig. 9.22.

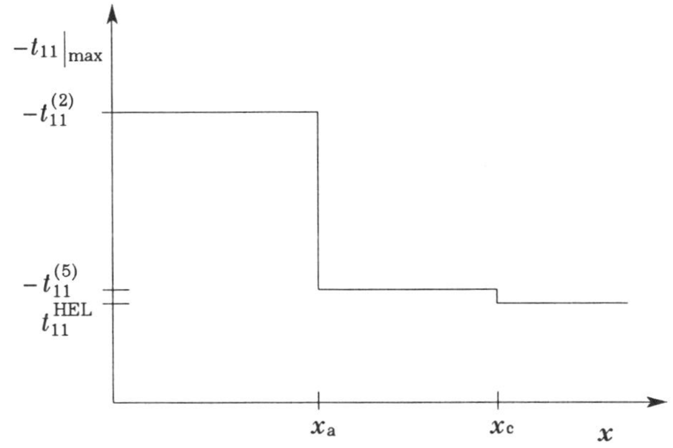

Figure 9.26. Attenuation curve for a pulse in stress range E. The points x_a and x_c are defined in Fig. 9.22.

9.6.5. Attenuation of a Pulse of Amplitude in Stress Range F

The case of the previous section concerned the small stress interval in which the compressive stress produced by the first wave interaction (state 5, as shown in Fig. 9.23) exceeds the Hugoniot elastic limit, but the longitudinal stress in all subsequent states falls below this level. In this section, we consider the next higher stress range, range F in Table 9.2. As we have noted, a pulse attenuation analysis requires consideration of increasingly many interactions as cases of larger peak compressive stresses are addressed. These problems are usually solved numerically using finite-element or finite-difference codes incorporating artificial viscosity, but these methods do not offer the degree of insight into the nature of the solution that can be obtained by use of computer-based characteristics methods or closed-form analysis. An alternative approach is graphical analysis. It lends itself to development of insight, is quickly and easily performed, and often provides numerical results of accuracy comparable to that obtainable by evaluating closed-form solutions (especially when the latter involve heavy cancellation). The solution to a problem in stress range F, as obtained solely by graphical means, is presented in Figs. 9.27–9.31.

The upper limit on $-t_{11}^{(2)}$ defining range F is established by the requirement that the $t_{11}-v_1$ Hugoniot for right-propagating elastic decompression shocks centered on state 5 intersect the v_1 axis at v_1^R. The requirement is actually imposed to ensure proper treatment of the solution in a region of the x–t plane not analyzed.

The results shown in the following five figures were obtained using a drawing program on a small personal computer but could equally well have been obtained by hand drafting. The easiest way to proceed seems to be to plot the basic hysteresis loops in both the stress–particle velocity and stress-strain planes (side-by-side with the same stress scale) using the analytical results for states 1–4 before proceeding solely by graphical means. Although it is essential to know whether a transition will be elastic or plastic before it can be analyzed in detail, this determination can be made without plotting the transverse stress states.

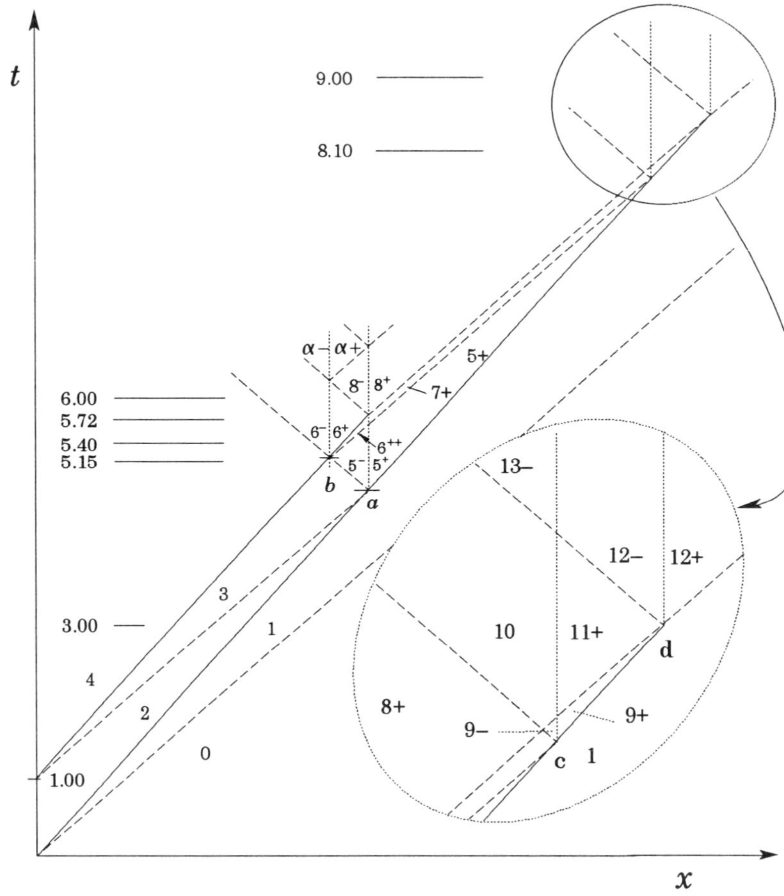

Figure 9.27. Space–time diagram for pulse propagation in range F. The notational conventions are as on the previous x–t diagrams and times marked correspond to plots of waveforms in Fig. 9.30. Several very weak interactions that occur beginning with the formation of the region labeled α have not been analyzed. These interactions do not have a significant effect on the waveform and do not affect the attenuation curve (steel).

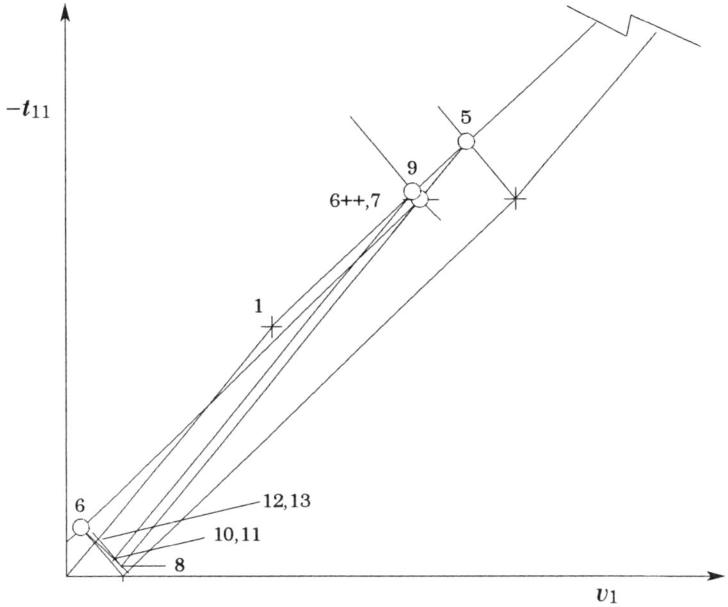

Figure 9.28. Low-stress portions of the stress–particle velocity Hugoniot curve for a pulse of amplitude $-t_{11}^{(2)}$ falling in the stress range F. The example is drawn for a –6.25-GPa pulse propagating in steel.

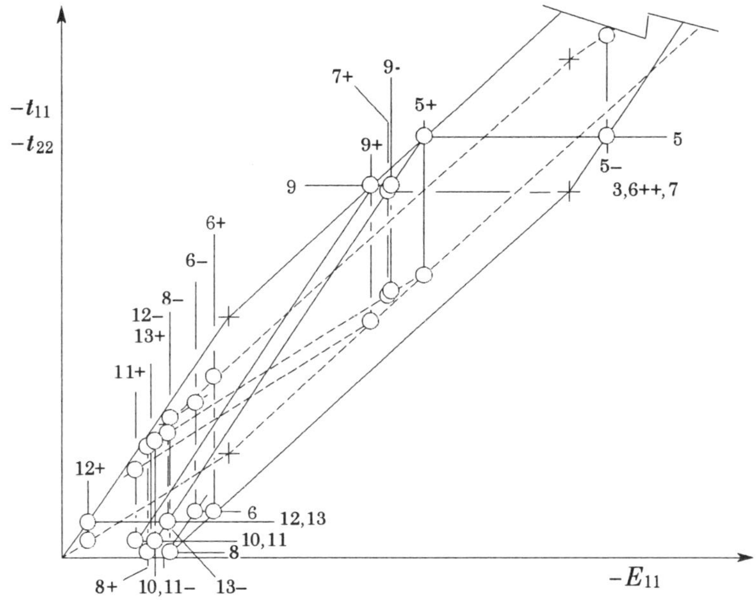

Figure 9.29. Low-stress portions of the stress-strain Hugoniot curves for a pulse of amplitude $-t_{11}^{(2)}$ falling in the stress range F. The example is drawn for a stress pulse of amplitude –6.25 GPa propagating in steel.

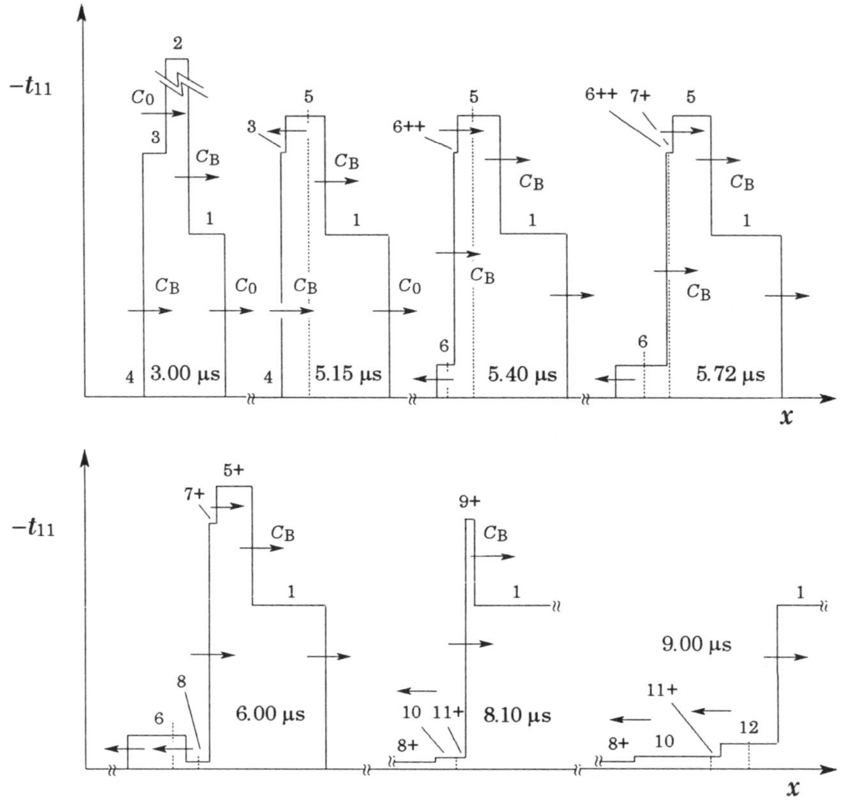

Figure 9.30. Waveforms for a pulse having an initial amplitude of –6.25 GPa and duration of 1 μs in stress range F in steel at the times shown on the figure. Numerals relate portions of the waveform to the corresponding region of the x–t diagram of Fig. 9.27.

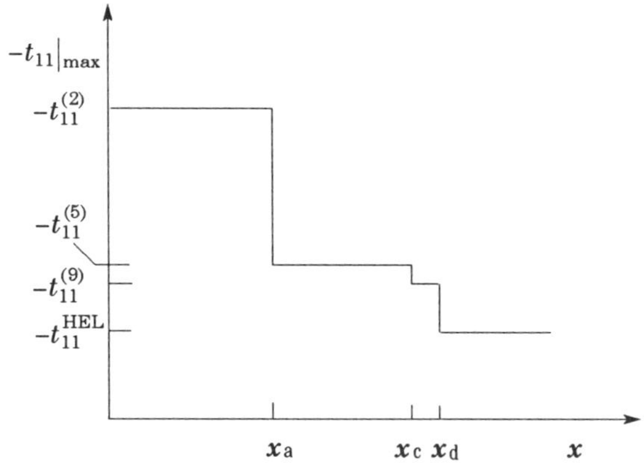

Figure 9.31. Attenuation curve for a pulse in stress range F. The points x_a, x_c, and x_d are defined in Fig. 9.27.

9.6.6. Attenuation of Higher-Amplitude Pulses

The methods outlined in previous sections suffice for analysis of propagation and attenuation of a pulse of any amplitude. Unfortunately, the analyses become increasingly awkward as attention is turned to pulses of ever-greater amplitude. This is because of the multiplicity of stress ranges and the larger number of interactions that must be considered. If a complete understanding of the process of is required, there is no alternative to analysis of the sort that has been demonstrated. However, an approximation to the attenuation curve be obtained very easily, as shown in this section.

An indication of the overall trend of the attenuation process can be obtained by examination of the x–t diagram for stress range F (Fig. 9.27). We see on this diagram that the shock interactions are clustered in zones near the points at which elastic waves overtake the plastic compression wave. In ranges B–D, attenuation to the elastic precursor amplitude is completed at the first interaction point. In ranges E and F, attenuation in the amount

$$\Delta t_{11} = \frac{4C_{\mathrm{B}}}{C_0 + C_{\mathrm{B}}} t_{11}^{\mathrm{HEL}} \tag{9.61}$$

occurs at the first interaction point, with a second attenuation step to the elastic precursor amplitude occurring in the second interaction zone (i.e., at a propagation distance of about $2x_{\mathrm{a}}$).

To see how this observation might be generalized to pulses of still greater amplitude, we begin by presenting the graphical solution for a pulse of compressive stress amplitude –10 GPa propagating in steel. Examination of Figs. 9.32–9.35 shows that there are three interaction zones in the stress range considered. Attenuation in the amount given by Eq. 9.61 occurs in each of the first two interaction zones, with the final steps to the precursor amplitude taking place in the third interaction zone. The attenuation that occurs in the first interaction zone takes place in a single step. The attenuation increment occurring in the second interaction zone occurs over two steps, whereas that occurring in the third interaction zone is accomplished in three steps.

The trend suggested by the analyses presented is that attenuation occurs in interactions formed at each reverberation of one or a packet of elastic shocks between the main plastic compression and decompression shocks. As the number of reverberations increases (as it does with increasing peak stress), the number of elastic waves in the reverberating wave packet increases, leading to more attenuation steps in each zone.

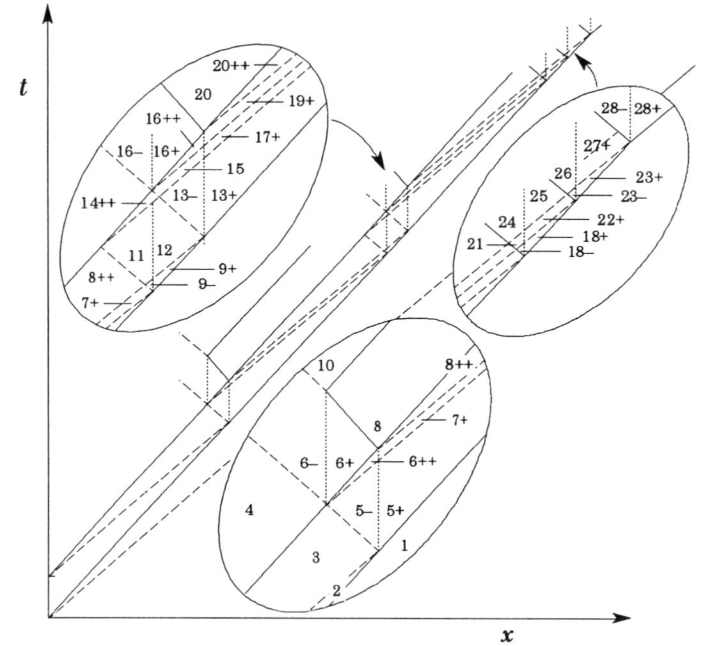

Figure 9.32. The x–t diagram for formation and attenuation of a stress pulse of 1 μs duration propagating in steel.

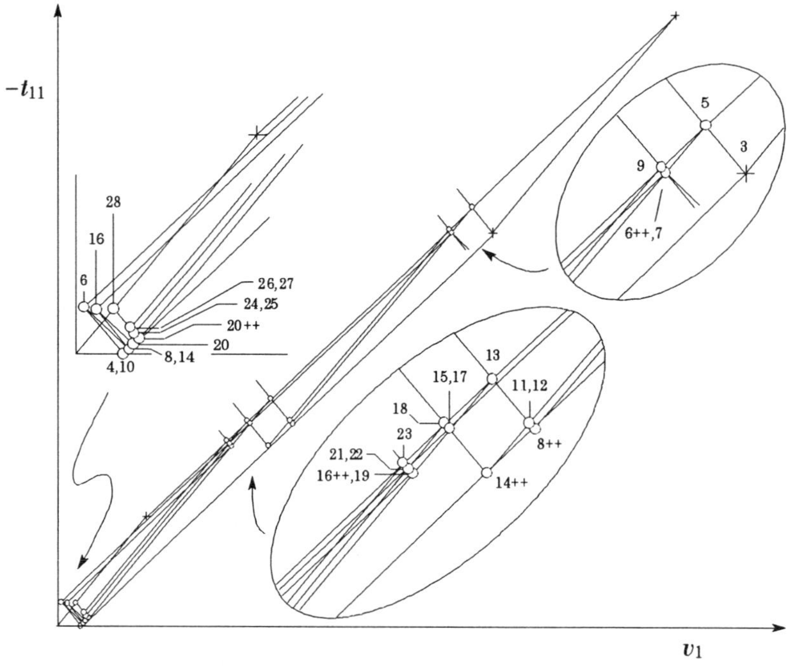

Figure 9.33. The t_{11}–v_1 Hugoniot for a –10-GPa pulse propagating in steel.

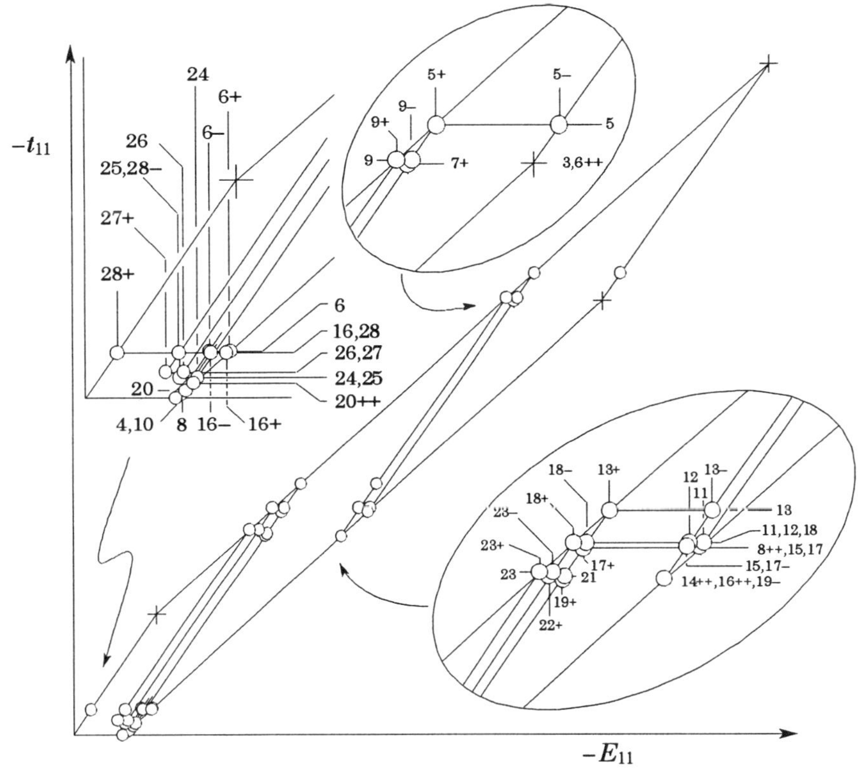

Figure 9.34. Longitudinal stress-strain Hugoniots corresponding to the stress–particle velocity Hugoniots of Fig. 9.33.

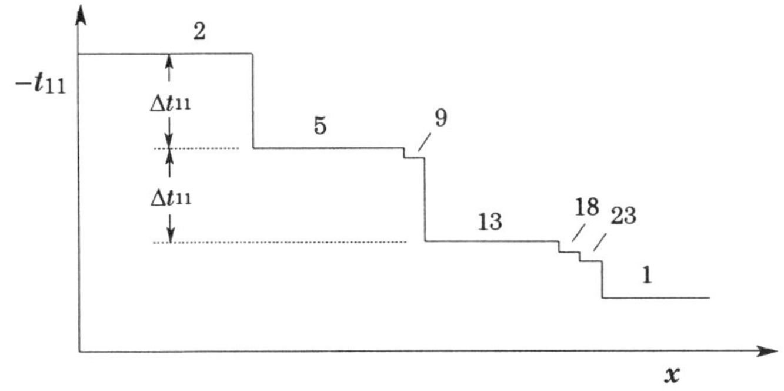

Figure 9.35. Attenuation curve for a stress pulse of amplitude –10 GPa in steel. Numerals refer to regions on the x–t diagram of Fig. 9.32 and Δt_{11} is as given by Eq. 9.61.

9.7. Summary and Conclusions

In this chapter we have exhibited solutions to the simplest problems of propagation and attenuation of uniaxial compression pulses in an ideal elastoplastic solid. The propagation and attenuation process is complex, involving many small shock interactions. Each of these interactions influences the propagated waveform, but the attenuation curve is independent of many of these interactions, and a simple approximation of its form was discussed.

It was shown that a given solution to this problem is valid only within a rather narrow range of peak stress and that a comprehensive treatment requires identification of these ranges and calculation of the solution within each range. Although the number of ranges to be considered is infinite, the theory used in the analyses presented is limited to small strains. In the case of the –10-GPa shock in steel, the peak compressive strain (calculated by the linear equation) is somewhat over 5 %, a value 13 % lower than the same strain calculated by a nonlinear measure. The plastic wavespeed at the peak stress will exceed C_B by some 8 %. These observations suggest that this example is at the upper limit of the range over which the theory is useful and that analyses covering some small number of ranges would suffice for any practical application.

The history of inelastic deformation in zones where shock interactions are concentrated is complicated, and many of the interactions produce a contact surface at which the plastic strain and residual transverse stress components are discontinuous. This must produce effects on material microstructures that may play an important role in interpreting metallurgical observations of recovered samples.

Most analyses of elastoplastic waves are performed numerically and study of the material in this chapter makes it easy to see why this is so. Numerical methods permit solution of problems involving continuous loading and taking account of effects such as hardening, nonlinear elastic response, and strain-rate dependence that have not been considered in this chapter. It is important to realize, however, that intelligent application of numerical methods requires an understanding of the nature of the shock interactions occurring.

9.A. Appendix: Field Values for Pulse Attenuation in Stress Range C

The formulas given are valid for peak stress, $-t_{11}^{(2)}$, in range C, $2t_{11}^{\mathrm{HEL}} < -t_{11}^{(2)} \le [(C_0+3C_B)/(2C_B)]t_{11}^{\mathrm{HEL}}$. Numerical values given for the field variables are for the case in which a compression shock of amplitude $t_{11}^{(2)} = -3.7$ GPa, followed by decompression to $t_{11}^{(2)} = 0$, is introduced into a steel characterized by the material parameters $\rho = 7870$ kg/m^3, $C_0 = 5.982$ km/s, $C_B = 4.729$ km/s, and $Y = 1$ GPa.

Region 1: (see Sec. 9.5.1)

$$t_{11}^{(1)} = -t_{11}^{\text{HEL}} = -1.78 \text{ GPa}, \quad v_1^{(1)} = v_1^{\text{HEL}} = 0.0377 \text{ km/s}$$
$$E_{11}^{(1)} = -E_{11}^{\text{HEL}} = -0.00631, \quad t_{22}^{(1)} = -t_{22}^{\text{HEL}} = -0.78 \text{ GPa} \tag{9.A1}$$

Region 2: (see Sec. 9.5.1)

$$t_{11}^{(2)} = -3.7 \text{ GPa}$$
$$v_1^{(2)} = \frac{1}{\rho C_B}\left\{-t_{11}^{(2)} - \left(\frac{C_0 - C_B}{C_0}\right)t_{11}^{\text{HEL}}\right\} = 0.0894 \text{ km/s}$$
$$E_{11}^{(2)} = \frac{1}{\rho C_B^2}\left\{ t_{11}^{(2)} + \left(\frac{C_0^2 - C_B^2}{C_0^2}\right)t_{11}^{\text{HEL}}\right\} = -0.0172 \tag{9.A2}$$
$$t_{22}^{(2)} = t_{11}^{(2)} + \frac{3C_0^2 - 3C_B^2}{2C_0^2}t_{11}^{\text{HEL}} = -2.70 \text{ GPa}$$

Region 3:

$$t_{11}^{(3)} = t_{11}^{(2)} + 2t_{11}^{\text{HEL}} = -0.15 \text{ GPa}$$
$$v_1^{(3)} = v_1^{(2)} - 2v_1^{\text{HEL}} = \frac{-1}{\rho C_B}\left[t_{11}^{(2)} + \left(\frac{C_0 + C_B}{C_0}\right)t_{11}^{\text{HEL}}\right] = 0.0140 \text{ km/s} \tag{9.A3}$$
$$E_{11}^{(3)} = E_{11}^{(2)} + 2E_{11}^{\text{HEL}} = \frac{1}{\rho C_B^2}\left[t_{11}^{(2)} + \left(\frac{C_0^2 + C_B^2}{C_0^2}\right)t_{11}^{\text{HEL}}\right] = -0.00463$$
$$t_{22}^{(3)} = t_{22}^{(2)} + 2t_{22}^{\text{HEL}} = t_{11}^{(3)} - Y = t_{11}^{(2)} + \frac{C_0^2 + 3C_B^2}{2C_0^2}t_{11}^{\text{HEL}} = -1.15 \text{ GPa}$$

Region 4:

$$t_{11}^{(4)} = 0, \quad v_1^{(4)} = v_1^{\text{R}} = \frac{C_0 - C_B}{\rho C_0 C_B}t_{11}^{\text{HEL}} = 0.00999 \text{ km/s}$$
$$E_{11}^{(4)} = -E_{11}^{\text{R}} = -\frac{C_0^2 - C_B^2}{\rho C_0^2 C_B^2}t_{11}^{\text{HEL}} = -0.00379 \tag{9.A4}$$
$$t_{22}^{(4)} = -t_{22}^{\text{R}} = \frac{-3C_0^2 + 3C_B^2}{2C_0^2}t_{11}^{\text{HEL}} = -Y = -1.00 \text{ GPa}$$

Region 5:

$$t_{11}^{(5)} = \frac{C_0 + C_B}{2C_B}t_{11}^{(2)} + \frac{C_0 + 3C_B}{2C_B}t_{11}^{\text{HEL}} = -0.40 \text{ GPa} \tag{9.A5}$$

$$v_1^{(5)} = -\frac{1}{\rho C_0} t_{11}^{(5)} = 0.00856 \text{ km/s}$$

$$E_{11}^{(5+)} = -\frac{1}{C_0} v_1^{(5)} = -0.00143 \qquad (9.A5, cont.)$$

$$E_{11}^{(5-)} = \frac{2C_0^2 + C_0 C_B - C_B^2}{2\rho C_0^2 C_B^2} t_{11}^{(2)} + \frac{2C_0^2 + C_0 C_B + C_B^2}{2\rho C_0^2 C_B^2} t_{11}^{\text{HEL}} = -0.00553$$

$$t_{22}^{(5-)} = \frac{-C_0^3 + 5C_0^2 C_B + 3C_0 C_B^2 - 3C_B^3}{4C_0^2 C_B} t_{11}^{(2)} + \frac{-C_0^3 + 3C_0^2 C_B + 3C_0 C_B^2 + 3C_B^3}{4C_0^2 C_B} t_{11}^{\text{HEL}} = -1.26 \text{ GPa}$$

$$t_{22}^{(5+)} = \frac{-C_0^3 - C_0^2 C_B + 3C_0 C_B^2 + 3C_B^3}{4C_0^2 C_B} t_{11}^{(2)} + \frac{-C_0^3 - 3C_0^2 C_B + 3C_0 C_B^2 + 9C_B^3}{4C_0^2 C_B} t_{11}^{\text{HEL}} = -0.18 \text{ GPa}$$

Region 6:

$$t_{11}^{(6)} = -\frac{C_0 - C_B}{2C_B} t_{11}^{\text{HEL}} = -0.24 \text{ GPa}$$

$$v_1^{(6)} = \frac{C_0 - C_B}{2\rho C_0 C_B} t_{11}^{\text{HEL}} = -0.00500 \text{ km/s} \qquad (9.A6)$$

$$E_{11}^{(6-)} = -\frac{2C_0^2 + C_0 C_B - 3C_B^2}{2\rho C_0^2 C_B^2} t_{11}^{\text{HEL}} = -0.00462$$

$$E_{11}^{(6+)} = \frac{C_0^2 - C_B^2}{\rho C_0^2 C_B^2} t_{11}^{(2)} + \frac{2C_0^2 - C_0 C_B - C_B^2}{2\rho C_0^2 C_B^2} t_{11}^{\text{HEL}} = -0.00494$$

$$t_{11}^{(6-)} = \frac{C_0^3 - 7C_0^2 C_B - 3C_0 C_B^2 + 9C_B^3}{4C_0^2 C_B} t_{11}^{\text{HEL}} = -1.10 \text{ GPa}$$

$$t_{11}^{(6+)} = \frac{3(C_0^2 - C_B^2)}{2C_0^2} t_{11}^{(2)} + \frac{C_0^3 + 5C_0^2 C_B - 3C_0 C_B^2 - 3C_B^3}{4C_0^2 C_B} t_{11}^{\text{HEL}} = -1.18 \text{ GPa}$$

Region 7:

$$t_{11}^{(7)} = t_{11}^{(6)}, \quad v_1^{(7)} = v_1^{(6)}, \quad E_{11}^{(7-)} = E_{11}^{(6+)}$$

$$E_{11}^{(7+)} = -\frac{C_0 - C_B}{2\rho C_0^2 C_B} t_{11}^{\text{HEL}} = -0.000836 \qquad (9.A7)$$

$$t_{22}^{(7+)} = -\frac{C_0 - C_B}{4C_0^2 C_B}(3C_B^2 - C_0^2)\, t_{11}^{\mathrm{HEL}} = -0.10 \text{ GPa} \qquad (9.A7, cont)$$

9.B. Appendix: Field Values for Pulse Attenuation in Range D

The formulas given are valid for peak stress, $-t_{11}^{(2)}$, in range D, $[(C_0 + 3C_B)/(2C_B)]t_{11}^{\mathrm{HEL}} < -t_{11}^{(2)} \le [(C_0 + 5C_B)/(C_0 + C_B)]t_{11}^{\mathrm{HEL}}$. The numerical values given for the field variables are for the case in which a compression shock of amplitude $t_{11}^{(2)} = -4.5$ GPa, followed by decompression to $t_{11}^{(2)} = 0$, is introduced into a steel characterized by the material parameters $\rho = 7870 \text{kg/m}^3$, $C_0 = 5.982 \text{km/s}$, $C_B = 4.729 \text{km/s}$, and $Y = 1 \text{GPa}$.

Region 1: Results for this region are as given by Eq. 9.A1.

Region 2: Formulas for this region are as given by Eq. 9.A2, but the numerical values change to those given.

$$t_{11}^{(2)} = -4.5 \text{ GPa}, \quad v_1^{(2)} = 0.111 \text{ km/s}$$
$$E_{11}^{(2)} = -0.0218, \quad t_{22}^{(2)} = -3.50 \text{ GPa} \qquad (9.B1)$$

Region 3: Formulas for this region are as given by Eq. 9.A3, but the numerical values change to those given.

$$t_{11}^{(3)} = -0.95 \text{ GPa}, \quad v_1^{(3)} = 0.0355 \text{ km/s}$$
$$E_{11}^{(3)} =, -0.00917, \quad t_{22}^{(3)} =, -1.95 \text{ GPa} \qquad (9.B2)$$

Region 4: Results for this region are as given by Eq. 9.A4.

Region 5: Formulas for this region are as given by Eq. 9.A5, but the numerical values change to those given.

$$t_{11}^{(5)} = -1.31 \text{ GPa}, \quad v_1^{(5)} = 0.0278 \text{ km/s}$$
$$E_{11}^{(5-)} = -0.0105, \quad E_{11}^{(5+)} = -0.00465 \qquad (9.B3)$$
$$t_{22}^{(5-)} = -2.107 \text{ GPa}, \quad t_{22}^{(5+)} = -0.57 \text{ GPa}$$

Region 6:

$$t_{11}^{(6++)} = t_{11}^{(3)}, \quad E_{11}^{(6++)} = E_{11}^{(3)}, \quad v_1^{(6++)} = -\frac{1}{\rho C_0}\left[t_{11}^{(2)} + 2t_{11}^{\mathrm{HEL}}\right] = 0.201 \text{ km/s}$$
$$t_{11}^{(6)} = \frac{C_0 - C_B}{C_0 + C_B}\left[t_{11}^{(2)} + t_{11}^{\mathrm{HEL}}\right] = -0.32 \text{ GPa} \qquad (9.B4)$$

$$
\begin{aligned}
v_1^{(6)} &= \frac{C_0 - C_B}{\rho C_0 (C_0 + C_B)} t_{11}^{(2)} + \frac{C_0^2 + C_0 C_B - 2C_B^2}{\rho C_0 C_B (C_0 + C_B)} t_{11}^{HEL} = 0.00323 \text{ km/s} \\
t_{22}^{(6-)} &= \frac{-C_0^3 + C_0^2 C_B + 3C_0 C_B^2 - 3C_B^3}{2C_0^2 (C_0 + C_B)} t_{11}^{(2)} - \frac{2C_0^2 + C_0 C_B - 3C_B^2}{C_0 (C_0 + C_B)} t_{11}^{HEL} = -1.14 \text{ GPa} \\
t_{22}^{(6+)} &= \frac{C_0 - C_B}{C_0 + C_B} t_{11}^{(2)} - \frac{C_0^3 + 5C_0^2 C_B - 3C_0 C_B^2 - 3C_B^3}{2C_0^2 (C_0 + C_B)} t_{11}^{HEL} = -1.32 \text{ GPa} \\
E_1^{(6-)} &= \frac{C_0 - C_B}{\rho C_0^2 (C_0 + C_B)} t_{11}^{(2)} - \frac{C_0^2 + C_0 C_B - 2C_B^2}{\rho C_0 C_B^2 (C_0 + C_B)} t_{11}^{HEL} = -0.00492 \\
E_1^{(6+)} &= \frac{C_0 - C_B}{\rho C_B^2 (C_0 + C_B)} t_{11}^{(2)} - \frac{2C_0^2 - C_0 C_B - C_B^2}{\rho C_0^2 C_B (C_0 + C_B)} t_{11}^{HEL} = -0.00559
\end{aligned}
\qquad (9.B4, cont)
$$

Region 7:

$$
\begin{aligned}
& t_{11}^{(7+)} = t_{11}^{(6++)} = t_{11}^{(3)}, \quad v_1^{(7+)} = v_1^{(6++)} \\
& E_{11}^{(7+)} = \frac{1}{\rho C_0^2}\left[t_{11}^{(2)} + 2t_{11}^{HEL}\right] = -0.00337 \\
& t_{22}^{(7+)} = \frac{3C_B^2 - C_0^2}{2C_0^2}\left[t_{11}^{(2)} + 2t_{11}^{HEL}\right] = -0.415 \text{ GPa}
\end{aligned}
\qquad (9.B5)
$$

Region 8:

$$
\begin{aligned}
t_{11}^{(8)} &= \frac{(C_0 - C_B)^2}{(C_0 + C_B)^2} t_{11}^{(2)} - \frac{2C_B (C_0 - C_B)}{(C_0 + C_B)^2} t_{11}^{HEL} = -0.245 \text{ GPa} \\
v_1^{(8)} &= -\frac{(C_0 - C_B)^2}{\rho C_0 (C_0 + C_B)^2} t_{11}^{(2)} + \frac{2C_B (C_0 - C_B)}{\rho C_0 (C_0 + C_B)^2} t_{11}^{HEL} = 0.00521 \text{ km/s} \\
E_{11}^{(8-)} &= \frac{(C_0 - C_B)^2}{\rho C_B^2 (C_0 + C_B)^2} t_{11}^{(2)} - \frac{(C_0 - C_B)(C_0^3 + 5C_0^2 C_B + 3C_0 C_B^2 + C_B^3)}{\rho C_0^2 C_B^2 (C_0 + C_B)^2} t_{11}^{HEL} \\
&= -0.00518 \\
E_{11}^{(8+)} &= -\frac{1}{C_0} v_1^{(8)} = -0.000871 \\
t_{22}^{(8-)} &= \frac{(C_0 - C_B)^2}{(C_0 + C_B)^2} t_{11}^{(2)} - \frac{(C_0 - C_B)(3C_0^3 + 13C_0^2 C_B + 9C_0 C_B^2 + 3C_B^3)}{2C_0^2 (C_0 + C_B)^2} t_{11}^{HEL} \\
&= -1.24 \text{ GPa} \\
t_{22}^{(8+)} &= \frac{3C_0^2 - C_B^2}{2C_0^2} \frac{(C_0 - C_B)}{(C_0 + C_B)^2}\left[(C_0 - C_B) t_{11}^{(2)} - 2C_B t_{11}^{HEL}\right] = -0.107 \text{ GPa}
\end{aligned}
\qquad (9.B6)
$$

Region 9:

$$
t_{11}^{(9)} = t_{11}^{(8)}, \quad v_1^{(9)} = v_1^{(8)}, \quad E_{11}^{(9+)} = E_{11}^{(8-)} \qquad (9.B7)
$$

$$E_{11}^{(9-)} = \frac{-(C_0 - C_B)^2}{\rho C_0^2 (C_0 + C_B)^2} t_{11}^{(2)} - \frac{C_0^3 + 3C_0^2 C_B + C_0 C_B^2 - 5C_B^3}{\rho C_0 C_B^2 (C_0 + C_B)^2} t_{11}^{HEL} \quad \text{(9.B7, cont.)}$$

$$= -0.00459$$

$$t_{22}^{(9-)} = \frac{-(C_0 - C_B)^2 (C_0^2 + 3C_B^2)}{2C_0^2 (C_0 + C_B)^2} t_{11}^{(2)} - \frac{(C_0 - C_B)(3C_0^2 + 10C_0 C_B + 15C_B^2)}{2C_0 (C_0 + C_B)^2} t_{11}^{HEL}$$

$$= -1.09 \text{ GPa}$$

9.C. Appendix: Field Values for Pulse Attenuation in Range E

The formulas given are valid for peak stress, $-t_{11}^{(2)}$, in range E, $[(C_0 + 5C_B)/(2C_B)]t_{11}^{HEL} < -t_{11}^{(2)} \le 3t_{11}^{HEL}$. The numerical values are for the case in which a compression shock of amplitude $t_{11}^{(2)} = -5.15$ GPa, followed by decompression to $t_{11}^{(2)} = 0$, is introduced into a steel characterized by the parameters $\rho = 7870$ kg/m^3, $C_0 = 5.982$ km/s, $C_B = 4.729$ km/s, and $Y = 1$ GPa.

Region 1: Results for this region are as given by Eq. 9.A1.

Region 2: Formulas for this region are as given by Eq. 9.A2, but the numerical values change to those given.

$$t_{11}^{(2)} = -5.15 \text{ GPa}, \quad v_1^{(2)} = 0.128 \text{ km/s}$$
$$E_{11}^{(2)} = -0.0255, \quad t_{22}^{(2)} = -4.15 \text{ GPa} \quad \text{(9.C1)}$$

Region 3: Formulas for this region are as given by Eq. 9.A3, but the numerical values change to those given.

$$t_{11}^{(3)} = -1.60 \text{ GPa}, \quad v_1^{(3)} = 0.0526 \text{ km/s}$$
$$E_{11}^{(3)} = -0.0129, \quad t_{22}^{(3)} = -2.60 \text{ GPa} \quad \text{(9.C2)}$$

Region 4: Results for this region are as given by Eq. 9.A4.

Region 5:

$$t_{11}^{(5)} = t_{11}^{(3)} - 2\frac{C_0 - C_B}{C_0 + C_B} t_{11}^{HEL} = t_{11}^{(2)} + \frac{4C_B}{C_0 + C_B} t_{11}^{HEL} = -2.01 \text{ GPa}$$

$$v_1^{(5)} = v_1^{(3)} - 2\frac{C_0 - C_B}{C_0 + C_B} v^{HEL} \quad \text{(9.C3)}$$

$$= \frac{-1}{\rho C_B} \left\{ t_{11}^{(2)} + \frac{C_0^2 + 4C_0 C_B - C_B^2}{C_0 (C_0 + C_B)} t_{11}^{HEL} \right\} = 0.0441 \text{ km/s}$$

$$E_{11}^{(5-)} = E_{11}^{(2)} + \frac{4C_B}{(C_0 + C_B)} E_{11}^{HEL}$$

$$= \frac{1}{\rho C_B^2} \left\{ t_{11}^{(2)} + \left[\frac{C_0^3 + C_0^2 C_B - C_0 C_B^2 + 3C_B^3}{C_0^2 (C_0 + C_B)} \right] t_{11}^{HEL} \right\} = -0.0143 \quad \text{(9.C3, cont.)}$$

$$E_{11}^{(5+)} = E_{11}^{(2)} + \frac{4C_0^2}{C_B(C_0 + C_B)} E_{11}^{HEL}$$

$$= \frac{1}{\rho C_B^2} \left\{ t_{11}^{(2)} + \left[\frac{C_0^3 + 5C_0^2 C_B - C_0 C_B^2 - C_B^3}{C_0^2 (C_0 + C_B)} \right] t_{11}^{HEL} \right\} = -0.00767$$

$$t_{22}^{(5-)} = t_{11}^{(2)} + \frac{3C_0^3 - C_0^2 C_B - 3C_0 C_B^2 + 9C_B^3}{2C_0^2 (C_0 + C_B)} t_{11}^{HEL} = -2.78 \text{ GPa}$$

$$t_{22}^{(5+)} = t_{11}^{(2)} + \frac{3C_0^3 + 11C_0^2 C_B - 3C_0 C_B^2 - 3C_B^3}{2C_0^2 (C_0 + C_B)} t_{11}^{HEL} = -1.01 \text{ GPa}$$

Region 6:

$$t_{11}^{(6++)} = t_{11}^{(3)}, \quad v_1^{(6++)} = v_1^{(3)} - 4\frac{C_0 - C_B}{C_0 + C_B} v_1^{HEL} = 0.0350 \text{ km/s}$$

$$E_{11}^{(6++)} = E_{11}^{(3)}, \quad t_{22}^{(6++)} = t_{22}^{(3)}$$

$$t_{11}^{(6)} = -4\frac{C_B(C_0 - C_B)}{(C_0 + C_B)^2} t_{11}^{HEL} = -0.37 \text{ GPa}$$

$$v^{(6)} = \frac{C_0^3 + C_0^2 C_B - 5C_0 C_B^2 + 3C_B^3}{C_B(C_0 + C_B)^2} v_1^{HEL} = 0.00220 \text{ km/s} \quad \text{(9.C4)}$$

$$t_{22}^{(6+)} = \frac{-3C_0^4 - 14C_0^3 C_B + 8C_0^2 C_B^2 + 6C_0 C_B^3 + 3C_B^4}{2C_0^2 (C_0 + C_B)^2} t_{11}^{HEL} = -1.37 \text{ GPa}$$

$$t_{22}^{(6-)} = \frac{-3C_0^4 - 2C_0^3 C_B - 4C_0^2 C_B^2 - 6C_0 C_B^3 + 15C_B^4}{2C_0^2 (C_0 + C_B)^2} t_{11}^{HEL} = -1.16 \text{ GPa}$$

$$E_{11}^{(6-)} = \frac{-C_0^4 - 2C_0^3 C_B - 2C_0 C_B^3 + 5C_B^4}{C_B^2 (C_0 + C_B)^2} E_{11}^{HEL} = -0.00509$$

$$E_{11}^{(6+)} = \frac{-C_0^4 - 6C_0^3 C_B + 4C_0^2 C_B^2 + 2C_0 C_B^3 + C_B^4}{C_B^2 (C_0 + C_B)^2} E_{11}^{HEL} = -0.00587$$

Region 7:

$$t_{11}^{(7+)} = t_{11}^{(6++)} = t_{11}^{(3)}, \quad v_1^{(7+)} = v_1^{(6++)} \quad \text{(9.C5)}$$

$$E_{11}^{(7+)} = E_{11}^{(2)} + \frac{4C_0 - 2C_B}{C_B} E_{11}^{HEL} = -0.00619$$

$$t_{22}^{(7-)} = t_{22}^{(3)}, \quad t_{22}^{(7+)} = t_{11}^{(2)} + \frac{C_0^2 + 12C_0 C_B - 9C_B^2}{2C_0^2} t_{11}^{HEL} = -0.83 \text{ GPa} \qquad (9.C5, cont)$$

Region 8:

$$t_{11}^{(8)} = -\frac{C_0 - C_B}{C_0 + C_B} t_{11}^{(2)} - \frac{2C_0^3 + 10C_0^2 C_B - 10C_0 C_B^2 - 2C_B^3}{(C_0 + C_B)^3} t_{11}^{HEL} = -0.22 \text{ GPa} \qquad (9.C6)$$

$$v_1^{(8)} = -\frac{C_0 - C_B}{\rho C_B (C_0 + C_B)} t_{11}^{(2)} - \frac{C_0^4 + 4C_0^3 C_B - 6C_0^2 C_B^2 + 4C_0 C_B^3 - 3C_B^4}{\rho C_0 C_B (C_0 + C_B)^3} t_{11}^{HEL} = 0.00607 \text{ km/s}$$

$$E_{11}^{(8-)} = -\frac{C_0 - C_B}{\rho C_B^2 (C_0 + C_B)} t_{11}^{(2)} - \frac{3C_0^5 + 13C_0^4 C_B - 8C_0^3 C_B^2 - 4C_0^2 C_B^3 - 3C_0 C_B^4 - C_B^5}{\rho C_0^2 C_B^2 (C_0 + C_B)^3} t_{11}^{HEL} = -0.00505$$

$$E_{11}^{(8+)} = \frac{C_0^2 + C_0 C_B - 2C_B^2}{\rho C_0 C_B^2 (C_0 + C_B)} t_{11}^{(2)} + \frac{C_0^5 + 7C_0^4 C_B + 8C_0^3 C_B^2 - 12C_0^2 C_B^3 - C_0 C_B^4 - 3C_B^5}{\rho C_0^2 C_B^2 (C_0 + C_B)^3} t_{11}^{HEL} = -0.00130$$

$$t_{22}^{(8-)} = -\frac{C_0 - C_B}{C_0 + C_B} t_{11}^{(2)} - \frac{7C_0^5 + 29C_0^4 C_B - 14C_0^3 C_B^2 - 10C_0^2 C_B^3 - 9C_0 C_B^4 - 3C_B^5}{2C_0^2 (C_0 + C_B)^3} t_{11}^{HEL} = -1.22 \text{ GPa}$$

$$t_{22}^{(8+)} = \frac{4C_0^2 + 2C_0 C_B - 6C_B^2}{2C_0 (C_0 + C_B)} t_{11}^{(2)} + \frac{5C_0^5 + 31C_0^4 C_B + 14C_0^3 C_B^2 - 38C_0^2 C_B^3 - 3C_0 C_B^4 - 9C_B^5}{2C_0^2 (C_0 + C_B)^3} t_{11}^{HEL} = -0.23 \text{ GPa}$$

Region 9:

$$t_{11}^{(9)} = \frac{C_0 + C_B}{2C_B} t_{11}^{(2)} + \frac{C_0^2 + 8C_0 C_B - C_B^2}{2C_B (C_0 + C_B)} t_{11}^{\text{HEL}} = -1.63\ \text{GPa}$$

$$v_1^{(9)} = -\frac{1}{\rho C_0} t_{11}^{(9)} = 0.346\ \text{km/s}$$

$$E_{11}^{(9-)} = \frac{(C_0 + C_B)(2C_0 - C_B)}{2\rho C_0^2 C_B^2} t_{11}^{(2)} + \frac{2C_0^3 + 11C_0^2 C_B + 6C_0 C_B^2 - 11C_B^3}{2\rho C_0^2 C_B^2 (C_0 + C_B)} t_{11}^{\text{HEL}} = -0.00629$$

$$E_{11}^{(9+)} = \frac{1}{\rho C_0^2} t_{11}^{(9)} = -0.00578 \tag{9.C7}$$

$$t_{22}^{(9+)} = \frac{3C_B^2 - C_0^2}{2C_0^2} t_{11}^{(9)} = -0.71\ \text{GPa}$$

$$t_{22}^{(9-)} = \frac{-C_0^3 + 5C_0^2 C_B + 3C_0 C_B^2 - 3C_B^3}{4C_0^2 C_B} t_{11}^{(2)} + \frac{-C_0^4 - 2C_0^3 C_B + 34C_0^2 C_B^2 + 18C_0 C_B^3 - 33C_B^4}{4C_0^2 C_B (C_0 + C_B)} t_{11}^{\text{HEL}} = -0.85\ \text{GPa}$$

Region 10:

$$t_{11}^{(10)} = \frac{(C_0 - C_B)^2}{2C_B (C_0 + C_B)} t_{11}^{(2)} + \frac{C_0^4 + 2C_0^3 C_B - 16C_0^2 C_B^2 + 14C_0 C_B^3 - C_B^4}{2C_B (C_0 + C_B)^3} t_{11}^{\text{HEL}} = -0.25\ \text{GPa} \tag{9.C8}$$

$$v_1^{(10)} = -\frac{1}{\rho C_0} t_{11}^{(10)} = 0.00540\ \text{km/s}$$

$$E_{11}^{(10)} = \frac{2C_0^3 + 3C_0^2 C_B - 4C_0 C_B^2 - C_B^3}{2\rho C_0^2 C_B^2 (C_0 + C_B)} t_{11}^{(2)} + \frac{2C_0^5 + 15C_0^4 C_B + 22C_0^3 C_B^2 - 20C_0^2 C_B^3 - 8C_0 C_B^4 - 11C_B^5}{2\rho C_0^2 C_B^2 (C_0 + C_B)^3} t_{11}^{\text{HEL}} = -0.00142$$

$$t_{22}^{(10)} = \frac{-C_0^4 + 8C_0^3 C_B + 8C_0^2 C_B^2 - 12C_0 C_B^3 - 3C_B^4}{4C_0^2 C_B (C_0 + C_B)} t_{11}^{(2)} + \frac{-C_0^6 + 4C_0^5 C_B + 61C_0^4 C_B^2 + 52C_0^3 C_B^3 - 59C_0^2 C_B^4 - 24C_0 C_B^5 - 33C_B^6}{4C_0^2 C_B (C_0 + C_B)^3} t_{11}^{\text{HEL}} = -0.24\ \text{GPa}$$

Region 11:

$$t_{11}^{(11)} = t_{11}^{(10)}, \quad v_1^{(11)} = v_1^{(10)}, \quad E_{11}^{(11+)} = \frac{1}{\rho C_0^2} t_{11}^{(10)}$$
$$t_{22}^{(11)} = \frac{3C_B^2 - C_0^2}{2C_0^2} t_{11}^{(10)} \tag{9.C9}$$

References

[1] Th. von Karman and P. Duwez, *J. Appl. Phys.* **21**, pp. 987–994 (1950).

[2] Kh.A. Rakhmatulin, *Prikl. Mat. Mekh.* **9**. p. 91 (1945).

[3] G.I. Taylor, in *The Scientific Papers of G.I. Taylor, Vol. 1, Mechanics of Solids* (ed. G.K. Batchelor), Cambridge University Press, London, p. 467 (1958).

[4] M.R. White and L.V. Griffis, *J. Appl. Mech.* **15**, p. 256 (1948); *Trans. ASME 69 J. Appl. Mech.* p. A-337 and *Trans. ASME 70 J. Appl. Mech.* p. A-337.

[5] L.W. Morland *Phil. Trans Roy. Soc. London,* **A251** p. 341 (1959).

[6] M.B. Boslough and J.R. Asay, in *High-Pressure Shock Compression of Solids* (ed. J.R. Asay and M. Shahinpoor), Springer-Verlag, New York, pp. 7–42 (1993)

Author Index

Subject Index